International Conference on Frontiers of Characterization and Metrology for Nanoelectronics 2017

Monterey, California, USA
21-23 March 2017

Editors:

E. M. Secula **D. G. Seiler**

ISBN: 978-1-5108-3911-3

Printed from e-media with permission by:

Curran Associates, Inc.
57 Morehouse Lane
Red Hook, NY 12571

Some format issues inherent in the e-media version may also appear in this print version.

Printed by Curran Associates, Inc. (2017)

Published by the National Institute of Standards and Technology
Publications Department
Gaithersburg, MD 20899
USA

Phone: (301) 975-2762
Fax: (301) 926-1630

inquiries@nist.gov

Additional copies of this publication are available from:

Curran Associates, Inc.
57 Morehouse Lane
Red Hook, NY 12571 USA
Phone: 845-758-0400
Fax: 845-758-2633
Email: curran@proceedings.com
Web: www.proceedings.com

2017 International Conference on Frontiers of Characterization and Metrology for Nanoelectronics

Welcome to the 2017 International Conference on Frontiers of Characterization and Metrology for Nanoelectronics (FCMN)! Our goal is to bring together scientists and engineers interested in all aspects of the characterization and measurement technology needed for nanoelectronic materials and device research, development, and manufacturing. All approaches are welcome: chemical, physical, electrical, magnetic, optical, in-situ, and real-time control and monitoring. The conference summarizes major issues and provides critical reviews of important semiconductor techniques needed as the semiconductor industry continues its move to silicon nanoelectronics and beyond. It is hoped that the invited talks, contributed posters, and informal discussions will be a stimulus to provide practical perspectives, breakthrough ideas for research and development, and a chance to explore collaborations and interactions on a world-wide basis.

We are pleased to have Dan Hutcheson, CEO, VLSI Research Inc., and Aaron Thean, VP, National University of Singapore, as keynote speakers for the event! Over thirty other invited talks will offer overviews in the sessions that follow. Posters will supplement these overviews with the latest metrology-based research results. These posters represent significant contributions to the latest developments in characterization and metrology technology, especially at the nanoscale.

The 2017 FCMN is the 11th in the series of conferences devoted to metrology frontiers for the semiconductor industry. It emphasizes the latest advances in characterization and metrology that will help shape the future of the nanoelectronics revolution. The proceedings for most of the previous conferences in the series were published as hardcover volumes by the American Institute of Physics, New York. Most of these proceedings as well as many archived presentation slides are available to view for free on-line at www.nist.gov/pml/engineering-physics-division/fcmn-publications-and-talks.

This year, the committee is excited to bring the FCMN to Monterey, California, the first time the conference has been held on the west coast! Monterey is a scenic California coastal city that rises from the pristine Monterey Bay to pine forested hillsides with sweeping bay views. We hope you enjoy your time here!

It is our sincere hope that you find this conference stimulating and enjoyable!

With best wishes from the Committee Co-Chairs,

David Seiler, NIST;
Alain Diebold, CNSE, SUNY Polytechnic Institute;
Zhiyong Ma, Intel; and
Bob McDonald, formerly of Intel (Treasurer)

Purpose and Goals

The FCMN brings together scientists and engineers interested in all aspects of the characterization technology needed for nanoelectronic materials and device research, development, and manufacturing. The conference summarizes major issues and provides critical reviews of important semiconductor techniques needed as the semiconductor industry continues its move to silicon nanoelectronics and beyond.

Contributed Posters

One of the major emphases of this conference is on the contributed posters. These extended poster abstracts selected by the committee represent significant contributions to the frontier, state-of-the-art materials, and device characterization.

Poster authors are responsible for setting up their displays, being present for the poster sessions on Tuesday and Wednesday afternoon, and removing their displays by the end of the lunch break on Thursday.

Poster Sessions

The poster sessions with complimentary snacks and beverages are scheduled for the end of Tuesday and Wednesday in San Carlos I and II at the Monterey Marriott.

Banquet

A dinner banquet will be held on Tuesday, March 21st, in the Ferrante's Bay View Room on the 10th floor of the Monterey Marriott.

Platinum Sponsors

Silver Sponsor

Exhibitors

- Bruker
- CAMECA Instruments, Inc.
- Frontier Semiconductor
- Intel
- NIST
- SAES Group
- Xenocs

Partners

- American Physical Society Physics
- American Vacuum Society
- National Institute of Standards and Technology
- National Science Foundation
- SEMI

Program at a Glance

	Morning	AM Sessions	PM Sessions	Evening
Tuesday Mar. 21st	**7:30** Registration / Attendee Check-in	**9:00** Conference Opening **9:15** Keynote Talks **11:15** TCAD	**11:45** Microscopy Metrology **3:45** Diagnostic Techniques	**5:15** Poster Session **6:45** Banquet
Wednesday Mar. 22nd	**8:00** Registration / Attendee Check-in	**8:30** Magnetics / Spintronics **10:30** Metrology for Patterning	**2:00** Interface and Gate Stack Metrology **3:30** Beyond CMOS Characterization and Metrology	**5:00** Poster Session
Thursday Mar. 23rd	**8:00** Registration / Attendee Check-in	**8:30** 7-10 NM Metrology and Defect Inspection **11:00** 2.5D/3D Packaging Metrology	**12:00** Emerging Metrology	

Monday, March 20

Reception and Registration

5:30 PM – 8:30 PM

Ferrante's Bay View Room, 10th Floor, Monterey Marriott

Tuesday, March 21

Registration and Breakfast

7:30 AM – 8:45 AM

Conference Opening

9:00 AM – 9:15 AM

Welcome and Introduction
David Seiler, NIST, Conference Chair

Keynote Talks

Session Chair: David Seiler, NIST

9:15 AM

10:00 AM

Coffee Break and Poster/Exhibit Viewing

10:45 AM – 11:15 AM

TCAD

Session Chair: Ajey Jacob, GlobalFoundries

11:15 AM

Microscopy

Session Chair: Alain Diebold, CNSE, SUNY Polytechnic Institute

11:45 AM

Lunch and Poster/Exhibit Viewing

12:15 PM – 1:45 PM

1:45 PM

2:15 PM

2:45 PM

Coffee Break and Poster/Exhibit Viewing

3:15 PM – 3:45 PM

Diagnostic Techniques

Session Chair: Deepak Goyal, Intel

3:45 PM

4:15 PM

4:45 PM

Poster Session (with Drinks and Hors d'oeuvres)

5:15 – 6:15 PM

Banquet at Hotel (Ferrante's Bay View Room)

6:45 PM

Wednesday, March 22

Registration and Breakfast

8:00 AM – 8:30 AM

Magnetics/Spintronics

Session Chairs: Scott List, SRC, and Ajey Jacob, GlobalFoundries

Coffee Break and Dedicated Poster/Exhibit Viewing Session
3:00 PM – 3:30 PM

Beyond CMOS Characterization and Metrology
Session Chair: Alain Diebold, CNSE, SUNY Polytechnic Institute

3:30 PM

4:00 PM

4:30 PM

Poster Session (with Drinks and Hors d'oeuvres)
5:00 – 6:00 PM

Thursday, March 23

Registration and Breakfast
8:00 AM – 8:30 AM

7-10 NM Metrology and Defect Inspection
Session Chair: Tuyen Tran, Intel

8:30 AM

9:00 AM

9:30 PM

Coffee Break and Poster/Exhibit Viewing
10:00 AM – 10:30 AM

10:30 AM

2.5D/3D Packaging Metrology

Session Chair: Ehrenfried Zschech, Fraunhofer IKTS Dresden

11:00 AM

11:30 AM

Emerging Metrology

Session Chairs: Wilfried Vandervorst, Imec, and Usha Varshney, NSF

12:00 PM

Lunch and Poster/Exhibit Viewing

12:30 PM – 2:00 PM

2:00 PM

2:30 PM

Coffee Break and Poster/Exhibit Viewing

3:00 PM – 3:30 PM

3:30 PM

4:00 PM

Conference End

4:30 PM

Poster Presentations

Semiconductor Metrology: Past Present and Future

G. Dan Hutcheson

VLSIresearch inc. 2290 North First Street, Suite 202, San Jose, CA 95131

INTRODUCTION

What comes next in Metrology as the industry enters the deep-nanometer, where nodes come in single-digit increments of nanometers? This presentation addresses this question with a historical timeline of how metrology and inspection have progressed historically in semiconductor manufacturing to show the momentum built up that will drive metrology into the future.

Early History

The early history of metrology, characterization, and inspection use in semiconductors was little different than in any other field. The most common tools in the fab were simple microscopes, while the most expensive were SEMs. These were industries all to themselves, serving virtually all segments of industry and science. Since handling of wafers was manual, there was no need to customize what was already available. But over time, the state of this would evolve into specialized tools, which are bundled under the term 'Process Diagnostics,' which is descriptive of what these applications would come to be focused on.

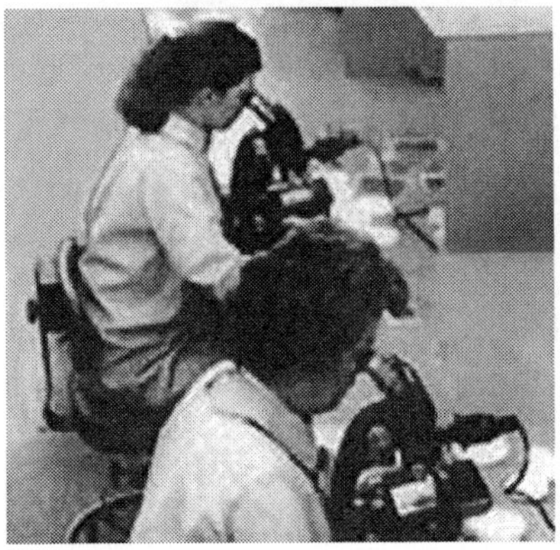

FIGURE 1. Wafer Inspection in the nineteen-sixties at Fairchild Semiconductor. Note that clean room attire was limited to a smock. Hair nets would be added by the seventies.

In these early years, typical die yields started in low single digits and would ramp to maturity levels of 40-50%, a far cry from today's 80-95-plus percent. And it would take years to get there. So once a design was released from manufacturing, operators inspected wafers by either counting defects on wafer surfaces under microscopes or looking for fringe patterns in reflected light to inspect the quality of thin film deposition and diffusion layers. SEMs were used more in process development and for failure analysis. Metrology tools were used mostly for incoming-inspection of raw wafers. This made the common microscope the most used in-line tool in semiconductor manufacturing.

At the time, the primary reason for inspection was to cull bad wafers that were either misprocessed (typically accounting for 15-20% of yield loss at the time) or had defect densities too high to justify further processing. Companies were concerned about maximizing utilization even in these days of fabs costing \$3-6M (\$19-38M in 2016 dollars). However, they generally tolerated defects rather than focusing on eliminating them.

The primary purpose of inspection was the same as today: To assure value, not add value. While this terminology was not used, its application can be seen in practice even in the early days. The economic purpose of culling bad wafers from production was to avoid the cost of adding further processing to wafers that were known to be bad, well before the finished die would be functionally tested.

The Eighties

With the dawn of the 1980s came the rise of Japan and the advent of the wafer stepper. The Japanese were intolerant of defects and saw this intolerance as a fundamental cultural advantage that could be turned to competitive advantage. Their ability to systematically increase yields would be a key game changer for this decade. At the same time, companies successfully mastering production with steppers found the starting point for yields was what had been the mature point for earlier device generations. The wafer stepper had a fundamental defect density advantage. But at the same time, companies that didn't master steppers could hardly yield at all, because a single defect on a reticle would repeat in every field, killing yield. This fact made the entire industry intolerant of defects … at least lithographic defects.

Because of this, a new philosophical approach emerged to compete with the concept that inspection was essential. Some process engineers believed it would be possible to make perfect wafers; if you did, there would be little need to inspect or measure anything. This would save cost and speed time-to-market.

In fact, a perfect wafer had been made in the late-seventies by Al Stein, a young fab-line manager at Texas Instruments, who would rise rapidly as the result of having made the first wafer with 100% die yield.[1] Taking this a step further, some believed if you made clean rooms clean enough and made tools with no process signatures, 100% yields could be delivered all the time. In fact, Mitsubishi would build a plant in Saijo they believed would be such a facility and a large part of Japan's semiconductor industry would follow.

Contrary to this was the rise of a belief that perfection would never be achieved, because the very process of scaling created new defects that had never been seen before. Hutcheson wrote, "The nature of VLSI manufacturing (Very Large Scale Integration) has exacerbated these issues. For example, a one micron defect might lie in any of over eight million points covering a four inch wafer. The magnitude of this problem was clearly pointed out by Gordon Moore, of Intel. He noted: '*Finding a defect … is like finding a matchstick in a field of hay.*'"[2]

So the rise in yields would prove a false prophet of perfection. The search for defects was actually becoming a bigger problem — one bigger than visual inspection through microscopes. Not only were there too many to count, it was also hard for humans to differentiate between killer and nuisance defects at anywhere near the timeframe needed. It was clear automated inspection was essential, and that solution would come from KLA-Tencor (then separate companies). KLA would develop the first brightfield tool, the 2020, on the foundation of its mask inspection technology. Tencor would take darkfield inspection capability from bare wafers to patterned ones. With these innovations, the market for wafer inspection tools began to soar.

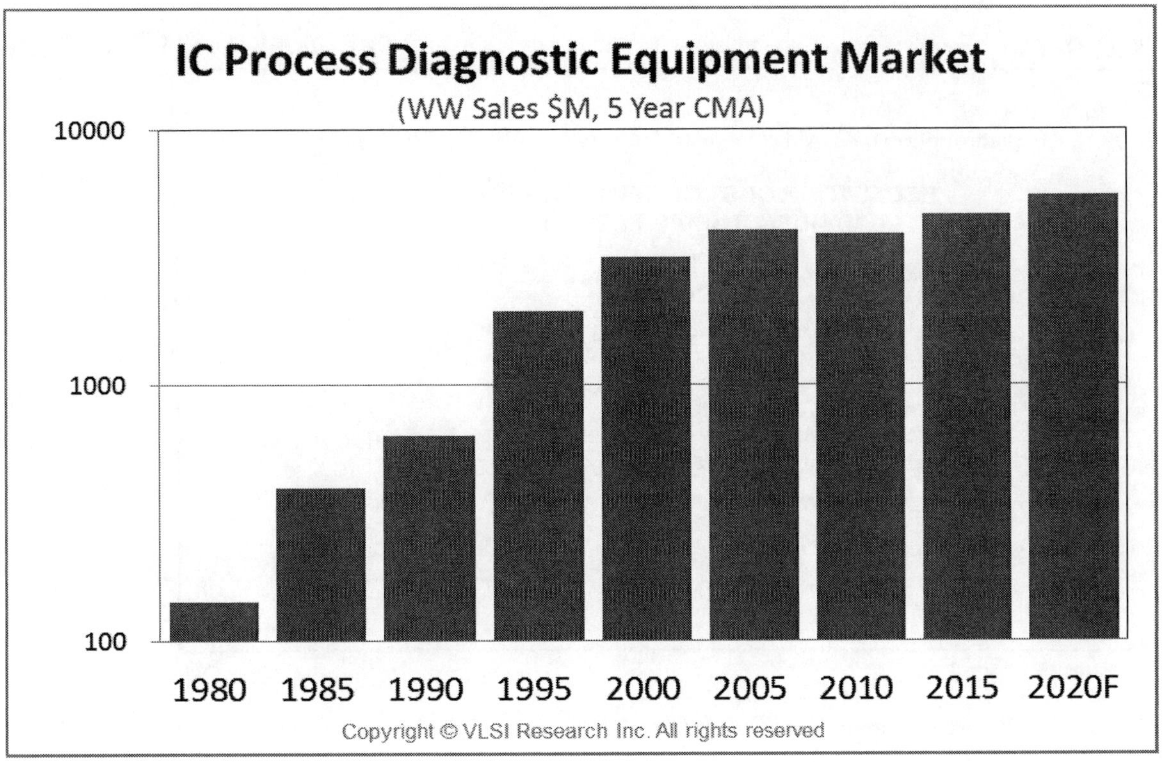

FIGURE 2. IC Process Diagnostic Equipment Market (WW Sales $M, 5 Year CMA)[3]

But there was one problem with Moore's simile: the search for defects was still a 2D process. But it was a vision of future problems. The ability to inspect across three-dimensions would become a significant problem three decades later.

Control over critical dimensions (CDs) and film thicknesses would also become a large problem as well. Both were proving critical to device performance. The latter would spark the introduction of multiple metrology technologies as well as new types of process tools that could deliver tighter uniformity across larger wafers. The coincidental development of performance binning at test meant the ability to capture higher revenues for die that had better electrical properties. The problem was how to control critical parameters before the devices had passed from the fab to test and assembly, thereby shifting the distribution to the more profitable side of the histogram. An important milestone was the adaptation of SEMs to measure CDs with Hitachi's EP-1035 CD Measurement Accessory. But like most of these tools, it was far from a solution.

The Nineties

KLA's 2020 would spark the yield management revolution in the nineties with a series of tools that came to be known under the nomenclature, "20XX." Hitachi would introduce S6000, which would spark a revolution not only in CD measurement, but also put measurement steps on the 'run card.' Once metrology was in-line, or on the 'run-card' as they said in those days, a large market was guaranteed.

With these events came new ways of thinking about how to wring more revenues and profits out of a fab as the cost of building one soared. By the mid-nineties, these costs would eclipse $1B. Fears were high they were tightly coupled to Moore's Law and that there was no stopping their rise.

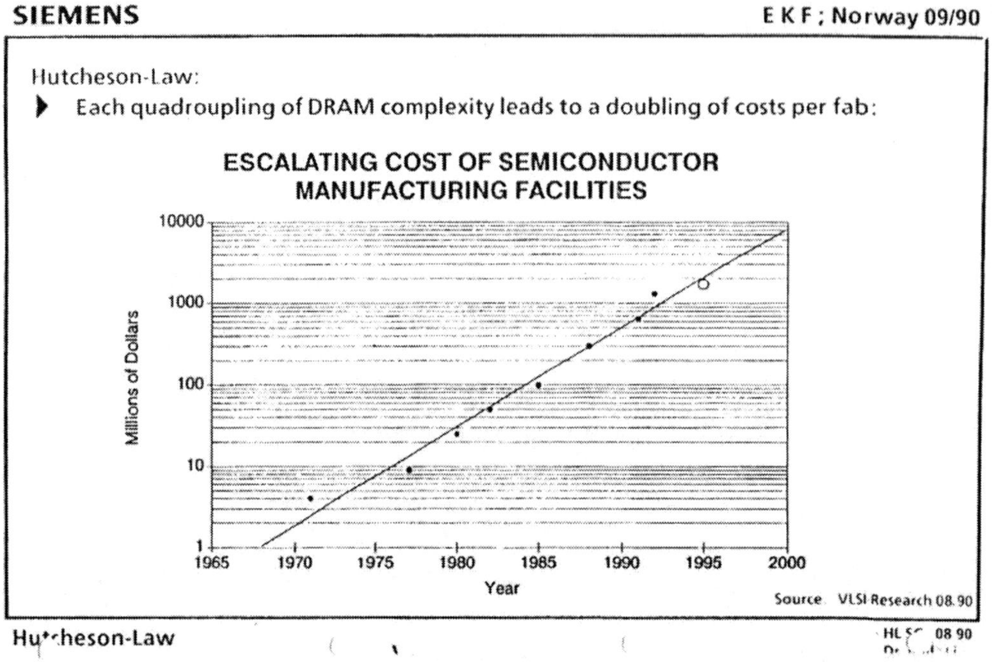

FIGURE 3. Fab cost escalation becomes a major fear in the early nineties, driving an intense need for process diagnostics.[4]

These fears were compounded by fears that optical would run out of steam as feature sizes got smaller. CD measurement had already made the jump from photons to electrons. Would this be true for in-line inspection as well? Even KLA was arguing this would be the case, placing a bet with its introduction of SEMSpec. E-beam inspection would have its place. But it was fundamentally limited by the need for a vacuum chamber, which made it slow, cumbersome, and costly. Like dark-field and bright-field, e-beam would prove complementary to optical, not a substitute for it.

High yields were now a given. Semiconductor companies were shifting from selling building-block commodities to intellectual property driven marketing with laser focus on end markets. Time-to-Market became important as a result and this fell back on fab managers in the form of Time-to-Yield. At the same time, while high yields were expected, they were not always given. So, fast recovery from yield excursions became essential. This was at a time when scaling was making data volume and variance soar. The concept of Time-to-Entitled Yield was introduced by KLA-Tencor and it quickly took off as a key management variable in their Yield Management. The term, 'Time is Money,' took on new meaning for fab managers around the world.

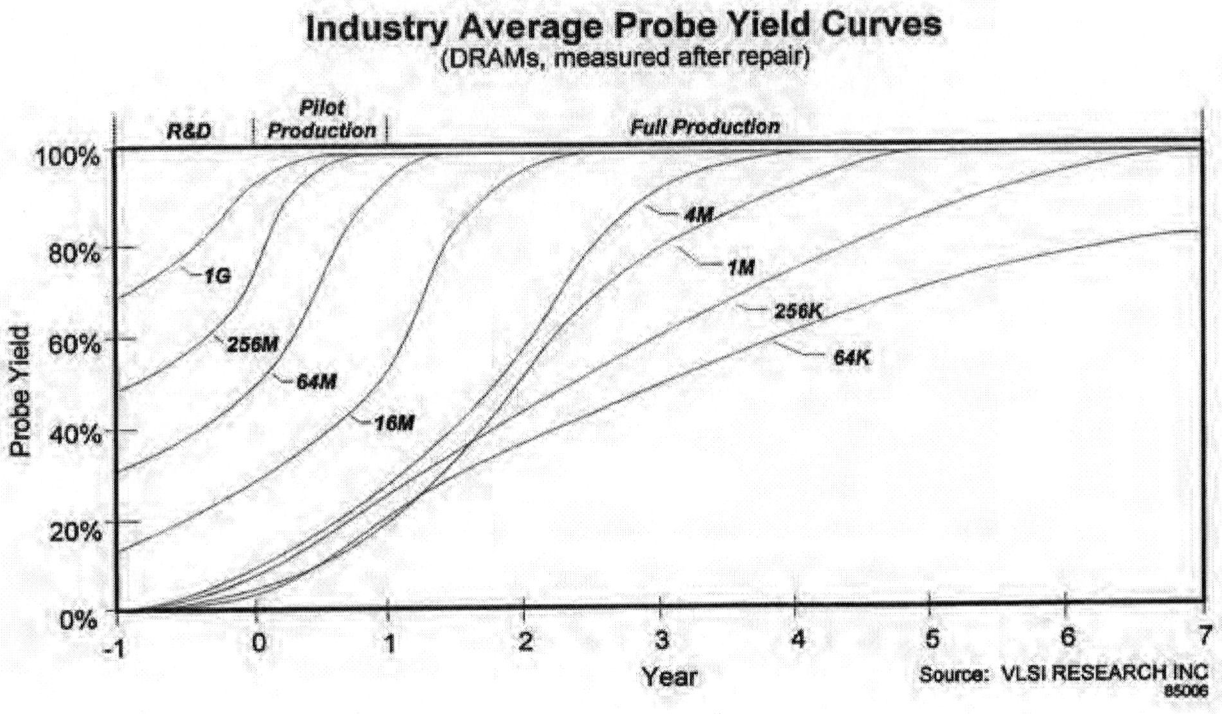

FIGURE 4. Time-to-Yield Acceleration from the eighties and into the nineties.[5]

The whole philosophy changed with the transition from manual to automatic defect classifications. It went from reactive to proactive. Yield models were no longer seen as stochastic. Now the approach assumed they were deterministic whack-a-mole games. Management's view of Process Diagnostics would shift from seeing it as a cost center to being a key profit center.

The concept of Yield Management was so important it toppled empires that didn't adopt it early. Moreover, it put an end to the idea a fab line could ever reliably produce perfect wafers. The first systematic application of Yield Management was at Samsung. This and Japan's late adoption could create fortunes and new silicon empires in Korea, while being a key part of destroying them in Japan.

The Two-Thousands

By the 2000s, all of this would be put together into networks in the fab, so that time-to-information could be brought from weeks and days to hours and minutes. This would be essential, as fab costs jumped by 3 times and would approach $5B by the end of the decade. So while they did slow, their growth was still a significant business issue that would drive process diagnostics.

One area of particular concern was lithography, where tool costs were soaring beyond what anyone could have imagined. So the issue was similar: How does a chip maker squeeze more value out of these tools? And it could not just be tighter CDs. A critical landmark came in late 2006, when ASML acquired Brion. They would mashup this diagnostic technology together with their scanner to create a whole new value concept in lithography. It would take almost 10 years, but in 2014 they pinned a name on this concept: Holistic Lithography.

But one thing had still not changed: the world of semiconductors was still planar CMOS with mostly two-dimensional process diagnostics. There were some 3-D mapping tools, such as AFMs, but there was little pressure to implement them in-line.

The Two-Thousand-Tens and Beyond

By the early 2010s, it had become clear that planar CMOS would not cut it electrically. Logic devices were drawing too much power, even with Intel's earlier introduction of HKMG technology. Performance gains were also dwindling with scaling. In 2011, Intel announced it would introduce finFET technology to production for 22nm devices. This made Intel the first to market with true 3D logic circuits. Meanwhile, NAND memories could not hold enough electrons to reliably store a readable charge. In 2013, Samsung would be first to announce the start of 3D NAND production.

The need to image, measure, and inspect in three-dimensions is becoming ever more critical. Today's devices need the equivalent of an MRI for chips to see in and around structures. So new approaches are needed, like KLA-Tencor's use of infrared for 3D defect inspection. Meanwhile, everything needs to get better to keep Moore's Law on track. But one thing is for sure; technology never stops nor tolerates any bounds for long. Central to this will be the consistent innovations that the metrology, characterization, and inspection community brings to market in the future.

REFERENCES

1. Conversations with Howard Moss of Texas Instruments and the author in the mid-eighties.
2. J. D. Hutcheson, *The VLSI Capital Equipment Outlook*, San Jose: VLSIresearch inc, vlsiresearch.com, 1982, Pg. 48 12.
3. *Chip Making Markets*, San Jose: VLSIresearch inc's database, vlsiresearch.com.
4. H. Reull, *Major Trends Shaping the Microelectronics of the* 90s, Keynote Address, SEMI ISS Europe proceedings, 1992.
5. G. D. Hutcheson, *Change in Chip Making and How It is Driving Process Diagnostics*, San Jose: VLSIresearch inc, vlsiresearch.com, June 1996, Pg. 9.

KEYWORDS

Metrology, Inspection, IC Process Diagnostic Equipment Market Trends, Inspection Methodology, Semiconductor Industry

CMOS to Beyond CMOS: From Advanced Structures to Complex Material Systems

Aaron Thean, Ph.D.

Department of Electronical & Computer Engineering

National University of Singapore

Email: Aaron.Thean@nus.edu.sg

As the semiconductor industry scales deep into the the sub-10nm technology nodes in the next decade, the need to maintain chip-level energy budget with increasing circuit density motivates the search for new materials and new transistor technologies. To increase functionality and performance, there is a trend to integrate different process technologies heterogeneously. This drives innovations from advanced structures processing to increasingly complex integration of material systems. There is now an important need for characterization and metrology technologies to catch up in order for these technologies to be realized for high-volume manufaturing. In this paper, we will review the advanced CMOS towards Beyond-CMOS technologies that are being explored today, and the implications on tomorrow's metrology technologies.

The CMOS industry took a major departure from basic planar processing with the introduction of FinFET technology in 2011. The ability to control, monitor, and manufacture ultra-thin vertical channels (8-10nm thickness) for the FinFET technology marks a major advancement in CMOS process technology. When device features are well into the 10nm scale, the advanced structural features are not only difficult to fabricate and control, but diffcult to design and characterized. Especially for highly scaled devices with complex structures, there is now a greater need to correlate detailed modeling with structural and material analyses, beyond simple film-level metrology. For example, scatterometry-based techniques are becoming an integral in-line metrology for the manufacture of 3-D Fin structures. Recently, layer-by-layer Scannning Spreading Resistance microscopy (SSRM) to create a 2-D carrier mapping for a 3-D structure has been investigated to calibrate TCAD models, allow us to understand the dopants, electrodes and their interactions with nanoscale features of scaled 3-D devices [1]. As we scale beyond FinFETs to more aggressive non-planar structures like the stacked nanowires (Fig. 1(c-d)), the characterization task becomes even more complex. We expect a greater fusion of modeling and ultra-sensitive physical metrology analysis is needed to support the high-volume development and manufacturing of such new technologies.

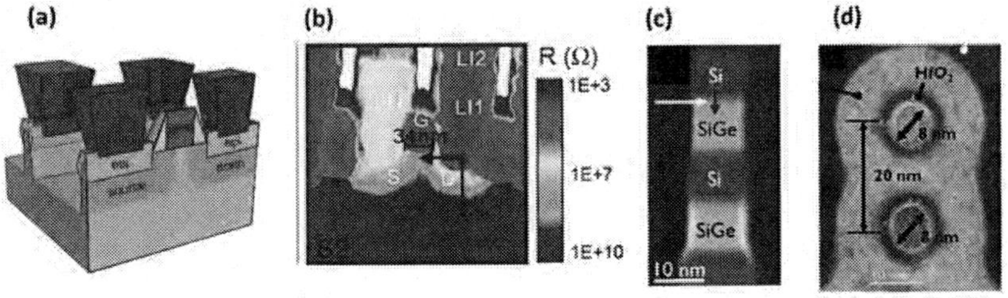

FIGURE 1. (a) Cross-section of scaled FinFET showing gate, electrodes, and junctions. (b) 2D SSRM resistance maps for FinFET sample. (c) Si-SiGe Supper lattice fin necessary to produce the Gate-All-Around Nanowires as should in (d). Sources: 1(a),(b) from Ref. 1, 1(c) and (d) from Ref. 2.

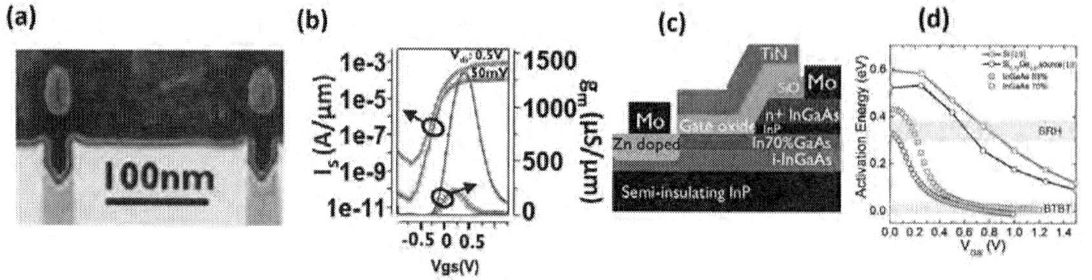

FIGURE 2. (a), (b) Successful scaled InGaAs –channel GAA device on 300mm Si wafer. (c) Simple InGaAs homojunction Tunnel FET (d) Activation energy of device distinguishing leakage components from Band-to-band and SRH transition, dependent on the quality of the material & device/material design. Sources 1(a),(b) from Ref. 3, 1(c) and (d) from Ref. 4.

The industry has started to explore beyond Silicon materials for transistors, there are enormous recent learning on how non-Si materials like III-V and Ge bring advantages over traditional silicon, as well as new challenges. Such devices typically involve the heterogeneous combination of materials (Fig. 2). Successful heteroepitaxy techniques have allowed us to achieved pseudomorphic structures in scaled dimensions (Fig. 2 (a)). Due to the challenging material combinations, defects become the major limitations. Moreover, emerging ultra-low voltage devices like TFETs are inevitably more sensitive to defects. Therefore, defect metrology and modeling gain importance. To support high-volume manufacturing at advanced CMOS-level densities, high-throughput in-line characterization techniques will need to be developed. This remains a major challenge today.

With the mastery of heterogeneous material integration, even greater opportunities are enabled to bring in non-tranditional semiconductor material to augment or expand CMOS. There are a variety of charge-based and non-charge based Beyond-CMOS based technologies being investigated actively world-wide. New classes of ultra-thin materials from Graphene to transition Metal Dichalcogenides and their 2-D Van Der Waal Heterostructures offer a rich gamut of interesting physical properties that can be harness for new functions. There is also much interest to integrate non-charge based computational systems like spintronics with CMOS. To introduce non-volatility memory-like capabilities to logic computatons are among some of the value propositions for these new approaches. The ability to characterize ultra-thin material properties becomes the key challenge. Techniques like micro Raman analysis for 2-D materials, Brillouin Light Scattering for Magnons, or Magneto-Optic Kerr Effect magnetometry for magnetic material are examples of lab techniques that may have to be scale up to wafer-level metrology.

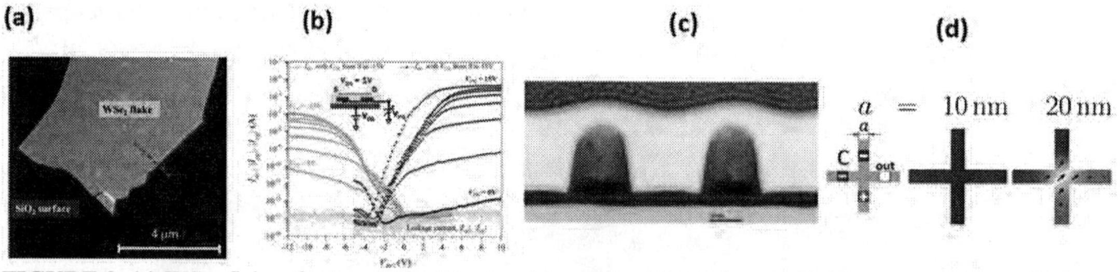

FIGURE 3. (a) WSe_2 flake of 7.5nm teck. (b) n & p bi-polarity of WSe_2 controlled by gate can be applied to enable new ambipolar logic XOR gates (c) XTEM of magnetic tunnel junctions on a MgO free layer for a Spin-Torque majority gate (STMG) (d) Micro-magnetic simulations of the domain switching of the 3-input STMG as a function of dimension. Sources: (a) & (b) Ref. 5, (c) and (d) from Refs. 6 and 7, respectively.

REFERENCES

[1] P. Eyben et. al. "*Accurate prediction of device performance in sub-10nm WFIN FinFETs using scalpel SSRM-based calibration of process simulations*,", 2016 International Conference on Simulation of Semiconductor Processes and Devices (SISPAD), pp. 287 – 290, 2016.

[2] H. Mertens et. al., "*Gate-all-around MOSFETs based on vertically stacked horizontal Si nanowires in a replacement metal gate process on bulk Si substrates*," 2016 IEEE Symposium on VLSI Technology, pp. 1–2, 2016.

[3] X. Zhou et. al., " *Scalability of InGaAs gate-all-around FET integrated on 300mm Si platform: Demonstration of channel width down to 7nm and Lg down to 36nm*," 2016 IEEE Symposium on VLSI Technology, pp 1-2, 2016

[4] A. Alian et. al. "*Record performance InGaAs homo-junction TFET with superior SS reliability over MOSFET*," 2015 IEEE International Electron Devices Meeting (IEDM), pp. 31.7.1 - 31.7.4, 2015.

[5] R. Giovanni et. al. "*Polarity control in WSe2 double-gate transistors*," SCIENTIFIC REPORTS Volume: 6 Article Number: 29448, Jul 2016

[6] I. Radu, A. Thean, "*Spintronic majority gates: A new paradigm for scaling*," SOLID STATE TECHNOLOGY Volume: 59 Issue: 3 pp. 12-15, May 2016

[7] A. Vaysset et. al. , "*Toward error-free scaled spin torque majority gates,*" AIP ADVANCES, Volume: 6 Issue: 6 Article Number: 065304, Jun 2016

KEYWORDS

Advanced CMOS, Beyond CMOS, FinFETs, Nanowires, III-V Tunnel FETs, 2-D Materials, Spintronics, Heterogeneous Integration and Metrology

Metrology Requirements and Challenges for Advanced FinFET Technology: Insight from TCAD Simulations

Peter Zeitzoff[b], Kerem Akarvardar[b], Jay Mody[a], Aniruddha Konar[c]

[a]GLOBALFOUNDRIES, 400 Stonebreak Ext, Malta, NY 12020; [b]GLOBALFOUNDRIES, 257 Fuller Road, Suite 3100, Albany, NY 12203;[c]GLOBALFOUNDRIES Engineering Private Limited, 9th Floor, Block N1 (Balsa), Manyata Embassy Business Park, Nagawara, Outer Ring Road, Bangalore - 560045, India

INTRODUCTION

For scaling to the 14nm technology node and below, the FinFET architecture (fully depleted transistors with a gate on 3 sides) has been preferentially adopted by the industry. This is because of the improved short channel control and device performance achievable with FinFETs for such scaled technologies [1-4]. FinFETs utilize tall thin fins, and have a complicated 3-dimensional (3D) structure, particularly in the source-drain (S-D) region. Due to the small dimensions, the complicated 3D geometry, and the fully depleted operation, the technology requirements for scaled FinFET technologies go beyond what was needed for previous planar technologies, and the FinFET requirements are significantly more stringent. For a representative 7nm FinFET technology, with 15nm gate length and 48nm contacted polysilicon pitch, TCAD simulations are utilized here as a key aid in determining the fundamental technology requirements, which drive the metrology requirements.

FINFET STRUCTURE

Figure 1 shows several TCAD simulated cross-sections of a FinFET device. In the figure, the metal layers have been removed and replaced with infinitely thin metal contacts. The 3D cross-section in Figure 1a) is through the middle of the fin, so only half the fin width is shown in the 2D cross-section A in Fig. 1b). As can be seen in Figure 1b), the gate stack (the gate dielectric and the metal gate contact) wraps around the fin, and stops at the top of the STI. The active fin height is defined by the vertical extent of the gate stack.

FINFET METROLOGY NEEDS

TCAD process and device simulations were run for the FinFET pictured in Figure 1, and the results were analyzed for the impact on transistor electrical performance of variation in key parameters. An example is shown in Figure 2, where a TCAD analysis of the sensitivity to the fin width has been run (since the control of short channel effects is strongly dependent on fin width for FinFETs). In the figure, the effective capacitance, Ceff, and the effective current, Ieff, are plotted versus changes in fin width from the nominal; note that Ieff decreases fairly sharply while Ceff increases somewhat. Since circuit speed ~ 1/(Ceff*Vdd/Ieff_stg), it is clear that the circuit speed will be noticeably reduced, even for increases in fin width ~1nm. Hence, the technology needs to be capable of controlling the fin width to accuracies of 1nm or better. Based on similar analysis for other parameters, and based on general process integration knowledge, a list of key technology requirements for this FinFET technology as well as potential metrology approaches to measure the key parameters has been put been put together in Table 1 below. The measurement techniques are divided into destructive and Off-Line, which are particularly useful at the R&D stage, and In-Line and non-destructive, which are particularly useful for process control measurements in production. (Note: for each parameter, the needed metrology resolution, which is not included in the table, is typically notably smaller than the required control; e.g., fin width measurement resolution significantly <1nm is needed to ensure the required 1nm fin width control.)

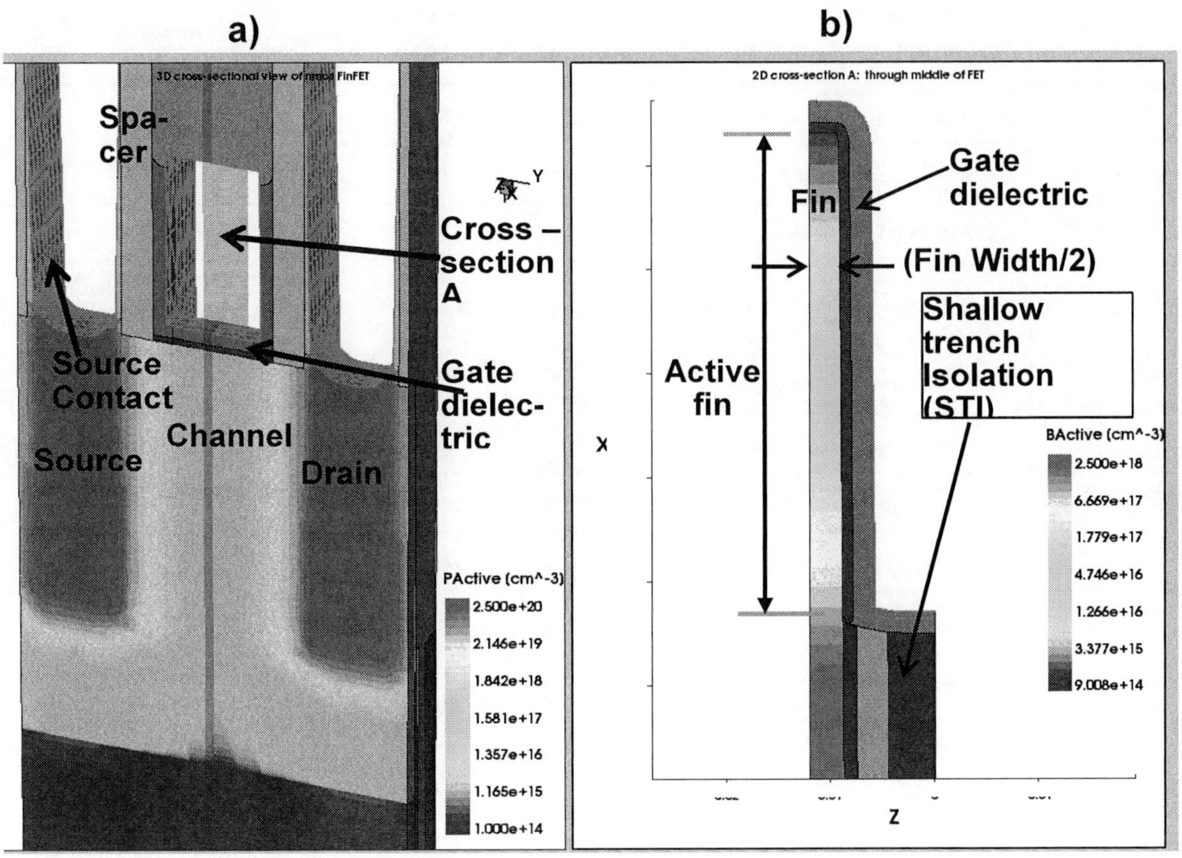

FIGURE 1. (a) 3D FinFET cross-section (b) 2D cross-section A from Figure 1a, showing a slice through the middle of the FET

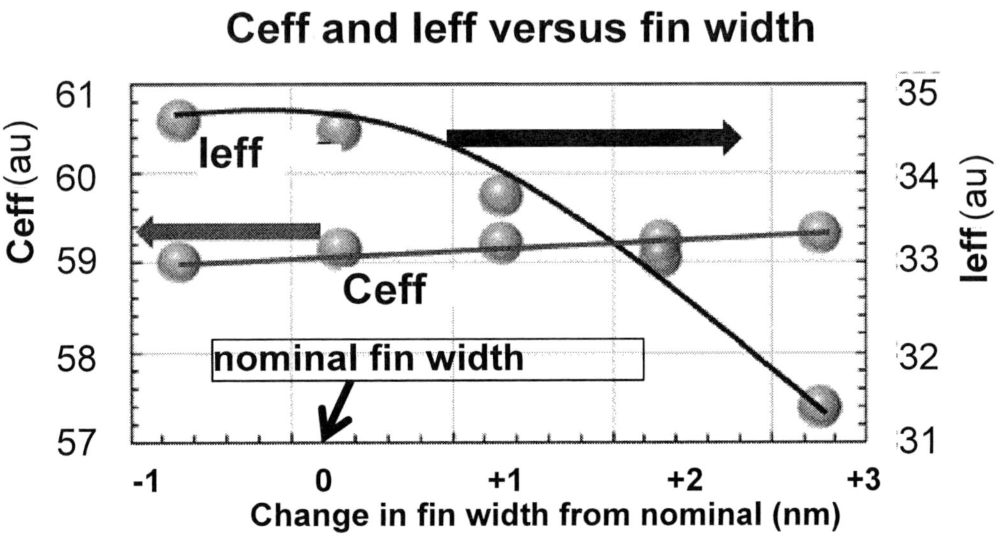

FIGURE 2. TCAD results for Ieff and Ceff versus change in fin width from the nominal

Parameter	Technology requirement	Destructive: potential approach	Non-destructive: potential approach	Comments
Fin width	Control to ≤1nm	TEM	Top-down SEM, OCD	Fin width CD critical to short channel effect control
Fin edge slope	≤2nm difference--top to bottom	TEM	OCD	Fin taper critical to short channel effect control
Spacer width, contact width	Control to ≤1nm	TEM	Top-down SEM, OCD	This controls transistor density
Metal gate length	Control to ≤1nm	TEM	Top-down SEM, OCD	Metal gate length critical to short channel effect control and transport
Fin pitch, Contacted Poly Pitch	Control to ≤2nm	TEM, SEM	Top-down SEM, OCD, XRD	These pitches control transistor density
Pitch walking	Control to ≤2nm	TEM, SEM	Top-down SEM, OCD, XRD	Pitch walking is unequal spacing between fins or poly lines
Gate metal to contact spacing	Control to ≤2nm	TEM	Top-down SEM, OCD	This will affect overlap capacitance,
STI width	Control to (2-4) nm	SEM	Top-down SEM, OCD	This will affect electrical isolation between transistors
Vertical doping profile in active fin	Doping level down to 1E17cm^-3	1.5D SIMS[5] or SCM or SSRM	Transistor electrical test	Critical to controlling mobility, transport in the transistor
Vertical doping profile below active fin	Doping level down to 1E17cm^-3	1.5D SIMS[5] or SCM or SSRM	Transistor electrical test	Adequate doping below channel is needed to control leakage below active fin
S-D lateral profile	Spatial control to ≤1nm	2D profile: SCM or SSRM	Transistor electrical test	Control of this profile is critical to short channel control and transport.
Active fin height: STI recess	Control to ≤1nm	SEM/TEM	OCD	This is critical to FinFET Ieff per unit area==>performance
In S-D: fin recess and STI recess	Control to ≤2nm	TEM	OCD	This is important to control of parasitic resistance
Gate dielectric: thickness of layers and uniformity along fin	Control to a few angstroms	TEM	XPS, OCD, Ellipsometry, Electrical: tinv	This is critical to FinFET short channel control, gate leakage, reliability, and mobility
Gate work function metal: thickness and uniformity along fin	Control to a few angstroms	TEM	XPS, OCD , Electrical: tinv	This is critical to FinFET threshold voltage and its variability

TABLE 1: Key technology requirements and potential metrology approaches

REFERENCES

1. R. Xie et al., IEDM, 2016..
2. S. Y. Yu et al., IEDM, pp. 224-227, 2013.
3. "Intel 14nm Technology," Intel.com, 2014.
4. K. Seo et al., VLSI Symposium, pp. 12-13, 2014.
5. J. Mody et al., J. Vac. Sci. and Tech. B, p. 1116, 2010.

KEYWORDS

Metrology, TCAD, FinFET, 7nm

Field mapping in semiconductor devices by transmission electron microscopy.

David Cooper[1,2], Nicolas Bernier[1,2], Jean-Luc Rouviere[1,2], Christoph Baumer[3] and Regina Dittman[3].

1. Université Grenoble Alpes, F-38000 Grenoble, France.

2. CEA, LETI, Minatec Campus, F-38054 Grenoble, France.

3. Peter Gruenberg Institute, Forschungszentrum Juelich GmbH and JARA-FIT, 52425 Juelich, Germany

INTRODUCTION

We will present experimental results from a range of different TEM-based techniques that have been used to measure the field that are present in semiconductor devices, such as the applied strain or from active dopant atoms. We will compare two different techniques that can be used for measuring the activity of dopant atoms with nm-scale spatial resolution. Results obtained by differential phase contrast (DPC) and off-axis electron holography will be compared. Deformation mapping has been performed using dark field electron holography and precession electron diffraction (PED), here we will briefly introduce the techniques and show experimental results applied to state-of-the-art silicon on insulator (SOI) devices that are currently under development at CEA LETI. Finally we will show experimental results of the in-situ biasing of resistive memory devices in the TEM. Off-axis electron holography and electron energy loss spectroscopy (EELS) have been applied to working devices in order to show the change in the distribution of oxygen vacancies as the devices are reversibly switched from a low resistance to a high resistance state.

DOPANT PROFILING BY OFF-AXIS ELECTRON HOLOGRAPHY AND DPC

The measurement of active dopants in semiconductor devices with nm-scale resolution is a difficult problem. Off-axis electron holography uses a biprism to interfere an electron wave that has passed through a specimen with a wave that has passed through only vacuum to form a hologram. The phase of the electrons can be recovered from the hologram from which information about the electromagnetic fields can be recovered with nm-scale spatial resolution [1]. Another approach which has recently received a lot of attention is DPC which uses a segmented detector to measure the deflection of a convergent beam as it is scanned across a region of interest containing an electric field [2]. These two different techniques have been applied both to simple p-n junctions and to bulk-type CMOS devices and SOI-devices. Figure 1(a) and (b) shows two examples of simple pMOS devices with different spacers (c) specimen preparation is an important step, and parallel-sided membranes are provided using back-side milling in a focused ion beam (FIB) tool. Figures 1(d) and (e) show quantitative maps of the electrostatic potential for the two different devices. The differences in electrical gate width is clear and this can be measured as shown in Figure 1(f). In this presentation we will compare the different approaches and highlight the advantages and disadvantages of each [3]. We will show that the problems that are encountered for dopant profiling are common to each technique and that there is no simple method that can be used for dopant profiling. We will also discuss the prospects of each technique to fulfill the requirements of the semiconductor industry for a method that can measure the active dopant concentration with nm-scale resolution.

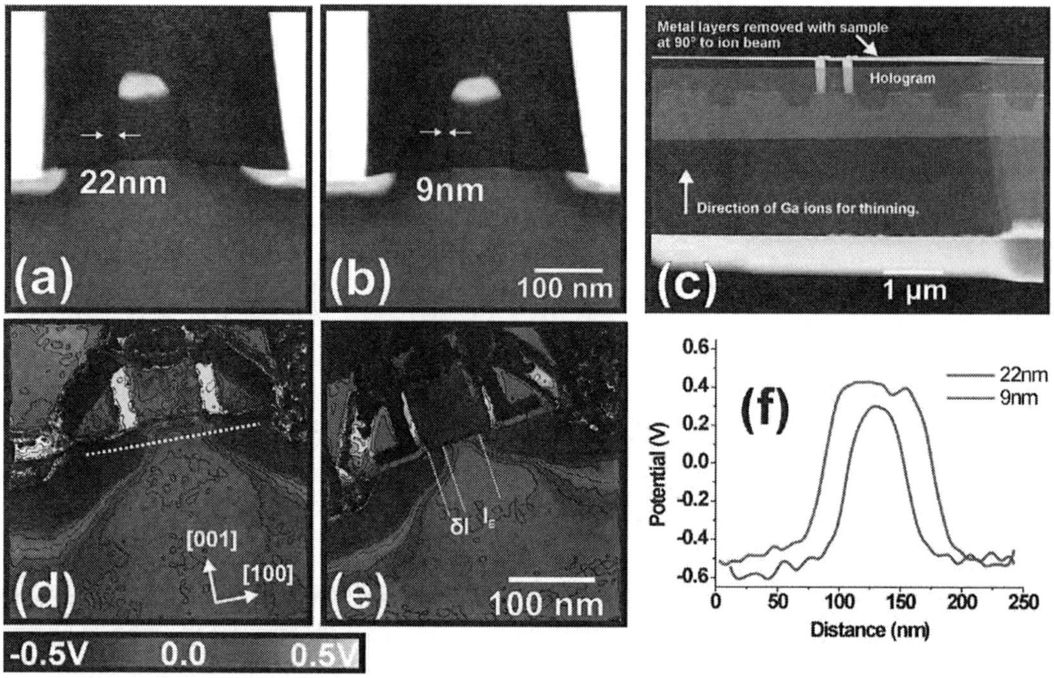

FIGURE 1. (a) and (b) show two p-MOS devices with different width spacers. (c) Electrically tested specimens have been prepared by back-side FIB milling. (d) and (e) show potential maps that have been acquired by off-axis electron holography. (f) Quantitative measurements of the electrostatic potential across the gate are shown.

STRAIN MAPPING BY PRECESSION ELECTRON DIFFRACTION

There has been a great deal of activity in the development of techniques that can be used to measure strain or deformation. In this presentation we will introduce dark field electron holography and show how deformation maps can be obtained by forming an interference pattern from a diffracted beam of interest [4]. We will also discuss precession electron diffraction (PED) which allows high quality diffraction patterns to be determined using a convergent electron beam [5]. By comparing the shift of the diffracted beams relative to a reference pattern, the deformation can be determined with up to 1 nm spatial resolution. In this presentation we will show how deformation mapping is now becoming routine in the development of semiconductor devices and will discuss the advantages and disadvantages of each technique [6].

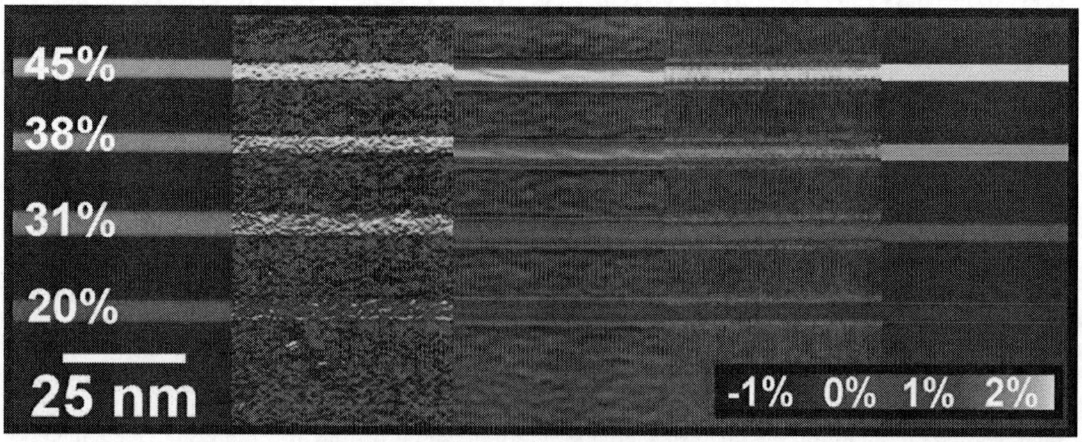

FIGURE 2. Shows a SiGe/Si test specimen with 10-nm-thick SiGe layers with different concentrations of Ge. Deformation maps for the growth direction showing the expansion of the lattice parameter relative to the unstrained substrate. The different maps from left to right have been acquired using geometrical phase analysis, dark field holography, precession electron diffraction and finite element simulations.

IN-SITU BIASING OF A RESISTIVE MEMORY DEVICE

We will discuss the need to make measurements of working semiconductor devices in-situ in the TEM [7]. When devices such as resistive memories (ReRAM) that have been switched ex-situ are separately observed in low resistance states (LRS) or high resistance states (HRS) in the TEM, it is not possible to be sure that specimen preparation or retention problems will lead to misleading results or even if the filament is present in the thin FIB-prepared TEM specimen. In order to understand better the switching mechanism in ReRAM devices, we have operated a device in-situ in the TEM and have applied atomic-resolution STEM, electron holography and EELS in order to measure the changes. In this experiment we have observed a simple STO device as this is considered to be the model system for ReRAM devices. Figure 3 shows how a device has been electrically operated to switch from a LRS and a HRS and that the movement of oxygen can be determined from changes in the fine structure of the Ti spectra which is sensitive to the Ti-O bonds. In this work we were able to limit the effects from artifacts such as from specimen heating and short circuits from FIB preparation and as such were able to reversibly operate the device to show that it is indeed he movement of oxygen under the Pt top electrode which is responsible for the different electrical properties of the device. We will finish the presentation by highlighting the problems that are present when performing in-situ electrical experiments in the TEM.

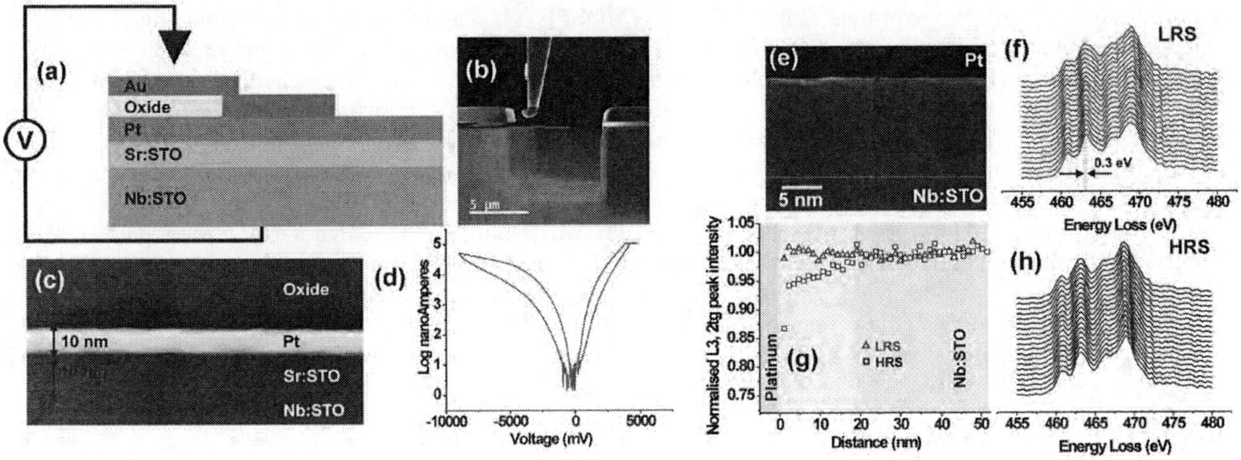

FIGURE 3. (a) Shows a schematic of the ReRAM device examined (b) a movable probe is placed onto the region of interest in-situ in the TEM. (c) A STEM image of the region of interest. (d) V/I response of the device as tested in the microscope. (e) An atomic resolution STEM image of the examined region of the device. (f) Ti spectra acquired from top to bottom acquired from the region indicated in (e) from a device switched into a LRS. (g) measured L3 2tg peak intensities of the Ti spectra which is sensitive to Ti-O bonds indicating a change in the oxygen conentration in the device switched to a LRS. (h) Ti spectra acquired from the same region of interest switched into a HRS.

REFERENCES

1. W-D. Rau et al, Phys. Rev. Lett. **82**, (1999) pp. 2614
2. N. Shibata et al, *Scientific Reports* **5**, (2015) pp. 10040
3. D. Cooper et al, *Semiconductor Science and Technology*, **28**, 2013, pp. 125013
4. M. Hytch at al, *Nature* **453**, (2008) pp. 1086
5. D. Cooper et al, *Nano Letters* **15**, (2015), pp. 5289
6. D. Cooper et al, Micron **80**, (2016) pp. 145
7. D-H. Kwon et al *Nature Nanotechnology* **5**, (2010) pp.148

KEYWORDS

Off-axis electron holography, Precession electron diffraction, Deformation Mapping, In-situ transmission electron microscopy

Quantitative Atom Probe Tomography of Complex Systems

A. D. Giddings

Taiwan Semiconductor Manufacturing Company, Ltd., 166, Park Ave. 2, Hsinchu Science Park, Hsinchu 30075, TAIWAN

INTRODUCTION

Metal-oxide-semiconductor field effect transistors (MOSFETs), which form the basic building blocks of logic and memory in modern electronics, are amongst the smallest man-made structures. The economics that has driven this situation is based on the principle that as the devices become smaller they not only have better electrical characteristics but are also cheaper to manufacture. To put the size in perspective, a virus typically has a diameter in the range of 20 to 400 nm while the pitch between transistors in a modern transistor device can be below 50 nm.

In order to achieve these levels of miniaturization, there have been necessary increases in structural and compositional complexity of the transistors. For example to increase the surface area that the gate acts over, modern transistor designs use a non-planar, fin-shaped channel, known as a finFET. In such devices, current paths will depend on the 3D distribution of dopants. Simultaneously to this, Si based gate and dielectrics have been replaced by a range of exotic materials with desirable properties such as lower resistances or better dielectric constants. A modern processor chip can be expected to contain over 20% of the naturally occurring elements of the periodic table.

The nano-scale dimensions, intricate 3D shapes and complex heterogeneous compositions of these devices makes them a uniquely difficult material for quantitative analysis.

ATOM PROBE TOMOGRAPHY

Atom probe tomography (APT)[1] is a nano-analysis technique that provides compositional measurements of nano-scale volumes. In principle it is the combination of a field ion microscope and a time-of-flight mass-spectrometer. In order to operate, the material to be studied is fabricated into a needle shape specimen with an apex diameter 50-100 nm that will be subjected to a high voltage. This will create an electric field at the needle apex sufficient that, with the aid of carefully controlled trigger event such as a laser pulse, single ions can be field evaporated from the surface and projected towards a position-sensitive detector. From the flight time of the ion, the chemical (or isotopic) species can be identified, and the original position in the specimen can be computed from back-projecting the impact position on the detector. As a result a 3D representation of the component atoms in a small volume can be produced.

As the analysis volume offered by APT has become appropriate for modern finFET devices, and the apparent need for 3D compositional measurements, it is no surprise that there has been considerable interest in applying APT for semiconductor applications. Indeed there are now several excellent examples of analysis of MOSFET and finFET-like structures in the literature.[2] However, in addition to playing the strengths of the technique, the complex nature of these materials also highlights a number of its drawbacks, both from the perspective of semiconductor device measurements and also from the perspective of using the tool in an industrial environment.

CHALLENGES

Sensitivity and Detection Limits

The sensitivity that can be achieved by APT is a function of the volume of measured, the efficiency of the detector and the uncertainty introduced by background noise. The field-of-view of the atom probe is sufficient to contain the entirety of particular device features, so even if the detector were perfect the sensitivity is limited by low number of atom counts intrinsic to the small device dimensions. Indeed because the correct operation of a computer chip requires millions of transistors working within certain tolerances, relevant bulk composition levels may not translate to meaningful discrete atom counts when considered on a single device. Furthermore, because of the statistical nature of mass-identification, a number of counts are required to get statistically significant qualitative detection. As such it may be fundamentally impossible to achieve required detection levels, at least with only a single device measurement.

Compositional Precision and Reproducibility

Although APT, in principal, has uniform sensitivity across the periodic table, there are, in fact, a number of issues in compositional determination caused by the APT measurement processes. Examples of these that affect key MOSFET materials include mass-spectral overlaps preventing unique ion identification ($^{14}N^{+}$ and $^{28}Si^{++}$), correlated evaporation (B in Si), and preferential evaporations (SiO_2 stoichiometry). Careful tuning of the operational parameter windows, such as laser power, can improve consistency of the results, but the heterogeneous nature of MOSFET structures means that these requirements may be mutually exclusive between different sub-volumes.

Heterogeneity and Spatial Accuracy

Because the 3D atom positions are determined by back projecting from the measured detector position onto the specimen apex, in order to get correct spatial position it is necessary to know the apex shape. However, in a heterogeneous system, different materials will have different evaporation fields, resulting in a non-uniform apex shape. This causes differential magnification of certain sub-volumes, leading to distorted shapes and poor dimensional fidelity, or worse actually overlapping signals from adjacent volumes.[3]

Yield and Time-to-Knowledge

For use in failure analysis applications, where there may be only one example of a bad device, a yield of 100% is necessary. However, a MOSFET specimen may contain a number of interfaces with weak mechanical strength or poor adhesion, seams & voids, or adjacent materials with widely different evaporation fields. Any of these could contribute to premature specimen failure, preventing complete data acquisition..

In semiconductor manufacturing time is a critical factor for metrology tools that support quality control or process monitoring. Although APT may offer unique compositional information, it also needs to be delivered fast enough in order to be useful. APT has three main issues here. Sample preparation is time consuming, and although in principle uses similar processes to TEM, it does not currently benefit from the high levels of automation that has been built up around TEM. Secondly, tool measurement times can be long, particularly for larger samples at low detection rates. Thirdly data analysis and interpretation takes considerable effort compared to SIMS and TEM.

REFERENCES

1. B. Gault et al., Atom Probe Microscopy, New York: Springer (2012)
1. H. Takamizawa *et al.*, *Appl. Phys. Lett.* **99**, 133502 (2011)
3. F. Vurpillot et al., *Appl. Phys. Lett.* **76**, 3127 (2000)

KEYWORDS

Semiconductor, MOSFET, finFET, atom probe tomography, failure analysis

Opportunities and Challenges for Lab-based Hybrid Metrology for Emerging Technologies

Markus Kuhn, Ying Zhou and Kevin Johnson

Intel Corporation, 2501 NW 229th Ave. Hillsboro, Oregon USA 97124
Email: markus.kuhn@intel.com

INTRODUCTION

Advanced CMOS technology development has evolved from the use of geometric scaling to the introduction of novel materials and non-planar architectures [1-4]. These inclusions have added significant complexity to all aspects of process development and also for the fab and lab metrologies that support the development and manufacturing of those processes. Furthermore, the exploration of novel memory and beyond CMOS technology options will require the development of metrologies targeted at measuring unique device properties (i.e. magnetism, polarization, spin, etc).

Much of the focus is on advancing dimensional metrology capabilities such as TEM, SEM, CDSAXS, etc. However, similar attention must be given to other device and material properties and the impact of increased process sensitivities and variations i.e. the measurement and control of the properties of atomically thin films and interfaces. Since lab-based methods are generally more suited to provide these enhanced capabilities, there is an increasing trend to use lab methods in a hybrid/near-fab approach. This requires lab metrology tools that provide improved efficiency, accelerated time to data, and high reliability, all at a lower cost of ownership. Velocity has become a key lab operational parameter.

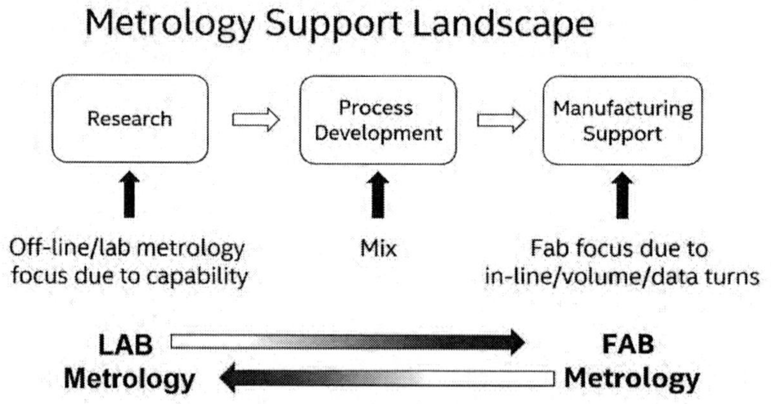

FIGURE 1. The synergies between lab and fab-based metrologies have evolved and the distinctions between them are becoming blurred [5].

HYBRID CHARACTERIZATION FOR CAPABILITY

The shift to highly complex, 3D device architectures and processing make it increasingly difficult for traditional lab techniques to provide unambiguous understanding of the material properties and interactions for these systems. Although TEM-based methods can provide more localized information, there are limits due to sensitivity, projection overlaps and statistics. Two approaches are highlighted that can close this gap.

1) Array-based methodologies. By the use of repeating array structures, large-spot techniques such as SIMS, XPS, XRD, etc. can be extended to provide statistical, local device measurements in a complementary fashion with TEM.
2) New localized capabilities. Atom probe tomography has been increasingly adopted by the semiconductor industry as a complementary local probe (with TEM) as it provides nanoscale 3D information, some of which is not accessible using TEM methods.

These alternatives along with TEM provide a potential hybrid lab framework for measuring, understanding and ultimately controlling complex, 3D devices.

Array-based Methods (Design for Metrology)

Array-structures can be utilized by many of the current large-spot techniques to extract local device properties with excellent statistics. This allows for the use of the strength of the individual techniques while not requiring the inherent spatial resolution to be scaled and the technique to be severely compromised in sensitivity, accuracy, precision, etc. An additional advantage is the ability to provide a statistically averaged view of the measured parameter without being skewed by local variations. As each technique can provide unique information, hybrid characterization is necessary to provide a fuller picture of the system of interest.

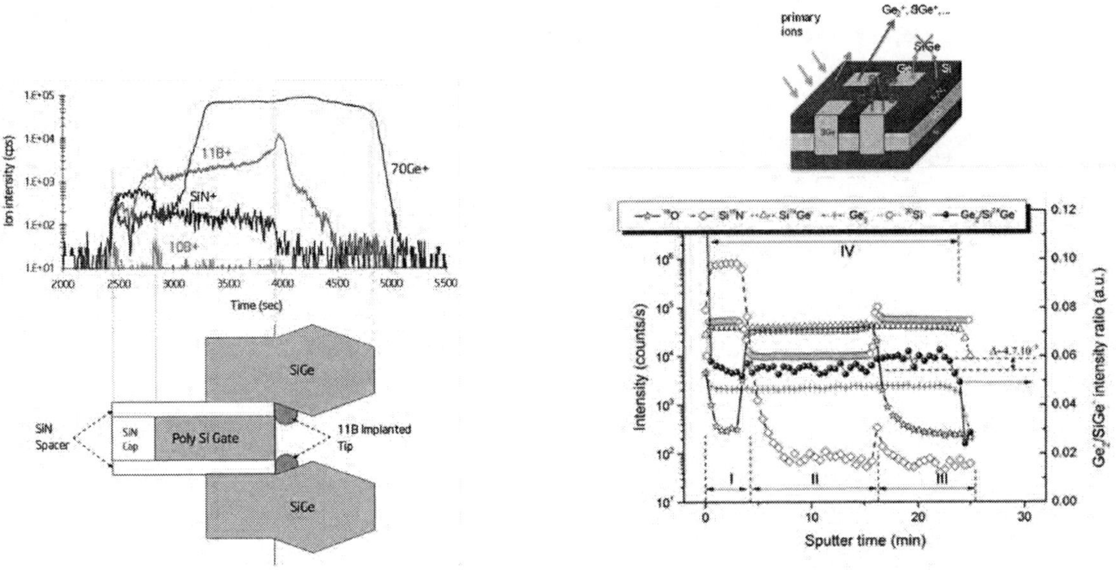

FIGURE 2. SIMS methods under development for measuring nanoscale device material parameters. 1.5D SIMS on left and Self-Focusing SIMS on the right [6].

Advancing Localized Methods

Atom Probe Tomography has emerged as a promising technique for providing unique three-dimensional compositional information of highly complex 3D nanostructures. There remain significant technical challenges in making the technique robust, distortions, trajectory overlaps, sample yield, etc., however, there has been progress in using the unique capabilities of this technique on various semiconductor-based nanostructures.

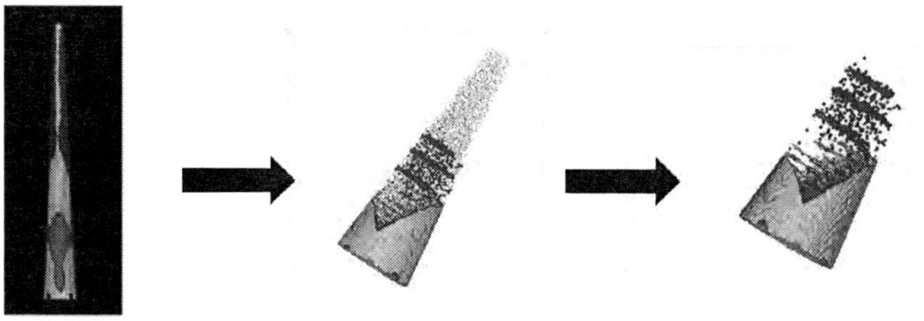

FIGURE 3. Atom Probe Tomography results highlighting localized 3D information from complex 3D nanostructures [5, 7].

HIGH VOLUME LAB METROLOGY THROUGH AUTOMATION

A key aspect to enabling lab-based or hybrid metrology is tool automation. Automation can provide the necessary time to data, efficiency, precision and low cost of ownership required to be an effective hybrid and near-fab metrology option. As an example, the increased demand for TEM dimensional analysis can be addressed through TEM automation. This output can then be coupled with in-fab metrologies such as OCD, AFM and CD-SEM to create a hybrid metrology solution.

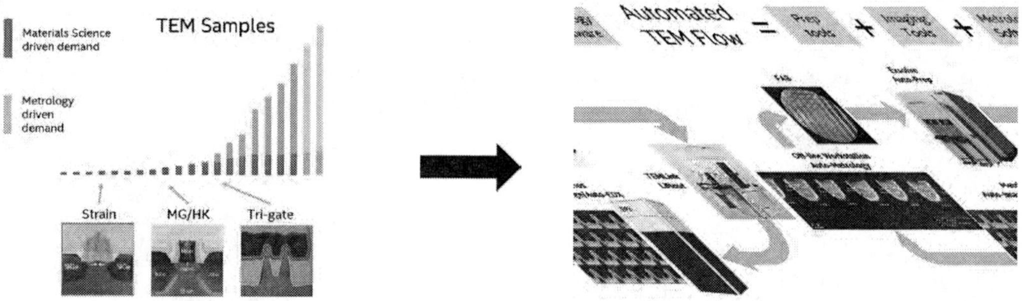

FIGURE 4. Technology driven increase in TEM sample output and the AutoTEM flow [5].

SUMMARY

The development of array-based capabilities along with improved lab automation provides a viable path for developing hybrid metrology solutions to enable the development and manufacturing of next-generation complex 3D nanostructured devices.

REFERENCES

1 T. Ghani, *Electron Devices Meeting, IEDM '03 Technical Digest. IEEE International*, 11.6.1 - 11.6.3 (2003).
2. K. Mistry, *Electron Devices Meeting, 2007. IEDM 2007. IEEE International*, 247-250 (2003).
3. C-H Jan, *Electron Devices Meeting, 2012. IEDM 2012. IEEE International*, 44-47 (2012).
4. S. Natarjan, *Electron Devices Meeting, 2008. IEDM 2008. IEEE International*, 1-3 (2008).
5. M. Kuhn and Z. Ma, *Metrology, Inspection, and Process Control for Microlithography XXX*, San Jose, 2016.
6. A. Budrevich and W. Vandervorst, *Metrology and Diagnostic Techniques for Nanoelectronics,* Pan Stanford, 2016, Ch 4.
7. Z. Ma and D. Seiler, *Metrology and Diagnostic Techniques for Nanoelectronics,* Pan Stanford, 2016, cover.

KEYWORDS

Lab-based metrology, hybrid metrology, AutoTEM, SIMS, Atom Probe Tomography

Hybrid Metrology and Machine Learning to Make a Virtual Fab from a Lab

J.Foucher, H.Gronqvist, A.Derville, A.Labrosse, J.Baderot, Y.Zimmermann, S.Martinez, G.Bernard

POLLEN Metrology, 9 rue du Rocher de Lorzier, 38430 Moirans - France

INTRODUCTION

The concept of multisensor data fusion is far from being new. Indeed, the use of a single source of information is a strong limitation for many applications. Humans and animals have evolved the capability to use multiple senses to improve their ability to survive [1]. Inspired by this ability of animal species, multisensor data fusion can also be used in modern technology. More specifically, multisensor data fusion in dimensional metrology can be defined as the process of combining data from several information sources (sensors) into a common representational format so that the metrological evaluation can benefit from all available sources of data [2]. In particular, this implies that measurement results can be determined, which could not – or only with lower accuracy – be determined solely on the basis of data from an individual source (sensor) only. Multisensor data fusion is formally defined as a "multi-level, multifaceted process handling the automatic detection, association, correlation, estimation, and combination of data and information from several sources" [3]. A generalization of this definition states that the data can be provided either by a single source or multiple sources [4, 5].

In this invited talk, we argue that the use of multisensor data fusion for metrology, called *hybrid metrology*, can significantly help in mastering nano-processes at the R&D level and their implementation in production lines.

1. STATE OF THE ART & THE HYBRID METROLOGY CHALLENGE

As can be seen from figure 1, there are non-negligible discrepancies between different measurement techniques, and we will focus on the case of dimensional metrology here. Having an accurate measure of a nano-object's dimensions is of utmost importance because it is well known that the properties of nanoparticles are size and shape dependent. Furthermore, in the semiconductor industry there is a race for the production of ever smaller electronic components. For these ever smaller components even the tiniest variations in the details of the design may impact the performance of the component. These are just a few examples of why there is a demand on the metrology community for highly accurate measurements at the nanoscale. By accuracy we also mean quantifying the variability of a measure. The lack of consensus concerning the appropriate measurands at each step of the manufacturing process often prohibits apples-to-apples comparisons, thus adding to uncertainties when inadequate or loosely defined measurands are used unwittingly.

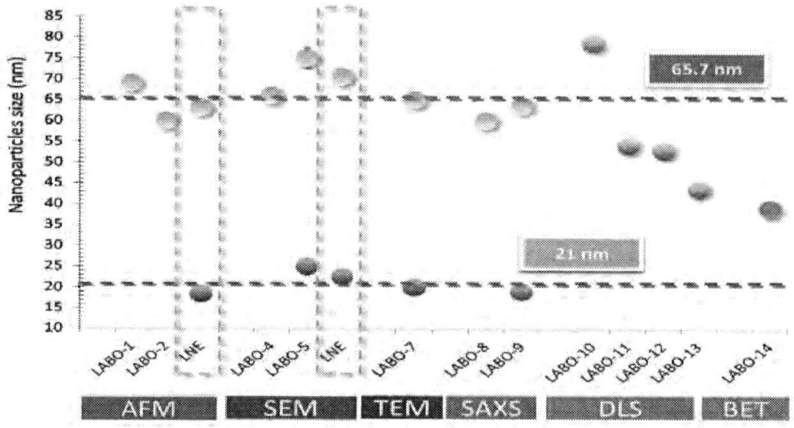

FIGURE 1. Measurements of nanoparticle size as a function of techniques used at different laboratories. Courtesy of Club Nanométrologie.

Most nanometrology analyses today are done in single-sensor mode (some partial data fusion is done, but only offline with the use of separate software for each data set). In addition, data analysis is often done manually or semi-automatically at best, with huge user influence on the results [6]. The hybrid metrology software platform that we have designed proposes automated one-click data analysis to solve problems in nanometrology. This software redefines the current state of the art to include multisensor data fusion and artificial intelligence in one platform. It thus provides a high degree of automation to R&D processes, enabling a rapid transition from lab to fab, cf. figure 2. The speed added to this transition notably implies reduced process costs. By design, the software is user friendly and simple to use - we have eliminated the prerequisite that the user have a PhD in order to master it. In an intuitive manner, the graphical user interface of the platform allows the user to define an entire manufacturing process from scratch, and analyse single or batch data at each step of this manufacturing process.

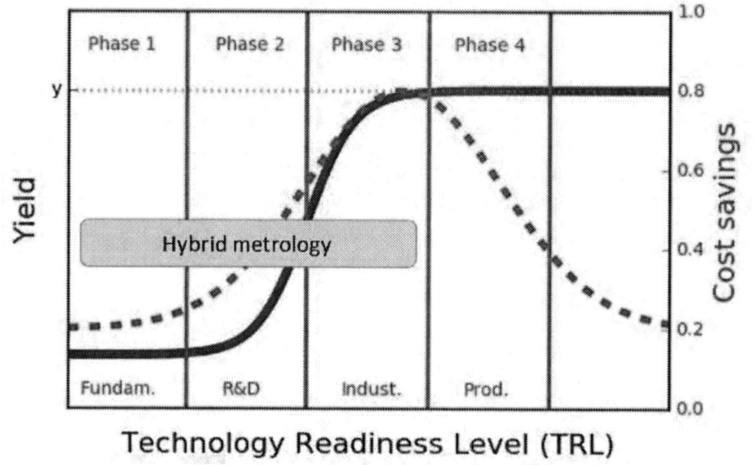

FIGURE 2. Production cycle. The positioning of hybrid metrology is in the capability to automatise data analysis in phases 1, 2 and 3. In this way, a virtual fab is made.

2. FROM LAB TO FAB

By providing the full infrastructure dedicated to the automation of data analysis and process control at the R&D level (figure 2), our software makes it possible for industry to rapidly pass from lab to fab. A unique feature of the provided services is the *transparency* of all methods and algorithms used. The algorithms developed by our expert team of physicists, mathematicians, computer scientists and software developers are characterized by their robustness and versatility when facing highly heterogeneous data from different sensors. This heterogeneity, often perceived as a difficulty in data analysis, can now justly be seen as a strength of the analysis.

We now describe with concrete examples how our solution provides speed and robustness to different problems that the community is facing every day in order to achieve its innovation goals.

2.1 Software infrastructure

First of all, we provide the infrastructure necessary for transforming a lab into a virtual fab, cf. figure 3. Our continuous integration server and agile development methods make sure that new solutions can be integrated quickly and in constant evaluation by all parties concerned. Some of the strengths of the development team include the know-how to analyze heterogeneous data, apply multisensor data fusion and machine learning to assist the concerned party in their decision making process.

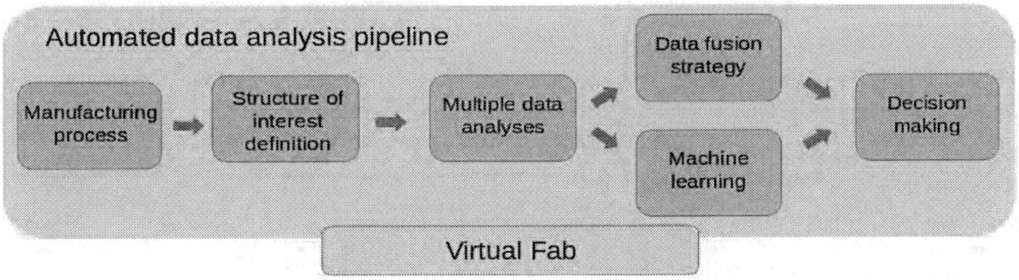

FIGURE 3. The making of a virtual fab.

2.2 Hybridization of direct and indirect techniques

Secondly, to provide a concrete example of data fusion, we propose the combination of imaging and nonimaging techniques. A typical case is the fusion of TEM with SAXS. It is well known that one cannot determine both the size and shape of particles from a scattering experiment. However, a priori information concerning shape from an imaging technique such as TEM can be fed into an algorithm analyzing SAXS data, giving rise to a size distribution, see figure 4.

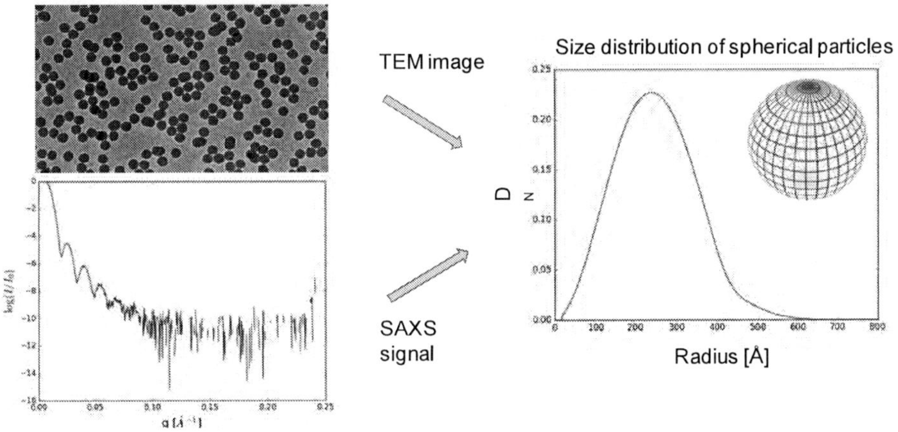

FIGURE 4. Data fusion of a direct (TEM) and an indirect (SAXS) measuring technique, giving rise to a new measurand, here the size distribution.

2.3 Machine learning for advanced node development

As a final example let us consider the VLSI, omnipresent in the semiconductor industry. By the application of pattern recognition algorithms combined with machine learning techniques we propose the automation of VLSI CD measurements and analysis of defects, see figure 5. The essence of this example is the use of machine learning as a completion step in hybrid metrology.

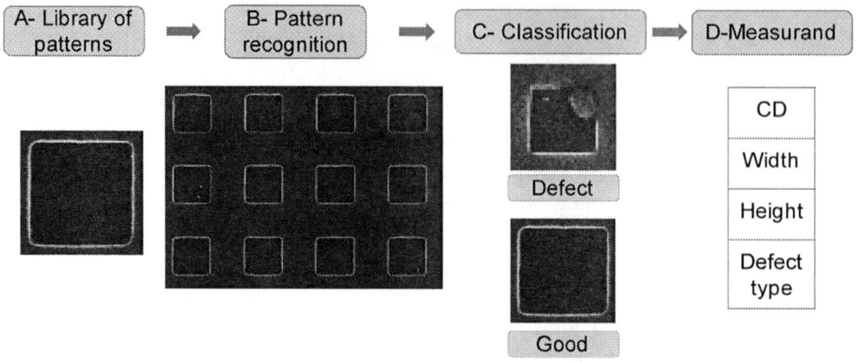

FIGURE 5. Machine learning sequence from library set-up on the left side to pattern recognition, classification and final measurements on the right side.

REFERENCES

[1] David L. Hall, An Introduction to Multisensor Data Fusion, Proceedings of the IEEE, Vol. 85, No. 1, January 1997

[2] A. Weckenmann, W. Jiang, K.-D. Sommer, U. Neuschaefer-Rube, J. Seewig, L. Shaw, T. Estler, Multisensor Data Fusion in Dimensional Metrology, CIRP Annals – Manufacturing Technology 58 (2009) 701-721

[3] F. E. White, Data Fusion Lexicon, Joint Directors of Laboratories, Technical Panel for C3, Data Fusion Sub-Panel, Naval Ocean Systems Center, San Diego 1991

[4]L. A. Klein, Sensor and Data Fusion Concepts and Applications, Second Ed., Society of Photo-Optical Instrumentation Engineers (SPIE), Bellingham, WA, 1991

[5] B. Khaleghi, A. Khamis, F. O. Karray, S. N. Razavi, Multisensor Data Fusion: A Review of the State-Of-The-Art, Information Fusion 14 (2013) 28-44

[6] D. Nečas, P. Klapetek, Study of user influence in routine SPM data processing, 2017 Meas. Sci. Technol. 28 034014

KEYWORDS

hybrid metrology, data fusion, machine learning, AFM, SEM, TEM

Fault Isolation in IoT Age - Turning Challenges into Chances

Christian Boit

Technische Universität Berlin, Einsteinufer 19 Sekr. E4, 10587 Berlin, Germany
Email: Christian.boit@tu-berlin.de

INTRODUCTION

The evolution of electronic devices for Integrated Circuits as we know it through Moore's law, is one of the fastest industrial development speeds – and still far too slow for the requirements of the age that is described as "Internet of Things" (IoT). Following the ITRS roadmap [3], we can expect a decrease of minimum feature size by a factor of 3 in the coming 10 years, bringing FinFETs to ca 2.5 to 5 nm nodes. The circuit debug and failure analysis (FA) requirements for this level of innovation can be managed with the optical interaction - based contactless fault isolation (CFI) techniques applied through chip backside when we leave near infra – red (NIR) regime. The big advantage of NIR is the relative transparency of silicon, so sample preparation is pretty convenient.

But as Fig. 1 is showing, we need to proceed to wavelengths about a factor of 2 smaller in order to achieve the resolution required for the technologies coming in the next decade. The substrate thickness is an issue then as it needs to be reduced to less than 10µm. Plus, looking at Tab. 1, we see there will not be much further margin, because the solid immersion lens that is substantial to those CFI performances, needs to have an index of refraction (n) close to the silicon number of 3.5. This works sufficiently well down to wavelengths of 550nm by using GaP material. Shorter wavelengths require more glass-like materials with a considerable loss in n. But the worst parameter of shorter wavelengths is the drastically decreasing transparency. In order to get resolution improvement

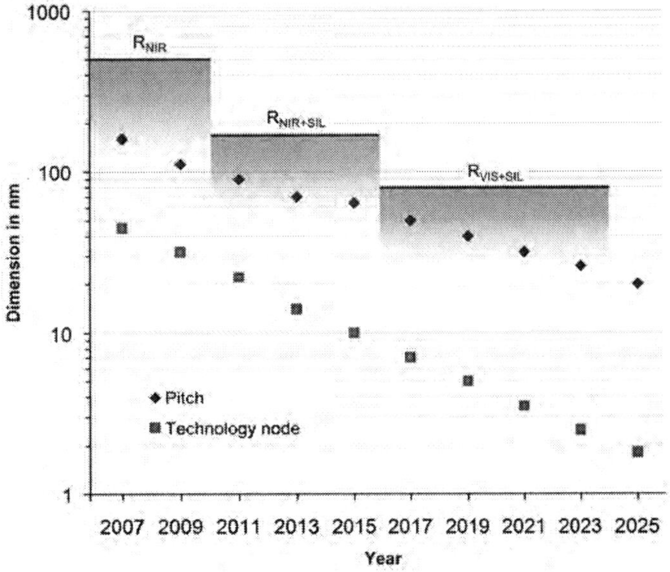

FIGURE 1. Technology generations with their FET pitch sizes according to ITRS roadmap [3]. Grey areas indicate fault isolating capability in real world IC debug with NIR and NIR+SIL based on [1] and adapted for VIS+SIL (after [6]).

Wavelength [nm]	SIL Material	n of SIL Mat.	Resolution [nm]	Absorption Depth in Si [μm]
1064	Si	3.5 [4]	152	< 300
650	GaP	3.3 [1]	98	3
550	GaP	3.4 [1]	81	1.5
440	SiC	2.7 [5]	81	0.13
330	C	2.5 [6]	66	0.012

Table 1: Wavelength of optical CFI technique, required SIL material and resolution (after [6])

by another factor of 2 over the 550nm visible light solution, device thickness can only be a few nanometers, which makes this a very unrealistic way to go.

FinFETS with such small geometries would provide the important advantage in shrinking size. But the gain of performance is by far not sufficient for the needs of the IoT age. Here, other technologies will join in. CFI concepts of the future will have to take those aspects into account as well.

CHALLENGES OF CHIP TECHNOLOGIES FOR IOT

The explosion of data collection, exchange and storage by IoT will require a dramatic decrease of the power necessary to operate these devices. The interconnection between all communicating parts will employ new concepts, materials and dimensions. IoT means very low power, radio frequency, sensors & actuators. All these aspects will have consequences for CFI that go far beyond the challenges of the past, being mostly imaging resolution and operating frequency with a given basic material of silicon.

Front End Challenges: Compound Semiconductors, 3D Active Devices

The devices ready for IoT will have to communicate with RFID functions. Operating frequencies will most likely go beyond 50GHz, maybe reaching THz regime. This can be done in silicon only with specific bipolar transistors like HBTs. But mainstream IC technology will take advantage of local epitaxial layers of compound semiconductors. This has already been established for FET strain in the past and will expand towards III-V materials for active devices. InSb is the most prominent representative for the direction of high frequency far beyond 100 GHz [2], due to high electron mobility (see Tab.2).

Property	Si	InSb
Lattice Constant [Å]	5.4	6.5
Dielectric Constant	11.7	17
Refractive Index	3.4	4.0
Eff. Electron Mass [m_o]	0.19	0.014
Eff. Hole Mass [m_o]	0.49	0.43
Band Gap [eV]	1.12	0.17
Intrinsic Carrier Conc. [cm^{-3}]	$1 \cdot 10^{10}$	$2 \cdot 10^{16}$
Electron Mobility [cm^2/Vs]	1400	77000
Hole Mobility [cm^2/Vs]	450	850
Breakdown Field [V/cm]	$3 \cdot 10^5$	10^3

Table2: Comparison of Basic and Electronic Properties of Si and InSb (After [4] and [5])

The small band gap allows for ultra low power operation, although high intrinsic carrier density is a limiting factor for off current. But with proper circuitry the stand by power consumption will be manageable.

In order to increase active device density, the next step after FinFETs seems to be Gate-All-Around Lateral and Vertical Nanowire FET Devices [7]. This technology is tightly connected to ultra thin chip to chip connections.

Back End Challenges: Ultra Thin Chip Layers, Interconnects By Photonics

Systems on Chip (SoCs) now realized with Through Silicon Vias (TSV) will become part of a regular Back End process like vias of interconnect levels as the semiconductor substrates can be processed to thicknesses smaller than 1µm. The main difference of interconnect technology will again be introduced by III-V material with direct semiconductor properties, perfectly suited as laser emitters [2]. Now, photonic signal transfer is possible inside the chip, using multi mode transmission for RF signals.

Security Challenges: Hardware Attack Risks

The drastically increased network density in IoT will make the system much more vulnerable to security attacks from software and hardware side. CFI techniques are extraordinarily successful to track signals from node to node. They can be used to attack sensitive data as well. The attack risk is an unprotected chip backside. Protection concepts are necessary [8]. Hardware Trojans are another substantial hardware risk that requires countermeasures.

VISION: TECHNOLOGIES FOR IOT - CHANCES TO CFI

All the technologies that will revolutionize chip functionality towards radio frequency inherit a lot of advantages for CFI: There is no IC tester operable at radio frequencies. Self test like BIST is the only chance to gain information about device performance. There will be a lot of programmable BIST opportunities that help fault isolation like scan paths today. Self-management, self healing, self configuration are keywords for supporting CFI in the future. Last but not least will the far infra red emission from the InSb devices provide a very helpful tool for CFI, including ultra thin SoCs. In addition, scanning probe techniques may help with interactions that are derivatives of nanostructuring. Electron beam interactions may help gaining some image resolution but huge currents are necessary to offer the required bandwidth. Only E-beam columns in nanoscale with acceleration voltages of less than 100V can reduce the electron energy injection and protect the devices from e-beam induced degradation. Complete new CFI techniques need to be explored in the near future. We better start today.

REFERENCES

1. M. Von Haartman et al., "Optical Fault Isolation and Nanoprobing Techniques for the 10 nm Technology Node and Beyond," *Proceedings of the 41st International Symposium for Testing and Failure Analysis, pp. 47-51, 2015.*

1. S. Datta et al., "85nm Gate Length Enhancement and Depletion mode InSb Quantum Well Transistorsfor Ultra High Speed and Very Low Power Digital Logic Applications", *IEEE IEDM 2005, 10-7803-9269-8/05/$20.00 (c) 2005 IEEE*

3. "ITRS - International Technology Roadmap for Semiconductors." [Online]. Available: http://www.itrs.net/reports.html.

4. http://www.ioffe.ru/SVA/NSM/Semicond/Si/

5. http://www.ioffe.ru/SVA/NSM/Semicond/InSb/

6. H. Lohrke et al.,"Contactless Fault Isolation for FinFET Technologies with Visible Light and GaP SIL", *Proc. ISTFA 2016, Fort Worth (TX), Nov. 6-10, ASM 2016*

7. Imec demonstrates Junction-less Gate-All-Around Lateral and Vertical Nanowire FET Devices, http://www2.imec.be/be_en/press/imec-news/vlsi-junctionless-device-iiiv.html

8. C. Helfmeier et al., "Cloning Physically Unclonable Functions", *Proc. 2013 IEEE International Symposium on Hardware-Oriented Security and Trust (HOST), 2013*

9. C. Boit et al., "From IC Debug to Hardware Security Risk: The Power of Backside Access and optical Interaction", Proc. IEEE IRPS 2016, July 18-20 Cloning Physically Unclonable Functions, *Proc. 2013 23rd IEEE International Symposium on the Physical an Failure Analysis of Integrated Circuits (IPFA), Marina Bay Sands, Singapore, July 18-21, 2016, ISBN 978-1-4673-8258-8, 2016*

KEYWORDS

Contactless Fault Isolation, Visible LVP, Technologies for IoT, Photonics in Semiconductors, Programmable BIST

Automated TEM Workflows for Process Control

Ozan Ugurlu, Hayley Johanesen, Anne Kenslea

Thermo Fisher Scientific, Materials & Structural Analysis (formerly FEI)
5350 NE Dawson Creek Drive Hillsboro, OR 97124

ABSTRACT

As the semiconductor industry moves from planar to 3D devices such as 3D NAND and FinFET, chip manufacturers are encountering multiple challenges. Process control for those 3D devices is one of the biggest challenges for both memory and logic manufacturers. Both metrology and wafer inspection are part of process control. Metrology is the science of measuring and analyzing small dimensions, while wafer inspection tests wafers for any kind of defects that affects the performance of the devices. Due to multi-patterning and EUV requirements, the 3D device manufacturing processes needs to be controlled within tight limits, which requires tighter metrology capabilities with fast turnaround times.

CD-SEM is one of the main in-line metrology solutions. However, during the last 4-5 years the demand for supplementing CD-SEM data with Transmission Electron Microscope (TEM) has exploded for both metrology and defects due to the small dimensions and complex structures used on the latest 3D technology nodes. Especially with 3D devices, the need for inspecting the cross section and not just the top surface as in CD-SEM has become one of the top priorities. Optical Critical Dimension (OCD) is another in-line metrology technique that is commonly used in the fabrication of devices. OCD is fairly fast and non-destructive, however it requires development of complex and time-consuming models. These models require significant amounts of TEM data to be accurate. This need highlights another area in the fab where fast, reliable and large quantities of TEM data are needed to reduce the total measurement uncertainty (TMU) [1]. Traditionally, TEM metrology and defect analysis are considered to be expensive, manual and time-consuming lab techniques with limited capacity. In order to perform TEM analysis, metrology or defect, one has to first prepare the sample (lamella) from the wafer via focused ion beam (FIB). Both sample preparation and TEM analysis used to be manual and labor intensive jobs that were owned by the lab. However, recent developments in FIB sample preparation and TEM analysis have made it possible to automate the entire workflow, making it more suitable for in-Fab and near-Fab solutions with fast turnaround times.

Automated FIB sample preparation and TEM analysis produce faster enhanced results from wafer to data while reducing human error and speeding up the process. In a recent study it has been demonstrated that automated TEM analysis provides 30% better metrology precision compared to manual image acquisition [2]. Since automation eliminates the need for manual operation and operators, the automated TEM workflows can be placed in the fab or near-fab and directly support the fab process control without taking the wafers out of the fab and following time-consuming lab procedures. The automation and high-throughput also make it possible to collect more data to improve the statistical analysis and reduce TMU.

In this study automated workflows for both metrology and defects are discussed, where the entire process from wafer to TEM data is done by full automation. Process variations in EUV patterning processes are studied, where different photo resists are used with both constant and varying parameters such as EUV exposure and focus, Table 1. Each wafer had a different photo resist and within the wafer the exposure dose in a row of dies increases from left to right with a fixed dose difference ΔE (mJ). The focus had been kept constant throughout both wafers.

TABLE 1. List of wafers and EUV process parameters

Wafer	Photo Resist	Exposure	Focus
Wafer1	EUV Resist 1	Increases from left to right	Uniform
Wafer2	EUV Resist 2	Increases from left to right	Uniform

As part of the automated metrology process control, 21 lamellae have been prepared from each wafer in a way to cover the exposure change from left to right, and also the uniform exposure and focus from top to bottom, Figure 1. All the information about the lamellae and downstream recipes for automated TEM imaging and metrology, have been assigned at the wafer level before lamellae prep via a factory host. Lamellae are prepared fully automatically (Figure 2a), plucked and transferred to TEM where they were automatically imaged. TEM images with multiple devices are measured automatically on an offline workstation. Fin width at three different locations, including fin height both from Si and SiOC interfaces as well as the pitch between devices have been measured via automated metrology on all of the TEM images. Dynamic precision of the automated workflow is evaluated using pooled variance and compared to process variation.

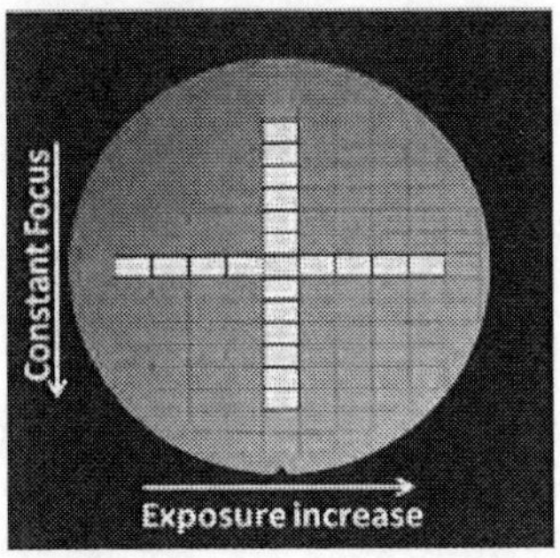

FIGURE 1. Wafer map for lamella prep used for both Wafer1 and Wafer2

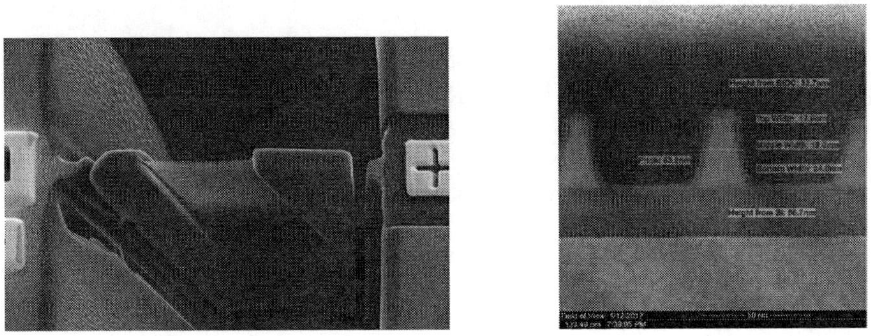

FIGURE 2. (a)Lamella after automated FIB sample prep on the wafer, (b) Automated Metrology on TEM images.

Automated workflow for defect analysis is also discussed in this study, where lamellae at various defect locations on the wafer are prepared automatically and sent to TEM for semi-automated defect redetection and analysis.

REFERENCES

1. H. Johanesen et al., "CD metrology for EUV lithography and etch" Advanced Semiconductor Manufacturing Conference (ASMC), 2015 26th Annual SEMI
2. O. Ugurlu et al., "High-volume process monitoring of FEOL 22nm FinFET structures using an automated STEM" SPIE 8681, Metrology, Inspection, and Process Control for Microlithography XXVII, 868107 (18 April 2013)

KEYWORDS

Process Control, Automated, Metrology, TEM, FAB, OCD.

Examination of Advanced Technologies in Characterization, Diagnostics, and Verification at Different Stages in the Manufacturing Lifecycle of Packaged IC Devices

Colin Ritchie[1], Scott West[1], Stuart Neches[1], Eiji Kato[2] and Masaichi Hashimoto[2]

1 Advantest America, Inc., 3061 Zanker Rd., San Jose, CA 95134 USA
2 Advantest Corporation, 2-48 Matsubara, Kamiayashi, Aoba-ku, Sendai 989-3124 Japan
stuart.neches@advantest.com

INTRODUCTION

This talk explores 3 applications of technology in the evaluation of IC packaged devices and systems at different stages in the manufacturing process: in the areas of Characterization, Diagnostics, and Verification/Reliability testing. Nondestructive mold thickness metrology using terahertz (THz) wave is examined for characterization of the moldcap and related manufacturing process in real time. Ultra-high resolution pulsed TDR/TDT is explored for diagnostic forensic analysis of packaged devices including complex 2.5D and 3D packages and interconnects. At the system level, verification of reliability and endurance is studied with respect to thermal consistency using Flash Memory-based SSD (Solid State Drive) SLT (System Level Test).

NONDESTRUCTIVE MOLD THICKNESS METROLOGY FOR CHARACTERIZATION OF MOBLE DEVICES

The worldwide embrace of smartphones, advanced consumer electronics and intelligent devices for Internet of Things (IoT) is driving demand for smaller, more highly integrated semiconductor devices. Semiconductor packages for this market must be compact, thin, and tough enough to stand up to mobile or ubiquitous usage. Accordingly, optimization and balancing of package durability and thinness is increasingly important. However, existing metrology methods are not suitable for volume production lines because they are slow, laborious and destructive and manufacturers have had to test package thickness by inspecting samples late in the production process. This makes it difficult to assess the quality of the entire production run and problematic to trace the root cause of any defects in the final product.

Advantest's TS9000 Mold Thickness Analysis (MTA) system (Fig. 1a) solves these issues by enabling non-destructive, rapid, repeatable, and highly accurate (± 3 μm) measurements of mold thickness, without the limitations of existing measurement methods. Furthermore, it can be deployed at various points in the assembly and packaging process, even immediately after the curing process, enabling earlier detection of production issues and improving product quality and yield. These innovative inspection capabilities are expected to contribute significantly to product quality improvement amid the trends towards smaller package sizes and its higher integration. As shown in Fig. 1b, in the mold thickness measurement with this system, pulsed THz waves with sub-picosecond duration are send to device under test (DUT) by THz emitter and specularly reflected waves are acquired by THz detector [1]. Since THz pulses are partially reflected at the package top surface, and partially at the mold-die interface due to Fresnel reflection, the reflected waveform from the DUT would have echo pulses like Fig.1c. The time difference between the echoes (Δt) and refractive index of the mold give mold thickness by calculation.

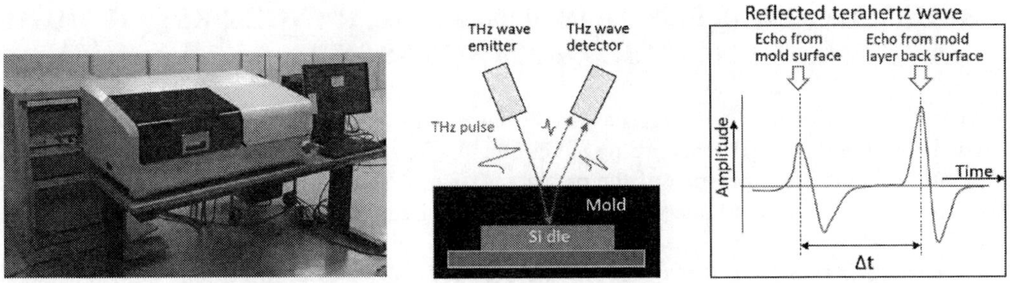

(a) (b) (c)

FIGURE 1. (a) TS9000 measurement unit (left side on the table), rack mounted controller electronics and optics (right), and controller PC (right side on the table). (b) Measurement configuration of overlaid mold thickness with pulsed THz radiation. (c) Example of reflected time-domain waveform from the DUT.

HIGH RESOLUTION TIME DOMAIN REFLECTOMETRY TOOL FOR DIAGNOSTIC FAILURE INVESTIGATION IN STATE-OF-THE-ART PACKAGES

Electronic device circuit fault analysis like open, short and high impedance is commonly performed by oscilloscope time domain reflectometry (TDR). However, as devices grow smaller and more highly integrated, the ability to locate fault with extreme spatial precision has become increasingly more important. The typical fault locations in cutting-edge semiconductor packages are shown in Fig. 2a. But existing TDR instruments have limited resolution, as the rise time of the TDR signal is facing the difficulty in shortening it much further, and this would be the risk that conventional TDR technologies will be inadequate to handle the requirements of fault isolation (FI) for highly integrated devices immediate future. Advantest's terahertz technology addresses these concerns and meets the need for high-resolution and long-haul measurement by utilizing ultrashort and low-jitter pulse generation and sampling technology.

Advantest's TS9000 TDR system relies on Advantest's market-proven TDR measurement technology to pinpoint and map circuit defects. The system is composed of an electro-optic sampling (EOS) system, TDR transceiver and a probe (Fig. 2b). The transceiver generates and detects ultrashort pulse signal with the help of femtosecond laser sources, of which the laser pulse width is less than <50 fs. Furthermore, an excellent jitter performance less than 30 fs is achieved [2]. The solution delivers circuit failure analysis with an extremely high spatial resolution of less than 5 μm, and a maximum measurement range of 300 mm, including for internal circuitry used in through-silicon via (TSVs) and interposers.

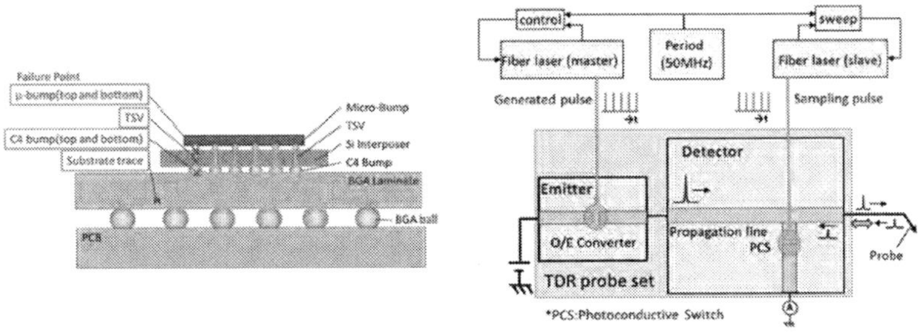

(a) (b)

FIGURE 2. (a) Advanced semiconductor package structure and typical fault locations. (b) Schematic diagram of the developed TDR transceiver and TDR signal generation/sampling scheme.

THERMAL CONSISTENCY DURING ENDURANCE TEST OF HIGH PERFORMANCE SSDS

The demand for high performance storage has driven the emergence and rapid growth of solid state drives with volumes to exceed 300M units by 2019, according to Gartner. Of key concern to SSD end-users in both the enterprise and consumer markets is the rated lifespan of the product. To ensure this lifespan, endurance testing must be done with long test times and elevated temperature to ensure statistical probably that the drives will meet lifespan specifications.

As SSD performance moves from that of single late 6G SATA devices to multi-lane 8G NVMe and 12G SAS, power consumption of drives increases to 25W or more per device. This creates a challenge during endurance testing to maintain thermal consistency across high power drives tested in parallel to achieve statistical confidence, while maintaining a low cost of test.

The MPT3000ENV test platform addresses this issue across a variety of SSD form factors and protocols, by maintaining a +/-5C thermal consistency for up to 256 25W devices, using a dual chamber design. Reliability demonstration testing (RDT) is performed, testing a sample of devices over time to prove that they will last over the desired lifetime. Typically, devices are run for 1,000 hours at high temperatures, simulating 3 months of constant read/write operations. Challenges in achieving the needed thermal consistency are explored.

REFERENCES

1. Wataru Momose et al., *Results in Pharm. Sci.* **2**, 29-37 (2012).
2. M. Hashimoto et al., 2016 IEEE 23rd Int. Symp. Phys. Failure Anal. Integ. Circ. (IPFA), Singapore, 238-240 (2016).

KEYWORDS

Molding, non-destructive, thickness, terahertz wave, time domain reflectometry, endurance test, thermal consistency

Overview of Spintronics and Nanoscale Magnetics

Kang L. Wang

ECE, MSE and Physics
University of California, Los Angeles, California 90095, USA

INTRODUCTION

Spintronics has emerged as an important area for research and applications. In addition to single spins, magnetism as a result of collective spin effect has been exploited for many applications, e.g., magnetic recording heads, magnetic sensors, and more recently nonvolatile magnetic random access memory (MRAM). In spintronics, spin Hall effect and spin orbit toque (SOT) have recently been advanced. For nanoscale magnetics, there is a surge of interest for MRAM as energy dissipation of scaled CMOS has become one of the most critical challenges in the continual of Moore's Law [1]. The nonvolatility of magnetics and spintronics devices can retain information even when power is removed, thus potentially eliminating the static power consumption. Recent advances in spintronics and nanoscaled magnetics have shown the potential of further increasing performance of memory with much reduced write energy dissipation. In addition, the magnetic/spintronic building block, i.e. magnetic tunnel junctions (MTJs), can be implemented in the back-end of line (BEOL) process, which is compatible with standard CMOS technology. Hence, the lowered static and dynamic power consumption offered by magnetics/spintronics presents great opportunities to realize hybrid CMOS-magnetic logic and memory systems, as well as novel computing architectures beyond CMOS. [2]

PHYSICS OF NANOSCALE MAGNETICS/SPINTRONICS

Today, spintronics technology, specifically spin transfer torque (STT) technology has been applied in limited production. Spin-transfer torque (STT) memory uses the electron current which carries the electron spin to exert a torque to switch the free magnetic layer of a magnetic tunnel junction. Alternatively, using electric field to control the switching yields a higher energy efficiency, i.e. voltage-controlled magnetic anisotropy (VCMA). In the past few decades, multiple other physical phenomena have also been shown to enable the use of electric field to control magnetism. This includes gate voltage control of dilute magnetic semiconductors (DMSs), and magneto-striction by using the strain (Fig. 1). Most of these effects are the result of relativistic spin orbit coupling (SOC). Energy efficiency memory using the effect of VCMA, referred to as magnetoelectric RAM or MeRAM, is shown to have orders of magnitude lower energy dissipation and higher density compared with spin transfer torque memory (STT-RAM). Riding on the success of STT-RAM, MeRAM which uses similar technology as STT-RAM can impact spintronics memory and logic applications for resolving this energy dissipation challenge. The technology readiness of MeRAM for implementation and manufacturing will be discussed.

In addition to ferromagnetic materials and devices, the emergent research is directed to exploit new effects for potential applications in the areas of topological insulators, skyrmions, anti-ferromagnetism (AFM), and their spin textures, etc. We will also discuss spin-orbit torque (SOT), which uses spin Hall effect to switch MTJ by passing current through a high SOC material, e.g., heavy metal or topological insulator. In particular, the interface of heterostructures has become a rich area for research, particularly in exploring the control mechanisms of magnetism, spin manipulation and transport. The understanding as of how the interfaces affect device performances will require various advanced metrology methods. Topological insulator will be used as an example to illustrate the interface effect, e.g., SOT enhancement as the result of Dirac fermions. Likewise, another real space topological object, skyrmions induced by the Dzyaloshinskii-Moriya interaction due to interfacial SOC will be discussed. This is a rich area and the demand of new characterization tools become more stringent as the structures are scaled down.

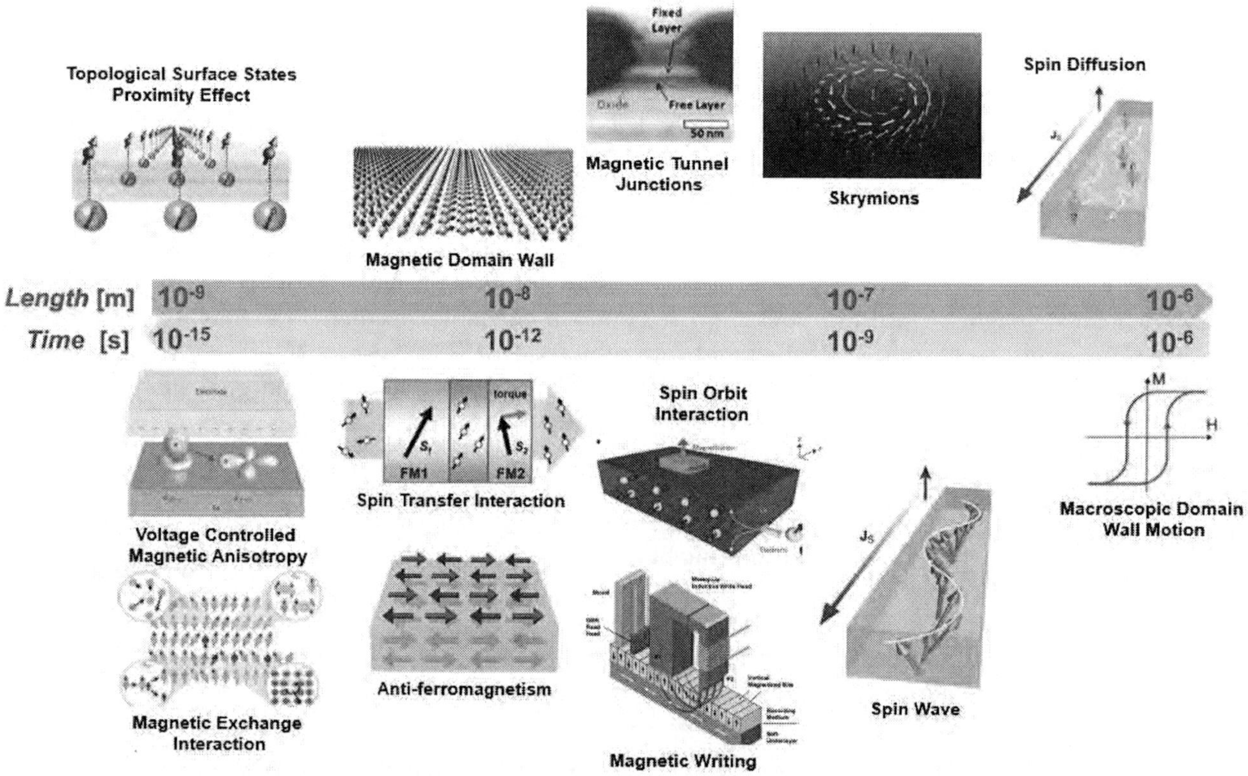

FIGURE 1. Various fundamental spin interaction mechanisms and spintronic devices in time and length scales. Revised from [3].

METROLOGY METHODS AND THEIR NEEDS

In general, the metrics for magnetic devices are energy of operation, speed of operation, and size of each element, among others. The interface SOC is responsible to perpendicular magnetic anisotropy (PMA) and VCMA. The dynamics of nanoscale spin textures is also important for full understanding of devices under external electrical stimuli. Fundamentally we need to access the atomic level details of spin textures such as domain walls and skyrmions. Likewise, AFM has been shown to have a potential for very high speed and high frequency applications but the details of the interface need further understanding. Different physics and phenomena have different time and length scales, hence metrology methods with different time and spatial resolutions are required for understanding physics and phenomena in different materials systems. Clearly materials properties such as chemical composition, chemical state and bonding, crystallinity, and strain are all relevant to various mechanisms in the control of magnetism and spintronics. All of these will require advanced instrumentation. Among them are high resolution instruments such as Lorentz transmission electron microscopy (TEM), spin-polarized scanning tunneling spectroscopy (STM), photoemission electron microscopy (PEEM), scanning near-field magneto-optic microscopy (SNOM), diamond magnetometry, magnetic force microscopy, scanning SQUID, etc. However, temporal resolutions are limited but both high spatial resolution and high temporal resolution are needed to understand the interfacial interaction of AFM. For practical applications, there is lack of rapid methods to characterize VCMA, PMA and DMI at the interface. The availability of these methods and tools will help provide fast feedback for material and structural development.

In conclusion, as the field of magnetics/spintronics expands its applications, further understanding of the physics of materials and structures at extreme spatial and temporal resolutions will be critical for continuing innovations. The characterization and metrology instruments have proved to be invaluable for the advancement of magnetics/spintronics research. New physics may also emerge with improved instrumentation.

References

1. N. S. Kim et al., “Leakage current: Moore’s law meets static power,” *Computer*, vol. 36, pp. 68–74, Dec. 2003.
2. K. L. Wang, X. Kou, P. Upadhyaya, Y. Fan, Q. Shao, G. Yu, and P. K. Amiri, *Proceedings of the IEEE* **104** (10), 1974 (2016).
3. “Ultrafast Magnetization Dynamics on the Nanoscale.” [Online]. Available: https://www.psi.ch/swissfel/MagnetismSideBar2/ScienceCaseMagnetism.pdf

KEYWORDS

Magnetics, spintronics, metrology, spatial resolution, temporal resolution

A short biography:

Kang L. Wang is currently a Distinguished Professor and holds Raytheon Chair Professor in Physical Science and Electronics in the University of California, Los Angeles (UCLA). He is affiliated with ECE, MSE and Physics. He received his BS degree from National Cheng Kung University (Taiwan) and his MS and PhD degrees from the Massachusetts Institute of Technology. He is a Fellow of the IEEE, and a member of the American Physical Society. He was a Guggenheim Fellow. He also served as Editor-in-Chief of IEEE TNANO, editor of Artech House, consulting editor for Spins, and Associate editor for Science Advances and other publications. His research areas include nanoscale physics and materials; molecular beam epitaxy; topological insulators; spintronics and devices, nonvolatile electronics and low dissipation devices; neurodynamics and neurotronics.

*The work was in part supported by ERFC-SHINES, FAME, C-Spin, CEGN and NSF

Spectroscopic Method To Measure Spin-Orbit Torque At Ferromagnet/Normal Metal Interfaces Intended For Magnetic Memory

Andrew J. Berger[1], Eric R. J. Edwards[1], Hans T. Nembach[1], Justin M. Shaw[1], Alexy D. Karenowska[2], Mathias Weiler[3,4], and Thomas J. Silva[1]

[1]*Quantum Electromagnetics Division, NIST, Boulder, CO 80305, USA*
[2]*Department of Physics, University of Oxford, Oxford, UK*
[3]*Walther-Meissner-Institut, Bavarian Academy of Sciences, Garching, Germany*
[4]*Physics Department, Technical University of Munich, Garching, Germany*

INTRODUCTION

Harnessing spin-charge conversion through current-driven spin torques [1, 2] and precession-driven charge currents [3-5] are promising for the development of scalable and efficient spintronic devices. This conversion occurs across ferromagnet/normal metal (FM/NM) interfaces with strong spin-orbit coupling (SOC), but details of the underlying physics are still much debated. These spin-orbitronic effects are short-range, making them highly advantageous for microelectronic applications that require device scaling to high densities such as nonvolatile memory and alternative state-variable logic [6]. SOC also underlies the interfacial Dzyaloshinskii-Moriya interaction (DMI) [7, 8]. It was recently proposed that a Rashba Hamiltonian operative at a FM/NM interface gives rise to both spin-orbit torques (SOT) and DMI, such that the presence of one effect implies the other [9]. Despite the complexity of interfacial spin interactions [10], this theory provides a simple, testable quantitative relation between the DMI and SOT. Here, we use a powerful new microwave spectroscopy method to quantify spin-charge conversion in FM/NM bilayers and demonstrate that the magnitude of the SOT is in good agreement with the theoretical prediction based on the previously measured value of DMI in identical bilayers [11].

Background

Damping-like torques due to the spin Hall effect (SHE) in NM layers such as Pt and β-Ta are well-studied and understood [6, 12, 13]. Field-like torques have also been measured for FM/NM interfaces in the forward configuration [2, 14, 15]. The Rashba-Edelstein effect (REE) [16, 17] has been invoked to explain such field-like torques, but the fundamental origin of the REE in metallic multilayers remains uncertain. For example, unlike Bi, Pt does not have an intrinsic Rashba splitting [18]. Furthermore, an inverse measurement of the field-like torque in $Ni_{80}Fe_{20}$/Pt has not yet been reported. Kim, Lee, Lee, and Stiles (KLLS) [9] have shed light on the issue by showing that SOT and the Dzyaloshinskii-Moriya interaction (DMI) at a FM/NM interface are both manifestations of an underlying Rashba Hamiltonian, and predict a straightforward relationship between the Rashba parameter α_R, interfacial DMI strength D_{DMI}^{int}, and the interfacial field-like SOT per DMI spin t_{fl}:

$$\alpha_R = \frac{\hbar^2}{2m_e}\left(\frac{D_{DMI}^{int}}{2A}\right) = \frac{\hbar^2}{2m_e}\left(\frac{t_{fl}}{v_s}\right) \tag{1}$$

where $\hbar$ is Planck's constant divided by 2π, m_e is the electron mass, A is the exchange stiffness, and v_s is the conduction electron spin velocity.

In this talk, we present simultaneous measurements of inverse field-like and damping-like torques in $Ni_{80}Fe_{20}$/Pt bilayers and find that Eq. (1) provides an accurate prediction for the magnitude and sign of the field-like SOT in a

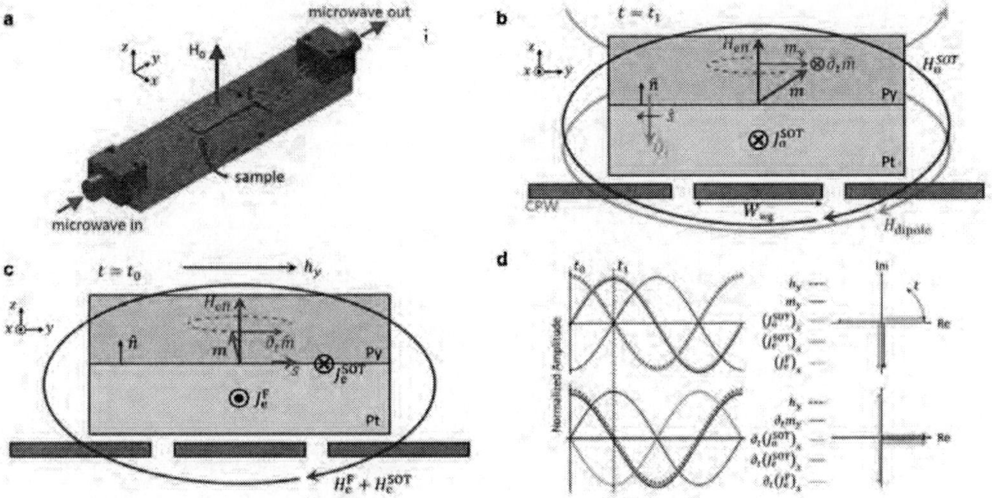

FIGURE 1. (a) Flip-chip sample on CPW, showing out-of-plane field H_0 and sample length l. Microwave driving field points primarily along $\hat{y}$ at the sample. (b) Schematic of the bilayer, with coordinate system, precessing magnetization $\boldsymbol{m}(t)$, odd symmetry SOT current $J_{\mathrm{o}}^{\mathrm{SOT}}$ (e.g., due to spin Hall effect), and the dynamic fields produced by each (H_{dipole} and $H_{\mathrm{o}}^{\mathrm{SOT}}$) at time t_1 when $\hat{m} = \langle 0, m_y, m_z \rangle$. Interface normal $\hat{n}$, spin orientation $\hat{s}$, and spin flow direction $\hat{Q}_{\hat{s}}$ due to spin pumping into the Pt are also shown. Bilayer is insulated from CPW using photoresist spacer layer (not shown). (c) Same as (b), except at time t_0 when $\hat{m} = \langle m_x, 0, m_z \rangle$. In this case, there is no spin Hall contribution to J_x, $J_{\mathrm{e}}^{\mathrm{F}}$, and $J_{\mathrm{e}}^{\mathrm{SOT}}$ (e.g., due to Rashba-Edelstein effect) are maximal along $\pm\hat{x}$, and the spin accumulation (with orientation $\hat{s}$) and $J_{\mathrm{e}}^{\mathrm{SOT}}$ are produced at the Py/Pt interface. (d) Amplitude of driving field h_y and different signal sources as a function of time (left), and viewed in the complex plane at time t_0 (right). Relative amplitudes not indicated. Note that currents and Py magnetization (top panel) are 90° shifted from the voltages they produce in the CPW ($\propto d\Phi/dt$, bottom panel).

sample system where the ratio $D_{\mathrm{DMI}}^{\mathrm{int}}/A$ has been measured over a wide range of FM thickness [11]. Our quantification of the inverse spin orbit torque (iSOT) relies on phase-sensitive microwave transmission spectroscopy and subsequent analysis of the complex inductance of the sample, which arises from both the FM magnetization dynamics and the superposition of all AC charge currents flowing in the sample. The determination of the phase of all AC charge currents generated in the FM/NM bilayers, combined with Onsager reciprocity for the specific phenomenology of these measurements, is a powerful advantage of this technique that allows us to accurately identify the processes that contribute to spin-charge conversion.

Method

We measured the ferromagnetic resonance (FMR) in metallic stacks consisting of substrate/Ta(1.5)/$Ni_{80}Fe_{20}$(3.5)/NM/Ta(3) and inverted stacks of substrate/Ta(1.5)/NM/$Ni_{80}Fe_{20}$(3.5)/Ta(3) (where the numbers in parentheses indicate thickness in nanometers). We focus on a Pt(6) NM layer due to its large intrinsic SOC, and use Cu(3.3) as a control material with nominally negligible COC [3, 19, 20]. The broadband, phase-sensitive FMR measurements utilize a coplanar waveguide (CPW) as both the excitation and detection transducer (see Fig. 1(a)). Any source of AC magnetic flux generated by the bilayer is inductively detected in the CPW. Therefore, the inductive load that the sample contributes to the CPW circuit consists of four terms: (1) The real-valued L_0 due to the oscillating magnetic dipolar fields produced by the resonating $Ni_{80}Fe_{20}$(Permalloy, Py) magnetization, (2) the Faraday-effect currents induced in the NM layer by the precessing Py magnetization, (3) currents produced by damping-like iSOT effects (e.g., spin pumping + iSHE), and (4) currents produced by field-like iSOT effects (e.g., REE). Currents in the NM that generate Oersted fields, which inductively couple to the CPW, produce the latter three inductances, which we collectively define as complex-valued L_{NM}. Since these currents are in response to a driving force (here, the time-derivative of the precessing magnetization, $\partial_t \hat{m}$), we quantify them with the effective conductivities $\sigma_{\mathrm{e}}^{\mathrm{F}}$, $\sigma_{\mathrm{o}}^{\mathrm{SOT}}$, and $\sigma_{\mathrm{e}}^{\mathrm{SOT}}$. Here, the superscripts indicate the source of the current as due to the Faraday effect or SOT, and the subscripts indicate "even" or "odd" with respect to time-

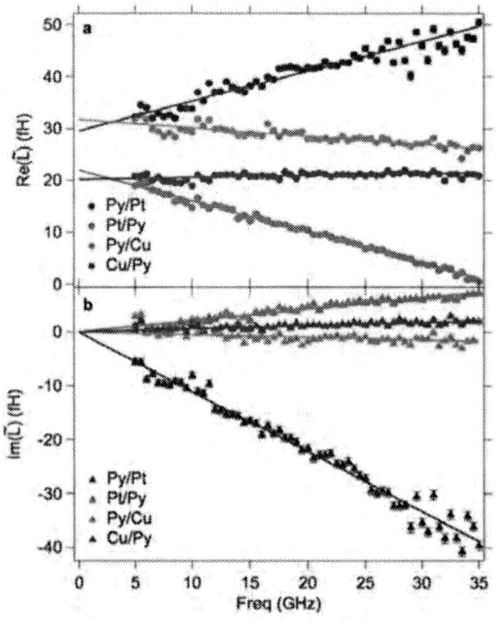

FIGURE 2. Frequency dependence of real and imaginary inductances extracted from S21 spectra (symbols) and fits to Eqs. 2 and 3 (lines). (a) $\mathrm{Re}(\tilde{L})$ for all samples with $l = 8$ mm. Zero-frequency intercept indicates the dipolar inductive coupling, while the linear slope reflects $\sigma_{\mathrm{o}}^{\mathrm{SOT}}$. (b) $\mathrm{Im}(\tilde{L})$ for all samples, as a function of frequency, where the linear slope includes contributions from both $\sigma_{\mathrm{e}}^{\mathrm{SOT}}$ and $\sigma_{\mathrm{e}}^{\mathrm{F}}$.

reversal, which determines the phase of the correspond ing inductance with respect to the driving microwave field. Importantly, while L_0 is independent of frequency, L_{NM} is linear in frequency, as the currents in the NM are driven by $\partial_t \hat{m}$. Hence, frequency-dependent measurements allow us to disentangle L_0 and L_{NM}.

Figure 1(b) and (c) show schematics of these four signal sources at two instants: when the dipolar/odd SOT effects are maximal (Fig. 1(b)) and when the Faraday/even SOT effects are maximal (Fig. 1(c)). Fig. 1(d) shows the time dependence of the signal-generating component of each source and its time derivative, the latter being proportional to induced voltage in the CPW. These plots show each SOT source has a distinct phase relationship to the driving field h_y, which we exploit below to determine their contributions separately.

We collected FMR spectra from 5 GHz to 35 GHz by microwave transmission S_{21} measurements with a vector network analyzer (VNA) with output power 0 dBm. Exemplary $\mathrm{Re}(\Delta S_{21})$ spectra are shown in Fig. 2. Phase and amplitude analysis of S_{21} vs. H_0 provide a measure of the complex inductance of the sample. Each raw spectrum has been normalized by the complex signal background, effectively de-embedding the sample contribution to the inductance from any (frequency-dependent) losses or phase shifts caused by the microwave cables and CPW.

Both Py/Cu and Cu/Py samples exhibit a mostly real inductance (symmetric Lorentzian dip for $\mathrm{Re}(\Delta S_{21})$ in Fig. 2(a) and (b)) with a magnitude largely independent of frequency, in accordance with $L_{\mathrm{NM}} \cong 0$. In contrast, the lineshape and magnitude of the Py/Pt and Pt/Py data in Fig. 2(c) and (d) exhibit a clear frequency dependence as expected for $L_{\mathrm{NM}} \neq 0$. In particular, the data for Py/Pt indicate that $\mathrm{Re}(L_{\mathrm{NM}}) > 0$, such that $\mathrm{Re}(L)$ increases with increasing f. Pt/Py evolves according to $\mathrm{Re}(L_{\mathrm{NM}}) < 0$ due to the stack inversion, leading to a decrease and eventual compensation of $\mathrm{Re}(L)$ at high f. The increasingly antisymmetric lineshape for both Py/Pt and Pt/Py reveals that the magnitude of $\mathrm{Im}(L)$ also increases with frequency, with a sign given by the stacking order.

Results

By normalizing the spectra in Fig. 2 to the magnetic susceptibility $\chi(\omega, H_0)$, we extract the complex inductance amplitude $\tilde{L} = L/\chi_{yy}(\omega, H_0)$. $\mathrm{Re}(\tilde{L})$ and $\mathrm{Im}(\tilde{L})$ are shown in Fig. 3 for all investigated bilayers with a length l of 8 mm. The behavior of these curves is described by:

$$\mathrm{Re}\left(\tilde{L}\right)=\frac{\mu_0 l}{4}\left[\frac{d_{\mathrm{FM}}}{W_{\mathrm{wg}}}\eta^2\left(z,W_{\mathrm{wg}}\right)+\mathrm{sgn}\left(\hat{z}\cdot\hat{n}\right)\eta\left(z,W_{\mathrm{wg}}\right)\frac{L_{12}\left(z,W_{\mathrm{wg}},l\right)}{\mu_0 \ell}\frac{\hbar\omega}{e}\sigma_{\mathrm{o}}^{\mathrm{SOT}}\right] \tag{2}$$

$$\mathrm{Im}\left(\tilde{L}\right)=\frac{\mu_0 l}{4}\left[\mathrm{sgn}\left(\hat{z}\cdot\hat{n}\right)\eta\left(z,W_{\mathrm{wg}}\right)\frac{L_{12}\left(z,W_{\mathrm{wg}},l\right)}{\mu_0 \ell}\frac{\hbar\omega}{e}\left(\sigma_{\mathrm{e}}^{\mathrm{SOT}}-\sigma_{\mathrm{e}}^{\mathrm{F}}\right)\right] \tag{3}$$

Briefly, $\mathrm{Re}\left(\tilde{L}\right)$ provides information about the dipolar inductance ($\tilde{L}_0$, zero-frequency intercept), and $\sigma_{\mathrm{o}}^{\mathrm{SOT}}$ (slope $\propto f$). Similarly, the slope of $\mathrm{Im}\left(\tilde{L}\right)$ reflects $\left(\sigma_{\mathrm{e}}^{\mathrm{SOT}}-\sigma_{\mathrm{e}}^{\mathrm{F}}\right)$. Immediately evident is the reversal of the slope for Py/Pt compared to Pt/Py, which is captured by the sgn function (where $\hat{n}$ is the FM/NM interface normal, pointing into the FM, and $\hat{z}$ is defined by the coordinate system in Fig. 1). This sign-reversal is consistent with the phenomenology expected for interface-symmetry sensitive effects, e.g., combined spin pumping and SHE, as well as REE. There is also a marked difference in the slope magnitude for Py/Pt and Pt/Py in panel (b), the implications of which are discussed below. After accounting for sample-to-sample variation in L_0 and measuring samples with a range of l to accurately normalize our results for sample length, we obtain the average values of $\sigma_{\mathrm{o}}^{\mathrm{SOT}}$ and $\left(\sigma_{\mathrm{e}}^{\mathrm{SOT}}-\sigma_{\mathrm{e}}^{\mathrm{F}}\right)$ shown in Table 1. The ratio of $\sigma_{\mathrm{o}}^{\mathrm{SOT}}/\sigma_{\mathrm{e}}^{\mathrm{SOT}}$ measured here is in quantitative agreement with the observation in Ref. [5]—using a complimentary technique and similar samples—that the dominant contribution to AC spin-charge currents in the Py/Pt samples opposes the Faraday effect currents.

TABLE 1. Effective conductivities and microscopic spin-charge conversion parameters (spin Hall angle θ_{S} and Rashba parameter α_{R}). All conductivities are expressed in units of $10^5\ \Omega^{-1}\ \mathrm{m}^{-1}$.

Sample	$\sigma_{\mathrm{o}}^{\mathrm{SOT}}$	$\sigma_{\mathrm{e}}^{\mathrm{SOT}}-\sigma_{\mathrm{e}}^{\mathrm{F}}$	θ_{SH}	α_{R} **(meV nm)**
Py/Pt	0.9 +/- 0.03	-1.9 +/- 0.2	0.12 +/- 0.04	-17 +/- 4
Pt/Py	1.38 +/- 0.04	-0.5 +/- 0.2	0.15 +/- 0.05	-4+/- 3
Py/Cu	-0.285 +/- 0.007	0.1 +/- 0.1		
Cu/Py	-0.07 +/- 0.03	0.1 +/- 0.2		

REFERENCES

1. L. Liu, T. Moriyama, *et al.*, *Phys. Rev. Lett.* **106**, 036601 (2011).
2. I. Mihai Miron, G. Gaudin, *et al.*, *Nat. Mater.* **9**, 230-234 (2010).
3. E. Saitoh, M. Ueda, *et al.*, *Appl. Phys. Lett.* **88**, 182509 (2006).
4. M.V. Costache, M. Sladkov, *et al.*, *Phys. Rev. Lett.* **97**, 216603 (2006).
5. M. Weiler, J.M. Shaw, *et al.*, *Phys. Rev. Lett.* **113**, 157204 (2014).
6. L. Liu, C.-F. Pai, *et al.*, *Science* **336**, 555 (2012).
7. I. Dzyaloshinsky, *J. Phys. Chem. Solids* **4**, (1958).
8. T. Moriya, *Phys. Rev.* **120**, 91-98 (1960).
9. K.-W. Kim, H.-W. Lee, *et al.*, *Phys. Rev. Lett.* **111**, 216601 (2013).
10. V.P. Amin and M.D. Stiles, *Phys. Rev. B* **94**, 104420 (2016).
11. H.T. Nembach, J.M. Shaw, *et al.*, *Nat. Phys.* **11**, 825-829 (2015).
12. F.D. Czeschka, L. Dreher, *et al.*, *Phys. Rev. Lett.* **107**, 046601 (2011).
13. M. Weiler, J.M. Shaw, *et al.*, *IEEE Mag. Lett.* **PP**, 3700104 (2014).
14. C.O. Avci, K. Garello, *et al.*, *Phys. Rev. B* **89**, 214419 (2014).
15. K. Garello, I.M. Miron, *et al.*, *Nat. Nano.* **8**, 587-593 (2013).
16. Y.A. Bychkov and E.I. Rashba, *JETP Lett.* **39**, 78-81 (1984).
17. V.M. Edelstein, *Solid State Commun.* **73**, 233-235 (1990).
18. S. Souma, A. Takayama, *et al.*, *Rev. Sci. Instrum.* **81**, 095101 (2010).
19. Y. Niimi, M. Morota, *et al.*, *Phys. Rev. Lett.* **106**, 126601 (2011).
20. J. Sinova, S.O. Valenzuela, *et al.*, *Rev. Mod. Phys.* **87**, 1213-1260 (2015).

KEYWORDS

Spintronics, magnetic memory, spin-orbit coupling, spin-orbit torque, Rashba parameter, MRAM, ferromagnetic resonance, FMR, spin Hall effect, Rashba-Edelstein effect, inductance, Dzyaloshinskii-Moriya interaction, DMI, coplanar waveguide.

Novel Magnetic Nanoprobes: Imaging of Magnetism, Current Flow, and MRI Signals with Nanometer Resolution

Christian Degen

Department of Physics, ETH Zurich, Otto Stern Weg 1, 8093 Zurich, Switzerland

INTRODUCTION

The ability to measure and analyze weak magnetic fields with high spatial resolution is of key importance to nanoscale materials science. For example, the imaging of stray fields from small ferromagnetic structures, such as magnetic domains, particles, and junctions is immensely helpful for the development of next-generation data storage media and spintronic devices. The imaging of stray fields further offers a means to study novel emerging materials and phases, including skyrmions, ferroelectrics, complex oxides, and topological insulators. In addition, nanoscale magnetic measurements can be used to map the two-dimensional current density in nanoscale conductors, with applications to semiconductor physics, integrated circuits, and thin photoactive films in energy research. Unfortunately, despite considerable effort, measurements of weak magnetic fields with nanometer spatial resolution are very challenging and are insufficiently supported by presently available technology [1].

Our group addresses this challenge by conceiving and implementing new scanning probe microscopies that can sensitively detect magnetic fields with high spatial resolution. The focus of this talk will be a technique known as "Nanoscale scanning diamond magnetometry" (NSDM). The NSDM is a new generation quantum metrology tool, which provides non-invasive, vectorial, quantitative measurements of electric currents and magnetic fields with 10 nm spatial resolution. The quantum nature of the diamond sensors additionally affords exquisite field sensitivity and a variety of schemes to measure fluctuating fields from the few Hz limit up to the GHz range. Its compatibility with different sample environments (ambient conditions, vacuum, liquids, biomaterials, cryogenic temperatures) make it a very versatile instrument that can serve a broad and diverse community in several areas of materials science, physics, chemistry, and biology.

PRESENTATION

This talk will give an introduction to nanoscale magnetic field measurements and their importance for modern science and technology. The basic principle of NSDM sensors will be discussed, and insight into the latest quantum sensing technology will be given. The talk will conclude with a gallery of emerging applications in nanoscale magnetic imaging, ranging from the characterization of magnetic materials, to the mapping of currents in patterned nanowire devices, to MRI tomography with nanometer spatial resolution.

A. Basic principle of Nanoscale Scanning Diamond Magnetometry

The basic principle of the Nanoscale Scanning Diamond Magnetometer is illustrated in Figure 1. A sharp diamond tip containing a single nitrogen-vacancy defect (NV center) is positioned over the sample under study. The local magnetic stray field B shifts the magnetic energy levels of the NV center by $2\gamma B$, where $\gamma = 2.8$ MHz/Gauss is the transduction factor (gyromagnetic ratio). The optical transitions are used to readout and re-pump the spin $|0\rangle$ state. The energy level shift is detected via a reduction of the optical fluorescence signal when microwaves are applied. Because NV centers are stable to within a few nm from the diamond surface, they can be brought very close to sources of magnetic field and permit very high imaging resolution. More information about the NSDM is found in Reviews [2, 3].

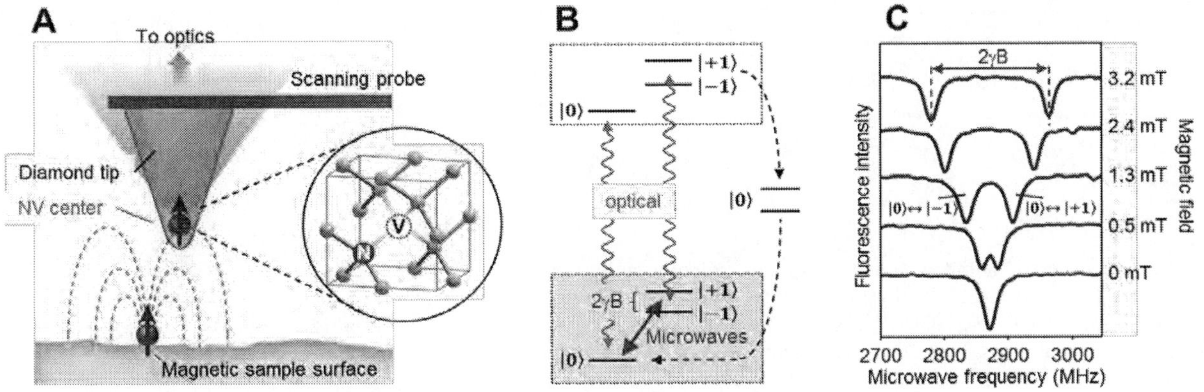

FIGURE 1. Basic principle of diamond-based magnetic sensing. (A) Scanning probe with an NV center at its apex. (B) Electronic and magnetic level structure of the NV center. A microwave transition (blue) is exploited for sensing. (C) Typical fluorescence response of an NV center when subjected to microwaves. The frequency of resonance peaks directly give the value of the local magnetic field.

B. Examples of Nanoscale Magnetic Imaging

The NSDM is a new tool that has been invented in 2008 [4], and that has seen a rapid development over the last years motivated by the many possible applications in physics, materials science and biology. Examples of recent research include the imaging of magnetic domains in disk drive media, the imaging of magnetic vortices, the study of domain wall motion, the study of spin waves, applications to helimagnetism in skyrmion materials, investigation of superconducting vortices, imaging of current density in micro- and nanostructures, and investigation of neuron activity and magnetotactic bacteria in biology [2, 3]. Some highlights of recent NSDM work in our and other laboratories are shown in Fig. 2.

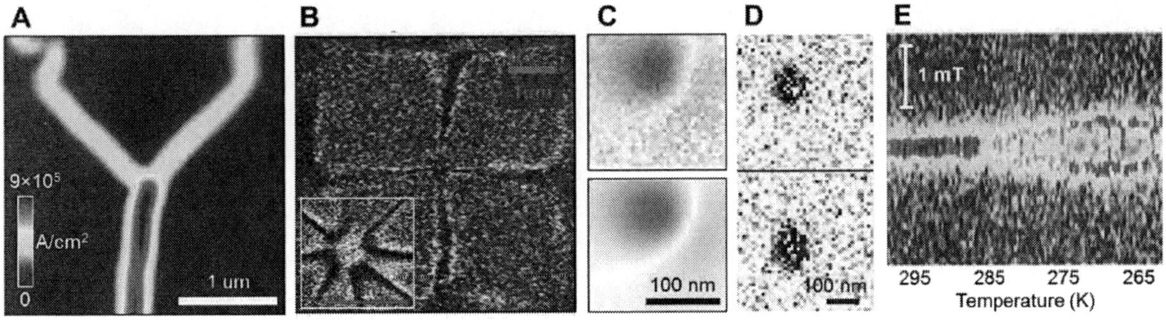

FIGURE 2. Examples of imaging nanomagnetism by NSDM: (A) Image of current density in a 50x50 nm^2 Pt nanowire [5], (B) Image of magnetic vortex in a 50 nm thin film of $Fe_{20}Ni_{80}$ [6]. (C) Image of a single FeCo nanoparticle (experiment and simulation) [7]. (D) Image of magnetic noise from single superparamagnetic magnetite nanoparticles [9]. (E) Local field at a FeGe surface above and below the helimagnetic phase transition [9].

REFERENCES

1. C. L. Degen, Nature Nanotech. 3, 643 (2008), and references therein.
2. R. Schirhagl et al., Annu. Rev. Phys. Chem. 65, 83 (2014).
3. L. Rondin et al., Rep. Prog. Phys. 77, 056503 (2014).
4. C. L. Degen, Appl. Phys. Lett. 92, 243111 (2008).
5. K. Chang, A. Eichler, and C. L. Degen, arXiv:1609.09644 (2016).
6. L. Rondin et al., Nat. Comms. 4, 2279 (2013).
7. K. Chang, unpublished.
8. D. Schmid-Lorch et al., Nano Lett. 15, 4942-4947 (2015).
9. A. Dussaux et al., Nature Communications 7, 12430 (2016).

KEYWORDS

Nanoscale imaging, Scanning probe microscopy, Magnetism, Quantum sensing, NV centers in diamond

Challenges in Nanotopography measurements at die level

C.Beitia

Univ. Grenoble Alpes, F-38000 Grenoble, France - CEA, LETI, MINATEC Campus, F-38054 Grenoble cedex 9, France

INTRODUCTION

In the last years there have been a lot of discussions about the end or Moore law and the role of the 3D integration approaches under this context. The actual trend show that both approaches will coexist and they will stand for few years waiting the appearance of some disruptive way to make computation with new processors or innovative ways to reduce cost of actual processes. In the meantime, from the process and from the metrology side, we still have some challenges to overcome to push even more the limits of the actual technological nodes. One of this challenge is in-die nanotopography measurements. Indeed because of miniaturization (More Moore, 3DIC) and complexity (More than More 3D Heterogeneous integration) process like wafer bonding and Chemical Mechanical Polishing are also getting crucial. Consequently in order to insure these process development and their appropriate yield, in-die nanotopography measurement is getting critical. The aim of the paper is to get an overview of challenges for in-die Nanotopography measurement with a focus on the control of CMP applications. Data acquisition, data fusion, data analysis challenges will be discussed. Another important topic, the metrology traceability and uncertainty evaluation will be discussed. Finally the challenge issue from the amount of data generated will be discussed.

DATA ACQUISITION CHALLENGE

When facing in-die measurements for Nanotopography characterization on wafers, the data acquisition over several orders of magnitudes scale is a challenge. Die data need to be acquire all over the wafer. Figure 1a shows the nanotopography domain in terms of amplitude (10^2 nm to nm range) and spatial wavelength (cm to μm). Figure 1b displays the height resolution and height range together with lateral resolution and dynamic range on x-y for different kind of instruments. Only those instruments that have some options for in-die and in-line metrology are presented. AFM has seen the appearance of fast acquisition modes and 300 mm wafer handling but still not fast and mature enough. In addition, fundamental problems about tip wear and probe-sample interaction need to be solved. Mechanical Stylus even if it is used in a routine way today (on scribe line) and it can cover higher spatial range it is too slow and its works in contact mode making it not acceptable as in-die measurements. The nanotopography measurement need to be fast enough; it needs to have the proper resolution to give enough in-die details and if intended to production in-line control it will need to be nondestructive at all. The selection of the proper instrument will be a compromise between those aspects. Figure 2b shows that optical measurements can be a solution. In this domain optical profilometry is the preferred choice, it can be fast, it acquires data a different scales with height varying sensitivity from sub-nanometer to hundreds of nanometer and as it is a non-contact method, it is in principle non-destructive. For this reason in the last years we have seen the emerging use of optical profiler for this kind of measurements [1, 2, 3]. Most of the optical profiler solutions available in the market are relying in some microscopy flavor: confocal; interferometry (Phase Shifting (PSI), Coherence Scanning (CSI), Shearing) even Digital Holography (DHM). The other approach is based on interferometry setups: Fizeau; Shearing or even 3D Deflectometry. The way these different technologies are implemented have an impact in the amount of information gathered, the data treatment and the limits of the associated metrology.

Confocal microscopy (CM) has seen the appearance in the semiconductor business in the last years. Hardware solutions have been developed to improve throughput and resolution to make it applicable in production line. For example, chromatic aberration is used to spread the different wavelength alone the optical axis of the microscope together with a spectrometer in front of the detector to avoid making Z scan [4].

Phase shifting microscopy (PSI) is well known and it is used in the semiconductor industry for MEMs and surface control [5]. In PSI a near monochromatic light source is used to create at interference image of the surface under study. This technique can achieve sub nanometer z resolution, limited only by system noise, however x-y resolution are limited by the Rayleigh diffraction limit yielding to a 200-300 nm lateral resolution limits.

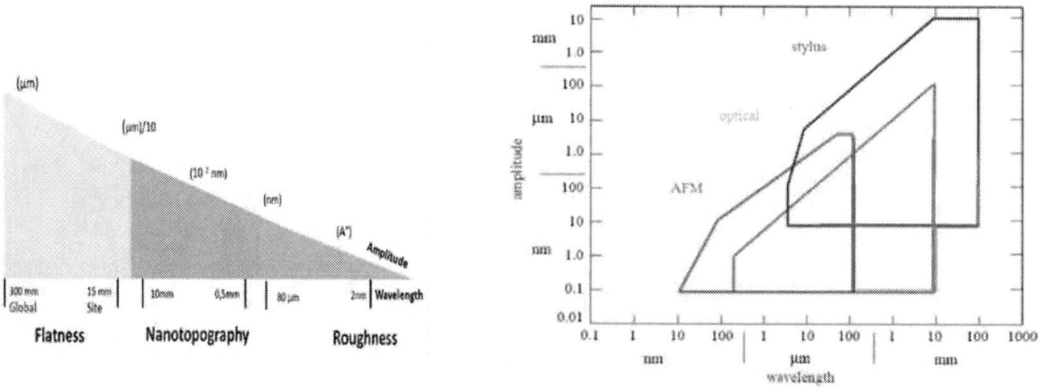

FIGURE 1. a) Nanotopography domain in spatial wavelength and amplitude; b) wavelength and amplitude of different surface topography instruments.

Coherence scanning interferometry (CSI) is another alternative way to assess nanotopography measurement [6]. In CSI visible light is used because of its incoherence is used to modulate the interference signal. Under this condition the height measurement is found by locating the peak of the modulation envelope of the interference signal, while scanning in the z direction. This mode has not the limitation of the wavelength ruler graduation but on the other side has less resolution in z (nanometer level). For the lateral resolution it presents the same limitation than CSI and all microscopies.

Digital Holography Microscopy (DHM) is also getting interest because of the flexibility and the potential gain on t-put since just few images should be acquired. However this technology is very dependent on the image reconstruction algorithms robustness and need to be explored as function of material sensibility e.g. dielectric layers [7]. For all the former microscopy based technologies a common weakness is the limited FOV of the microscope needing stitching solutions to improve the FOV size. This yields to the problems of the data fusion process and the methodology definition to have the maximum throughput and uncertainty propagation [8].

Fizeau [1] and shearing interferometer technology [9] has the advantage that the optical setup can be configured with a 300 mm optical field of view allowing full wafer measurements. In this case, the t-put problem is fully addressed and today there exists solutions that are available in the market. However the x-y resolution is reduced and today 100 µm lateral resolution is the best we can obtain.

The first challenge to all optical technology mentioned above is coming from the nature of the surface that need to be measured. Indeed they are deterministic structured surfaces by process and technology design having complex stack of material distribute in a non-homogenous way at nanometer scales on and underneath the top surface. Under this condition light arriving to the surface undergoes some phase change that will be reflecting as bias on the measured nanotopography. The usual way to avoid this problem is to deposit a very thin metal (few nm thick) conformal layer but even if this can be a solution for process development it is a real problem for in-line control. In this first challenge some work has been realized by signal modeling and combined metrology approaches e.g. PSI microscopy coupled with CSI or ellipsometry and CSI [10-11]. Although these approaches seems to work well for simple surfaces their performances on complex surface need to be tested and validated. Alternatives more generic solutions still need to be found.

From data acquisition point the best tool to acquire the whole nanotopography domain, will be a combination of two technologies. One acquiring information at Die level (obtained with/without stitching) at relative high t-put (Die/few minutes) with high resolution i.e. 200 nm lateral resolution or better and z resolution with sub nanometer sensitivity. The other will be a tool with low lateral x-y resolution but with a large field of view capable of making full wafer with high throughput (>30 Wafer/hours), with ~100 µm lateral resolution and z resolution at the nanometer level. Finally this combined metrology will need to deal with data fusion at the different data scales. Figure 2 shows a wafer level image (100 mm), a die level image at low resolution and the same die at high resolution and zoom of in-die structure [8]. This clearly show that the combined data has all information needed for process.

DATA ANALYSIS AND STORAGE

Once the nanotopography data has been acquired, the dataset size will be large. For instance in the high resolution size, a 1 cm^2 die nanotopography data at 3 μm resolution can need a hundred of individual images to stitch and produce in PSI/VSI about 64 millions of data points. Taking a minimal sampling of 3-5 dies per wafer that represents already a GB/wafer data volume. On the other hand in the low resolution interferometric data provided by Fizeau interferometer on a full wafer will require as well, about 1 GB/wafer. Thus two challenges are important: The analysis of a large set of data to get the pertinent information for process people and the storage of data and results of the analysis.

Concerning the analysis itself and parameter selection, this will depend in the application but also in the standardization level. As data is 3D in nature then 3D areal parameter are necessary. Thus lately several ways to monitor and characterize nanotopography has been proposed work still need to be done to avoid redundancy of correlation between the parameters [12]. It is also important to standardize them in order also to allow to build software gauge to check algorithms calculations accuracy and variations [13]. In all the cases the main objective is to have the pertinent information at the proper scale taking advantage of the massive statistic information provided for the 3D areal nanotopography measurements. Another important aspect will be the consistency of results at the different data scale in the frame of a combined data acquisition approach, overlap between high and low resolution will need to be consistent and transparent. This is evident in figure 2 comparison between low resolution image of the die level from a Fizeau interferometer and a high resolution PSI microscope on the same die. Comparison between them is not easy a clear methodology for comparison of this data need to be stablish.

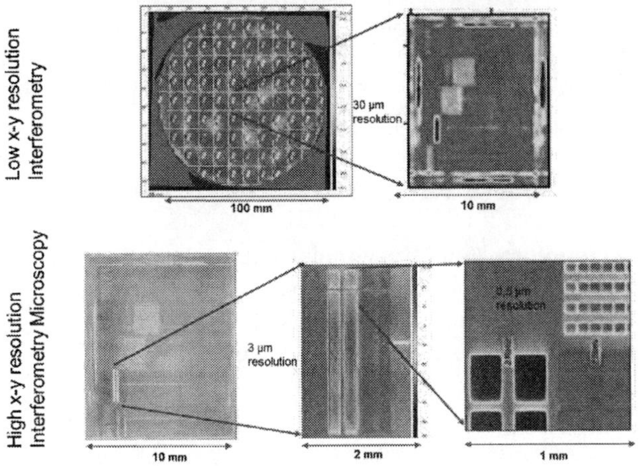

FIGURE 2. a) Low resolution Fizeau interferometer 100 mm scale data; b) Die level data extraction cm scale data c) high resolution PSI microscope Die level cm scale data d) in-die level μm scale data

Once the analysis has been done in the 3D surface images, the challenge of data storage is not trivial. It will be important to storage the raw data for potential post-treatment or just data traceability. The question about which data need to be kept and how it will be storage is key. In any case, the volume of the data collected at a GB/wafer rate is a concern and solution need to be provided in term of memory side optimization but as well, speed of information retrieval when needed. In order to keep the maximum of information the maximum amount of data should be saved, for this challenge approaches like compressing sensing at data saving might be an alternative [14]. It is worth to mention that this approach might help as well at data acquisition time.

TRACEABILITY AND UNCERTAINTY

Last but not the least there is the challenge of measurement traceability and uncertainty. In nanotopography measurements that means traceability and uncertainty of x-y and z axis for data acquisition but as well the traceability and uncertainty through the parameters calculations algorithms in the used software and final results.

Lateral axis can be more easily trace it because there are standards to trace the measurement chain to the meter definition. However the z or "height" scale (amplification coefficient) can be improved respect to usual methods. In semi the standard practice is to calibrate the z scale by measuring a traceable step height standard from National Metrology Institutes (e.g. NIST) or a certified lab (e;g. VLSI). For example nanometer level step heights standards from VLSI are available and are of two types: Quartz block Standards (~25 mm^2) and for the Semi industry 200 and 300 mm silicon wafer based standards, named "Autoload standards". However, the specified expanded uncertainty is about 0.6% and other contributors need to be added in the final results (operator variation, software). At the end the total uncertainty is in the percent range which is frustrating giving the high accuracy expected from interferometers since we are using light wavelength to measure. Another practical issue is that those standard are costly and require regular recalibration process so alternative methods will be welcome. Recently, Groot et al [15] has proposed an additional/complementary methodology using laser emission wavelength. Their work open the field for a better calibration approach to improve measurements traceability and its maintenance.

Concerning uncertainty the whole instrumental metrological characteristic need to be stablish before to try to quantify the uncertainty propagation in the final results. Today, work have been done establishing and standardizing methodologies to assess the instrument metrological characteristics (ISO25178) and standards references have been make available by NMI for performing this assessment. However surface are complex and general framework to quantify uncertainty of the final parameter including bandwith calibration contribution and software-algorithm need to be implemented.

SUMMARY

Nanotopography in-die measurements are getting crucial in the semi industry because of miniaturization but as well because they are important on alternative technological approaches where bonding and CMP are crucial like 3D heterogeneous and monolithic integration. In spite of the fact that some solution exists today, a lot of work still need to be done to insure a traceable measurement with the higher accuracy and lower uncertainty this is the main challenge in nanotopography. A general framework methodology to quantify and propagate the uncertainty from metrological instrumental characteristic up to the final calculated parameter need to be stablish. The abstract give an overview of the state of the art and the area where work need to be accomplish from Data acquisition up to uncertainty evaluation in order to deliver a complete solution to the industry.

ACKNOWLEDGEMENT

The author acknowledge the metrology, cmp, lithography and process integration teams for interesting discussions on this area from metrology point of view as well as the perspective of need and process requirement.

Some of the work that has yield to fruitful data and discussions yielding to subjects on the abstract has been issue from work in the **Polis European program**, Authors acknowledge its support.

REFERENCES

1. T.A. Brunner et al. *Optical Microlithography XXIX*, Proc. of SPIE, Vol. **9780** (2016).
2. F.Dettoni et al. *Microelectronic Engineering*, Elsevier, **113**, 105–108 (2014).
3 J. Qiu. et al. *ECS Journal of Solid State Science and Technology*, 155-161 (2015).
4 W. Lyda et al. *Meas. Sci. Technol.* **23** 054009 (2012).
5 K. Creath, Appl. Opt. **26** 2810– 6 (1987).
6 P. De Groot and L.Deck, *J. Mod. Opt.* **42** 389–401 (1995).
7 K Myung. et al, *SPIE reviews*, SPIE, vol **1** 018005-1-50. (2010).
8 Wang J. et al, *Surf. Topogr.: Metrol. Prop.* IOP Publishing, 023001 (2015).
9 J. Mileham et al. *2016 27th Annual SEMI Advanced Semiconductor Manufacturing Conference (ASMC)*, Saratoga Springs, NY, pp. 258-262 (2016).
10 X. Colonna de Lega and P. J. de Groot, *Optical Micro- and Nanometrology in Microsystems Technology*, Proc. of SPIE Vol. **6995**, 69950P, (2008)
11 A. Dubois, *Applied Optics*, Vol. **43**, No. 7 1503-1507 (2004)
12 Qi Q. et al, *Procedia CIRP*, Elsevier, **27**, 149 - 154 (2015)
13 P.M. Harris et al, *Meas. Sci. Technol.* Iop Publisher, **23** 105008 (11pp) (2012)
14 J. Ma, IEEE Transactions on instrumentations and measurements (2009.)
15 Groot P. et al, *Modeling Aspects in Optical Metrology V*, Proc. SPIE 9526952610 (2015)

KEYWORDS

Nanotopography, Traceability, Uncertainty, Interferometry

Critical Dimension Small Angle X-ray Scattering for the Semiconductor Industry

R. Joseph Kline and Daniel F. Sunday

National Institute of Standards and Technology
Gaithersburg, MD, 20899 USA

INTRODUCTION

The semiconductor industry continues to scale devices to ever smaller dimensions. It is likely that the 10 nm node will come into full production this year. As devices shrink, dimensional metrology becomes at the same time both more important and more difficult. Process control of the fabrication of these nanoscale devices is critical to obtaining and maintaining high production yields. Manufacturers need instruments to non-destructively measure the size and shape of their nanostructures to determine if the process is working properly. Manufacturing currently uses a combination of optical critical dimension scatterometry (OCD) and critical dimension scanning electron microscopy (CD-SEM). OCD uses broadband light with wavelengths much larger than the feature sizes being measured. In addition to the fundamental diffraction limits, devices are getting so small that their optical properties (n and k) are no longer constant and are becoming a function of the feature size. These challenges are making it ever more difficult to get sufficient information on the complex three dimensional (3D) nanostructures being fabricated for next generation devices. CD-SEM is great for measuring line widths and pitches, but is challenged by 3D structures. Physics-based models and tilted beams can provide some 3D information, but complex 3D nanostructures are very challenging. Critical dimension small angle X-ray scattering (CDSAXS) is similar to OCD,[1] but uses short wavelength X-rays instead of visible light to overcome the diffraction limit. At X-ray wavelengths, the optical parameters are determined by the specific atoms present and are independent of feature size. We will discuss laboratory scale CDSAXS and give an update on its development as a next generation in-line dimensional metrology tool for the semiconductor industry.[2-4] We will also discuss applications of CDSAXS to next generation lithography including directed self assembly (DSA) and EUV resists.[5]

CDSAXS

CDSAXS is a variable-angle, transmission X-ray scattering measurement for the periodic nanostructure scatters the X-ray beam. Beam energies greater than 17 keV are required to get sufficient transmission through the silicon wafer. Since the pitch of the measured nanostructures is much larger than the wavelength of the X-rays, the scattering angles are quite small (thus small angle X-ray scattering) according to Bragg's Law (Eq. 1). CDSAXS is a single crystal diffraction measurement where the repeating nanostructure is the "atom" and the grating is the "lattice". It is analogous to protein crystallography. The scattering pattern is the Fourier transform of the electron density distribution. Since the X-rays are not coherent, the phase information is unknown and the scattering pattern cannot be inverse Fourier transformed to the electron density distribution. An iterative inverse approach is used to guess the electron density distribution, calculate the scattering pattern for the guess and compare it to the original scattering pattern. The electron density distribution is modulated until a satisfactory solution is obtained. Since it is an inverse approach, the solutions must be carefully examined for uniqueness.[2]

$$n\lambda = 2d\sin\theta \quad (Eq. 1)$$

CDSAXS has been demonstrated on a variety of periodic nanostructures including 1D line/space gratings and 2D contact holes or posts. CDSAXS has also been done on finFET structures and can solve multiple component systems like conformal layers on a line or gated-fins.[4] CDSAXS is also particularly sensitive to pitch walking.[2-3] Most demonstrations of CDSAXS have been done at synchrotron beamlines due to insufficient brightness in laboratory sources. For CDSAXS to become a high throughput, in-line measurement new compact X-ray sources with several order of magnitude increase in brightness are needed.

FIGURE 1. NIST CDSAXS system.

We will discuss the current status of CDSAXS and what aspects of the measurement are critical to the throughput. In particular, we will discuss compact X-ray sources[4] and a study of how much data is required to get satisfactory parameter uncertainty on 32 nm pitch line gratings patterned by spacer-assisted quadruple patterning (SAQP).[3] Due to the inverse nature of the fitting and the Fourier transform, there is not a simple connection between data signal to noise and the parameter uncertainty. It will be sample dependent and some parameters will require much better signal to noise and more incident angles than others. CDSAXS is transitioning from the research stage to having instruments installed in the laboratories of high volume manufacturers. NIST hosted a short course on CDSAXS last summer with a large attendance including representatives from high volume manufacturers, equipment suppliers, and X-ray component vendors. The development of compact X-ray sources will be the primary determining factor for when or if CDSAXS transitions from the laboratory to the fab floor.

REFERENCES

1. Jones, R. L.; Hu, T.; Lin, E. K.; Wu, W.-L.; Kolb, R.; Casa, D. M.; Bolton, P. J.; Barclay, G. G. Small Angle X-Ray Scattering for Sub-100 Nm Pattern Characterization. *Appl. Phys. Lett.* 2003, 83, 4059.
2. Sunday, D.F.; List, S; Chawla, J.S.; Kline, R.J.; Determination of shape and periodicity of nanostructures through variable angle transmission small angle x-ray scattering. *J. Appl. Cryst.* 48 (5), 2015, 1355-1363
3. Sunday, D.F.; List, S; Chawla, J.S.; Kline, R.J.; Evaluation of the effect of data quality on the profile uncertainty of critical dimension small angle x-ray scattering. *J. Micro/Nanolith. MEMS MOEMS.* 15(1), 2016, 014001
4. Kline, R.J.; Sunday, D.F.; Windover, D.; Bunday, B; X-ray Scattering Critical Dimensional Metrology Using Compact X-ray Source for Next Generation Semiconductor Devices. *J. Micro/Nanolith. MEMS MOEMS* (*in press*)
5. Sunday, D.F.; Hammond, M.R.; Wang, C.; Wu, W.L.; DeLongchamp, D.M.; Tjio, M.; Cheng, J.Y. Pitera, J.W.; Kline, R.J.; Determination of the Internal Morphology of Nanostructures Patterned by Directed Self Assembly. *ACS Nano*, 8 (8), 2014, 8426-8437

KEYWORDS

Dimensional metrology, critical dimensions, critical dimension small angle x-ray scattering, CDSAXS

OCD Metrology for Advanced Lithography

Bernd Bodermann[1)], Sebastian Heidenreich, Juergen Probst[2)], Victor Soltwisch[1)], Matthias Wurm[1)]

[1)] Physikalisch-Technische Bundesanstalt, Bundesallee 100, 38116 Braunschweig and Abbestraße 2-12, 10587 Berlin, Germany, [2)] Helmholtz-Zentrum Berlin für Materialien und Energie GmbH, Albert-Einstein-Str. 15, D-12489 Berlin, Germany

INTRODUCTION

Scatterometry is widely used for dimensional metrology in nanoelectronic manufacturing [1, 2], here often referred to as optical CD (OCD) metrology. However, currently applications are limited to relative measurements for process development and process control due to the lack of traceability [3, 4]. In addition, with current and future trends in advanced lithography substantial new challenges arise and have to be addressed [5]. These challenges include e.g. in-die measurements, multi-parameter measurements in complex applications such as high aspect ratio (HAR) features or FinFET structures, 3D metrology or stochastic structure parameters (roughness), to name but a few.

At PTB we develop and apply a broad variety of scatterometry methods from the NIR down to the EUV and X-Ray spectral regime for CD metrology on photo masks and structured wafers [6-11]. Here we present an overview about our methods and approaches including recent and future developments. In particular, traceability and tool matching for different OCD techniques is an issue as well as matching with other methods such as CD-SEM and CD-AFM. A popular and promising approach to improve the measurement uniqueness and uncertainty and to address the mentioned future challenges is hybrid metrology, combining either different scatterometry methods or OCD and microscopy metrology tools. Finally, approaching structure sizes of about 10 nm or less, the quest for the ultimate limits of optical scatterometry arises. With it the perspectives of OCD and related methods with respect to the emerging metrology issues in the semiconductor industry have to be considered and evaluated thoroughly.

STATE-OF-THE-ART, CURRENT AND FUTURE CHALLENGES

Different non-imaging OCD techniques such as reflectometry or ellipsometry are widely used because they are fast and non-destructive. For practical reasons in industrial applications essentially spectroscopic methods and a data base approach for the data analysis is commonly used. These techniques have proven to be reliable and very sensitive methods for current technology nodes.

However, as an integral method, OCD cannot provide local feature parameters. In addition special relatively large scatterometry test patterns (gratings) and complex modelling and data analysis are required. In some applications in the data analysis cross-correlations between different measurand parameters of interest may be an issue. Up to now OCD is not yet an absolute traceable method, but there has been different work in progress on this issue [12, 13]. Thus today OCD measurements still have to be referenced to CD-SEM tools. The matching of CD-SEM and OCD is not always perfect for different reasons. Nevertheless, OCD is the workhorse for in-line metrology and statistical relevant process monitoring. It is expected that OCD can reach the principal suitability for metrology of future manufacturing technologies at least down to the 8 nm node. However, OCD will be strongly challenged by these future applications and significant research is required to reach this suitability. For these nodes, OCD may run into sensitivity issues and the cross-correlation issue may become worse for enhanced structure complexity. To enhance the sensitivity with respect to different structure parameters, such as CD or side wall angle, there has been extensive research on the exploitation of all available polarisation information using e.g. Mueller Polarimetry [14, 15].

At PTB different scatterometry methods and tools have been developed and are used for OCD measurements on photomasks and wafers, including a DUV goniometric scatterometer [6-8], a spectroscopic ellipsometer and Mueller polarimeter (fig. 1), an EUV reflectometer/scatterometer and a gracing incidence small angle X-Ray scattering (GISAXS) tool [8-11]. For all measurement systems the same finite element based method is applied to model the measurements and the data analysis is based on sophisticated nonlinear optimisation procedures and is not relying on precomputed data bases and a look-up table approach. With it the data analysis is much more flexible and access is given to reliable uncertainty estimations for the price of much larger evaluation time requirements.
To establish scatterometry as a traceable and absolute metrological method for dimensional measurements and to improve the tool matching with alternative and complementing metrology tools such as CD-SEM or CD-AFM, recently at PTB we have developed suitable high quality scatterometry reference standard samples [16, 17]. Currently the grating dimensions cover a CD range between 25 nm and 100 nm and periods between 50 nm and 250 nm, but have the potential to be extendable to smaller periods well below 50 nm. For GISAXS reference measurements the standard also includes a 1 mm • 15 mm large grating. A traceable calibration of these standards is currently being developed applying and combining different scatterometric as well as imaging calibration methods.

FUTURE PERSPECTIVES: OPTICAL SCATTEROMETRY IN THE SUB 10 NM REGIME AND HYBRID METROLOGY

To enhance the sensitivity and to reduce the impact of parameter correlations in particular for future multi-parameter metrology applications the combination of different methods in hybrid metrology approaches is very promising [18]. Figure 1 shows an example combining spectroscopic and goniometric measurement data: By applying a sophisticated Bayesian data analysis measurement uncertainty values significantly below the uncertainties obtained with the individual methods have been achieved.

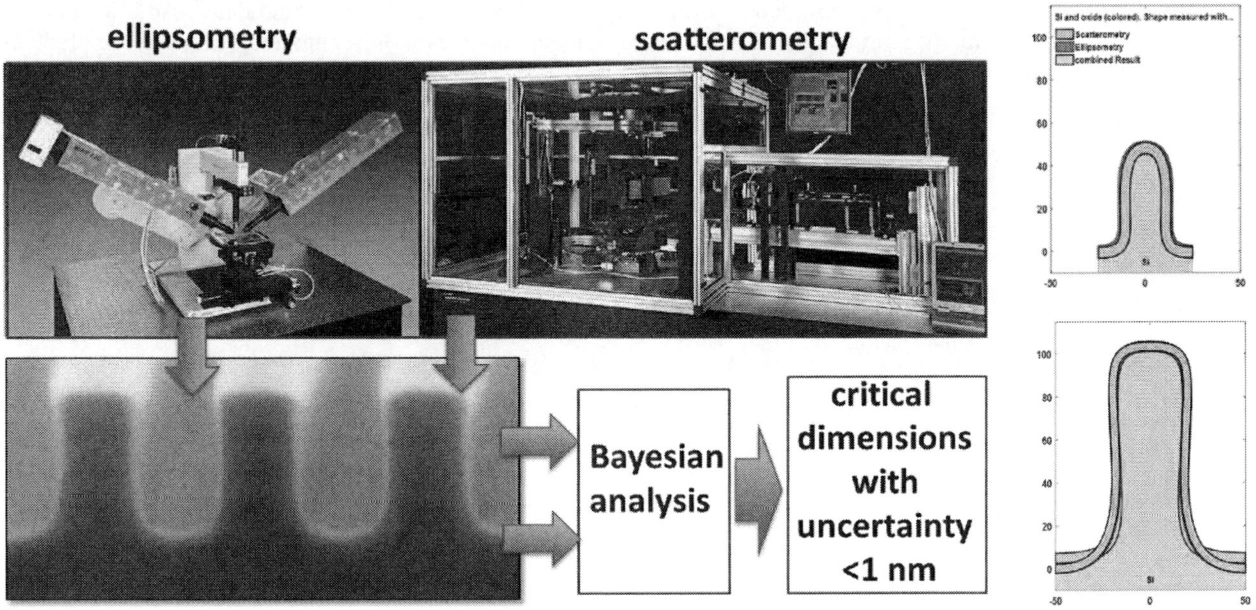

FIGURE 1. Left: scheme of a hybrid measurement approach: goniometric scatterometry and spectroscopic ellipsometry measurement data are combined in a Bayesian analysis to provide CD values with reduced measurement uncertainties; right: measurement results obtained with DUV (266 nm) reflectance measurements (blue), spectroscopic ellipsometry (red) and a hybrid analysis (yellow) of Si-gratings with nominal CD/pitch values of 25/50 nm (top) and 50/100 nm (bottom).

Recently we have investigated the performance of polarisation dependent goniometric reflectometry measurements on deep sub-wavelength grating structures down to the nm regime on the basis of numerical simulations [19]. Here no fundamental limit but an astonishing high sensitivity of this method has been observed. In fact even for grating periods as small 5 nm and a corresponding CD of only 2.5 nm a significant and easily measureable splitting of the two different reflectance curves measured using p-polarised light is observed (fig.2). So OCD metrology is a very promising method even for the sub 10 nm regime. Any practical limitations of this approach will be subject to further numerical and experimental investigations.

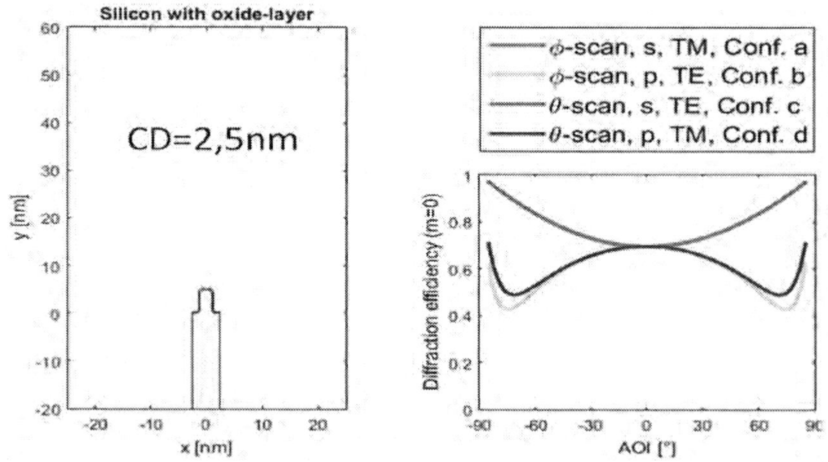

FIGURE 2. Numerical simulation of reflectance curves for a Si grating, period = structure height = 5 nm, CD = 2.5 nm

DISCUSSION AND OUTLOOK

Also in the future OCD (in the broader sense) is assumed to play an essential and probably even enhanced role for dimensional metrology for advanced lithography. However, to meet the dramatically increased future requirements novel enhanced methodologies will be required. Here in particular sophisticated methods such as Mueller ellipsometry and X-Ray scatterometry (GISAXS, CD-SAXS) are promising candidates. Additionally the combination of different methods in advanced hybrid metrology approaches will be inevitable. This will include as well the combination with imaging (CD-SEM) metrology.

REFERENCES

1. Thony P, Herisson D, Henry D, Severgnini E, Vasconi M *AIP Conf. Proc.* **683** 381-388 (2003)
2. Raymond C J, *AIP Conf. Proc.* **788** 394-402 (2005)
3. Heather J. Patrick, Thomas A. Germer, *Proc. SPIE* **6672**, 66720L, (2007)
4. T. A. Germer, H. J. Patrick, R. M. Silver, B. Bunday, *Proc. SPIE* **7272**, 72720T-1 - 72720T-11, (2009)
5. B. Bunday, T. A. Germer, V. Vartanian, A. Cordes, A. Cepler, C. Settens, *Proc. SPIE* **8681** 86813B, (2013)
6. Wurm M, Pilarski F and Bodermann B, *Rev. Sci. Instrum.* 81 023701, (2010)
7. M. Wurm, S. Bonifer, B. Bodermann, J. Richter, *Meas. Sci. Technol.* **22**, 094024-1 - 094024-9, (2011)
8. J. Wernecke, F. Scholze, M. Krumrey, Rev. Sci. Instrum. **83**, (2012) 103906-103916
9. Scholze F, Bodermann B, Burger S, Endres J, Haase A, Krumrey M, Laubis C, Soltwisch V, Ullrich A, Wernecke J , *Proc. SPIE* **9231** 92310M, (2014)
10. V. Soltwisch, J.Wernecke, A. Haase, J. Probst, M. Schoengen, M. Krumrey, F. Scholze, *Proc. SPIE* **9050** 9050-36, (2014)
11. V. Soltwisch, J.Wernecke, A. Haase , J. Probst, M. Schoengen, M. Krumrey, F. Scholze, J. Pomplun, S. Burger *Proc. SPIE* **9235**, 92351D, (2014)
12. Patrick H J, Germer T A *Proc. SPIE* **6672**, 66720L(2007)
13. Germer, T A, et al. *Proc. SPIE* **7272**, 72720T - 72720T, (2009)
14. M. Foldyna, A. De Martino, C. Licitra, J. Foucher *Proc. SPIE 7638* 763803-763803-10, (2010)
15. P. Vagos, Pedro, P. I. Rovira: US patent 8427645
16. Bodermann, Bernd, et al., *Proc SPIE* **9132**, 91320A-1-91320A-12, (2014)
17. Agocs, Emil, et al., *Proc SPIE* **9556**, 955610-1-955610-12, (2015)
18. A. Vaid, C. Osorio, J. Tsai, C. Bozdog, M. Sendelbach, E. Grubner, R. Koret, S. Wolfling, *Proc. SPIE* **9050**, 905009, (2014)
19. B. Bodermann,G. Ehret, J. Endres and M. Wurm, *Surf. Topogr.: Metrol. Prop.* **4** (2016) 024014
20. M. Wurm, J. Endres, J. Probst, M. Schoengen, A. Diener, B. Bodermann, submitted to Opt. Exp.

KEYWORDS

Scatterometry, OCD, reference standard, CD metrology, GISAXS, Mueller ellipsometry, hybrid metrology, sub-wavelength metrology

Mueller Matrix Spectroscopic Ellipsometry Based Scatterometry

Alain C. Diebold

SUNY Polytechnic Institute

INTRODUCTION

Typically, scatterometry refers to the use of use of spectroscopic ellipsometry to measure feature dimensions of arrays of patterned structures. It is also sometimes referred to as OCD or optical critical dimension measurement. Mueller Matrix spectroscopic ellipsometry (MMSE) provides a significant advance over traditional spectroscopic ellipsometry. Traditionally, spectroscopic ellipsometry provides the values of ψ and Δ at each wavelength where $\tan(\psi) = \frac{|r_p|}{|r_s|}$ and $\Delta = \delta_{rs} - \delta_{rp}$. Here, r_p and r_s are the reflectivity's of the p and s polarized light respectively, and δ_{rs} and δ_{rp} are the changes in polarization for the s and p polarized light respectively. ψ and Δ do not account for cross-polarized light scattering. The Mueller Matrix approach accounts for all forms of light scattering including cross-polarized scattering and depolarization. At each wavelength, 16 Mueller Matrix elements are measured. Rigorous Coupled Wave Analysis (RCWA) is used to solve Maxwell's equations, and RCWA simulations are used to simulate the Mueller Matrix spectra for a variety of feature dimensions and shapes. The closest fit between the simulated and measured spectra determines the average feature dimensions and shape of the measured array. Here we discuss the application of MMSE based scatterometry to the measurement of the feature dimensions of fin arrays, block copolymer line-space and contact hole arrays, lines etched from block copolymer patterning, Line Edge Roughness (LER) measurement, and the use of plasmons in metal cross-grating structures to enhance sensitivity to changes in line width for copper lines.

MMSE SCATTEROMETRY OF SILICON AND SILICON GERMANIUM FIN ARRAYS AND SILICON LINE SPACE ARRAYS

Measurement of the line width and feature shape of silicon and silicon – germanium alloy fin arrays challenge the measurement capability of CD-SEM and scatterometry.[1, 2] In addition to the sub 15 nm width, the sub 50 nm pitch of the fins and line edge roughness make scatterometry considerably more complicated. This section describes the application of MMSE-scatterometry to fin arrays. We also demonstrate the importance of the optimum azimuthal angle for maximum sensitivity to feature dimensions. Line edge roughness (LER) greatly complicates determination of line width. A method of including LER will be presented and is shown in Figure 1 below.

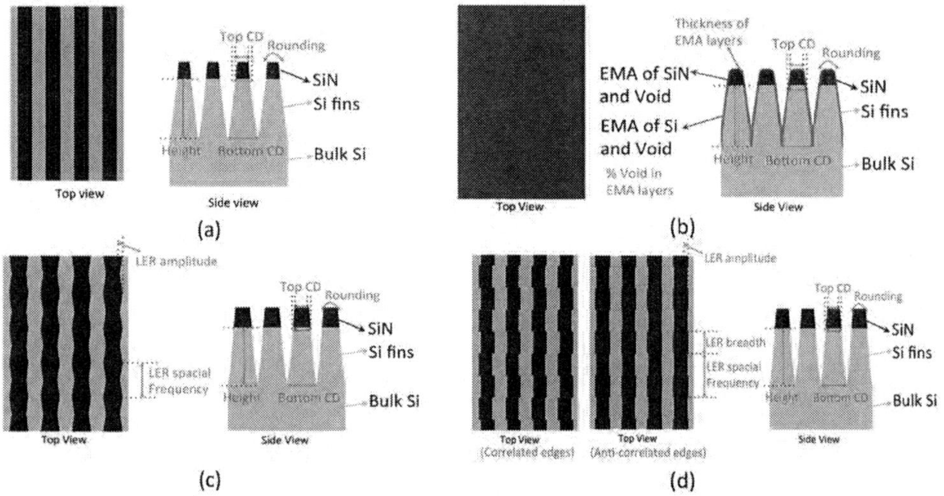

FIGURE 1. Different methods of including Line Edge Roughness are shown. 1.a shows idea lines with no LER. 1.b shows the use of and effective medium layer to model LER. 1.c shows a non-random sinusoidal line edge. This method proved to be the most successful 1.d shows a line segment off-set approach. [2]

MMSE SCATTEROMETRY OF BLOCK COPOLYMER ARRAYS

Directed Self Assembly of Block Copolymers is under consideration as a potential patterning method.[3, 4, 5] Arrays of line – space structures and contact hole structures were fabricated and subsequently imaged by CD-SEM. In this section we describe the measurement of feature dimensions and shape using scatterometry.

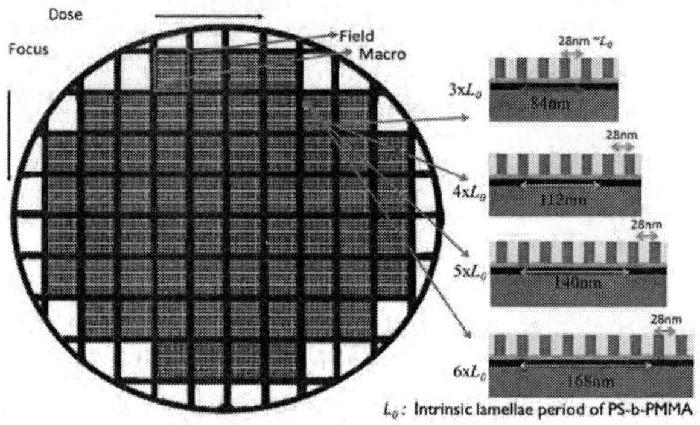

FIGURE 2. Line Space Block Copolymer arrays fabricated on a wafer with different focus-exposure conditions. The test areas contained sections (macros) with different guide structure pitches.[3]

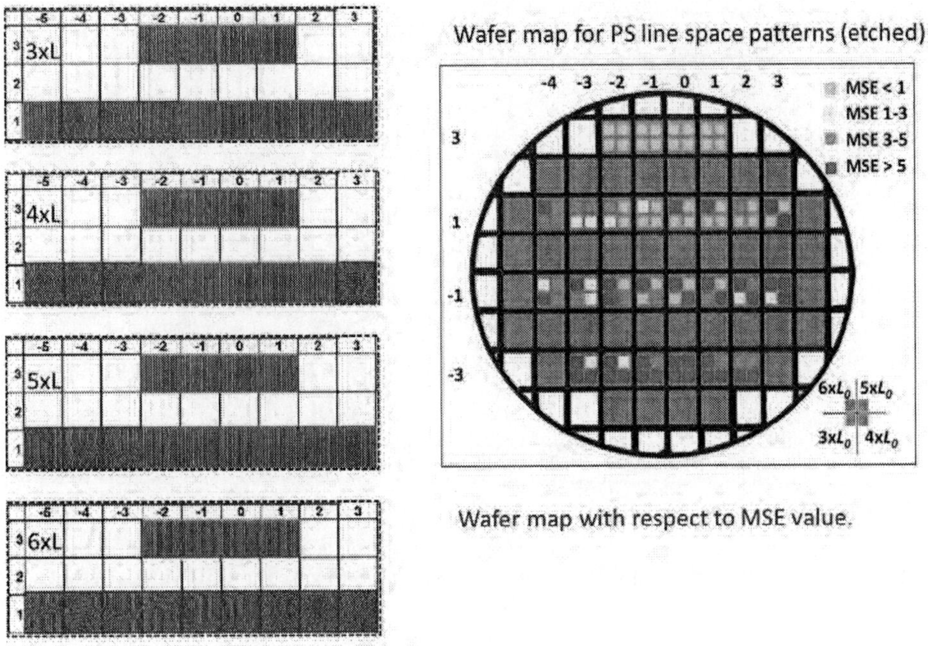

FIGURE 3. CD SEM images and results of optical model fitting for Line Space Block Copolymer arrays fabricated on a wafer with different focus-exposure conditions..[3]

MMSE SCATTEROMETRY OF PLASMONIC COPPER CROSSGRATING ARRAYS

One of the most challenging structures for scatterometry is a metal grating. The dimension of a metal grating that produces plasmon polaritons are much greater than those of state of the art copper lines.[6, 7] However, when a grating is crossed with a grating of fine copper lines, the changes in polarized reflectance greatly enhance the sensitivity of scatterometry to the linewidth of the copper lines. One of these structures is shown in Figure 4 below, and the enhanced sensitivity to copper line width for a cross grating in Figure 5.

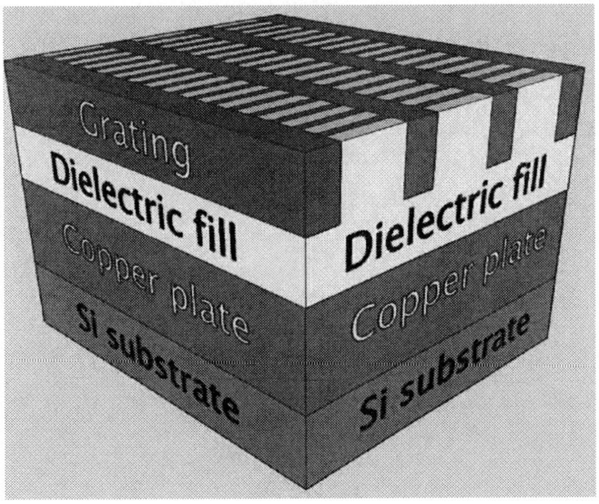

FIGURE 4. Copper Cross-Grating Structure showing a copper plate below the top cross-grating.[5, 6]

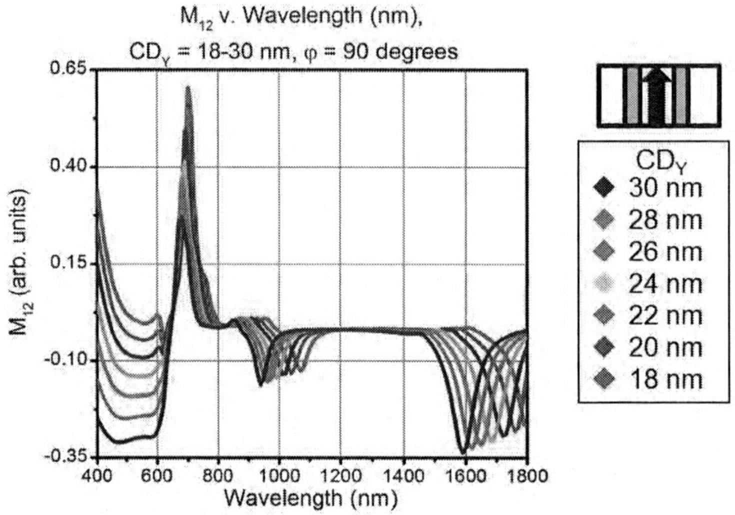

FIGURE 5. The M12 Mueller Matrix element for a copper cross-grating Structure for a CD values between 18 nm and 30 nm for the thin metal lines in the cross-grating.[5, 6]

ACKNOWLEDGEMENTS

The author gratefully acknowledges the work of the graduate students whose work is described in this paper: Vimal Kamineni, Raja Muthinti, Dhairya Dixit, and Sam O'Mullane. The author gratefully acknowledges the support of Brennan Peterson, Nick Keller, and Joe Race concerning scatterometry simulations, and the support of collaborators Erik Hosler, Vimal Kamineni, Jody Fronheiser, and Richard Farrell in supplying patterned wafers for this research. The author gratefully acknowledges funding support from Nanometrics, Inc. and the Semiconductor Research Corporation's Global Research Consortium.

REFERENCES

1. G.R. Muthinti, M. Medikonda, J. Fronheiser, V.K. Kamineni, B. Peterson, J. Race, and A.C. Diebold, Proc. SPIE 8681, Metrology, Inspection, and Process Control for Microlithography XXVII, 2013, 86810M.
2. D. Dixit, S.J. O'Mullane, S.Sunkoju, A.Gottipati, E.R. Hosler, V.K.Kamineni, M.Preil, N.Keller, J.Race, G. R. Muthinti, and A.C. Diebold, J. Micro/Nanolith. MEMS MOEMS **14**(3), 2015, 031208.
3. D. J. Dixit, V. Kamineni, R. Farrell, E. R. Hosler,, M. Preil, J. Race, B. Peterson, A. C. Diebold, Control of IC Patterning Variance. Part 1: Metrology, Process Monitoring, and Control of Critical Dimension in Journal of Micro/Nano Lithography, MEMS & MOEMS 14, (2015), 021102.
4. D.J. Dixit, A. Green, E.R. Hosler, V. Kamineni, M. Preil, N. Keller, J. Race, J.S. Chun, M. O'Sullivan , P. Khare, W. Montgomery, A.C. Diebold, Journal of Micro/Nano Lithography, MEMS & MOEMS 15(1), 2016, 014004.
5. S. O'Mullane, Nick Keller, J. Race, A.C. Diebold, Journal of Micro/Nano Lithography, MEMS & MOEMS 15, (2016), 044003.
6. S. O'Mullane, B. Peterson, J. Race, N. Keller, A.C. Diebold, Optics Express 22, (2014), 26247. Erratum Optics Express **24**, (2016) 3613.

KEYWORDS

Scatterometry, spectroscopic ellipsometry, critical dimension measurement

CHARACTERIZATION OF THE INTERLAYER OF BETWEEN A NANOLAMINATE GATE OXIDE AND SIGE(001)

K. Sardashti[1], E. Chagarov[1], I. Kwak[1], S. W. Park[1], S. Ueda[1], S. Madisetti[1], M. Yakimov[3], S. Oktyabrsky[3], and A. C. Kummel[1,2]

[1]*Materials Science and Engineering Department, University of California San Diego*
[2]*Department of Chemistry and Biochemistry, University of California San Diego*
[3]*SUNY Polytechnic Institute, Colleges of Nanoscale Science and Engineering, Albany, NY*

INTRODUCTION

SiGe is considered to be the leading channel material for p-FET devices at the 7-nm technology node. When high-k dielectrics such as HfO_2 are deposited on silicon, a few angstroms of SiO_2 are intentionally maintained between the Si channel and the HfO_2 to reduce interfacial defect density (D_{it}) and to reduce remote phonon scattering, which decreases channel mobility. Formation of a good interfacial layer between high-k gate oxides and SiGe interfaces is challenging since germanium suboxide (GeO_x, x<2 containing Ge^{+2}) is known to induce electronic defects, and it is nearly impossible to fully oxidize or nitride Ge to Ge^{+4} in the presence of Si since both O and N make stronger bonds to Si than Ge. An alternative approach is to form a subnanometer layer of amorphous SiO_x between the high-k dielectric and the SiGe channel via ALD or PVD. However, since SiO_x has a greater heat of formation than GeO_x, it is possible that just annealing of the interfacial $Si_xGe_yO_z$ oxide can result in formation of a pure SiO_x interface. In this report, multiple analytical techniques are used to investigate the formation and refinement of a Ge-free interfacial layer between gate oxides and SiGe. In-situ XPS is employed to show that after deposition of OH groups, annealing results in formation of a Si terminated surface. Angle-resolved XPS of buried oxide/SiGe interfaces is employed to show that ALD at high temperature can form a nearly pure SiO_x interlayer after sulfur treatment or pure SiON after NH_3 plasma nitridation. Finally, cross-sectional TEM-EDS is used to determine the composition and thickness of the SiO_x interface in a full SiGe nanolaminate gate stack structure.

RESULTS

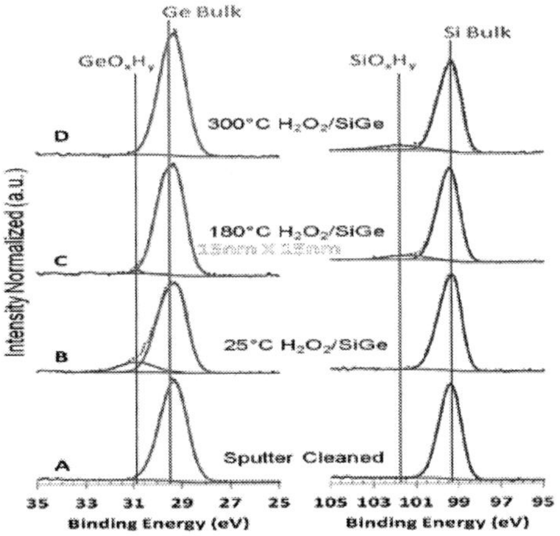

FIGURE 1: In-situ XPS in UHV of HOOH(g)/$Si_{05}Ge_{05}$(001). Dosing with HOOH(g) induces only a GeOx component. Annealing removes the GeOx and only a SiOx component is formed.[2]

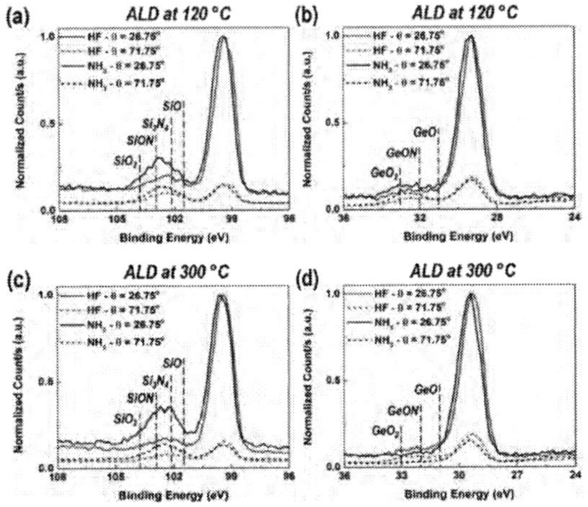

FIGURE 2: Angle Resolved XPS of HF(aq) vs NH_3 plasma passivated SiGe(001) with 8 cycles Al_2O_3 ALD at 120 °C vs 300 °C. Note: At 300 °C, there is no GeON component [5].

In-Situ XPS: To test the effect of annealing on formation of a Ge-free SiOx layer, the reaction of HOOH(g) with $Si_{0.5}Ge_{0.5}$(001) was studied with in-situ XPS. After sputter cleaning, the XPS (Fig 1) shows that the surface has only elemental Si and Ge peaks. After dosing with HOOH(g) at 25 °C, a high binding energy shoulder appears on the Ge peak consistent with GeOx formation since SiGe(001) is known to be terminated Ge atoms. However, upon annealing to 180 °C, the GeOx component is completely removed and a SiOx component is observed consistent with formation of a pure SiOx interface [2]. Similar results were also observed on SiGe(110) [3].

Angle Resolved XPS of Buried Interface: To determine if the process of forming a Ge-free interface could occur under typical ALD conditions, angle-resolved XPS was performed on buried interfaces. Two types of passivation were studied: $(NH_4)_2S$(aq) and NH_3 plasma [4,5]. Experiments were performed with both Al_2O_3 and HfO_2 oxides with consistent results. As shown in Fig 2, when Al_2O_3 ALD is performed at 120 °C on a NH_3 plasma passivated surfaces, there is both SiON and GeON formation. In contrast, when ALD is performed at 300 °C, there is no GeON component with NH_3 plasma passivation. For both ALD temperatures, the spectra at glancing angle have a larger SiON/Si ratio component consistent with a thin buried SiON interlayer.

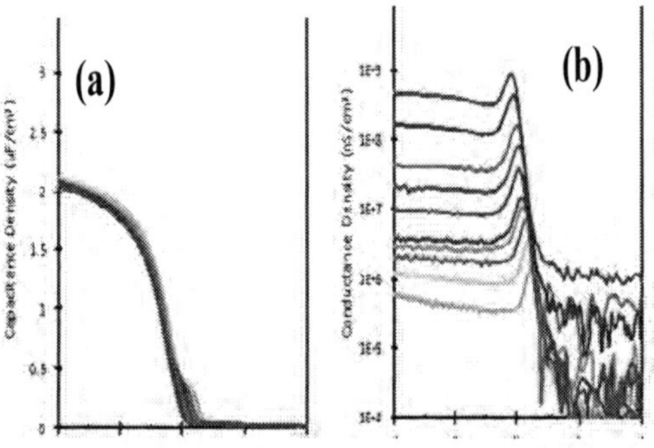

Figure 3: Electrical Characterization of Al_2O_3-HfO_2 alloy / $Si_{0.5}Ge_{05}$(001)/Si(001). (a) C-V shows D_{it} ~ $2x10^{12}$ /cm^2-eV using the conductance method. (b) G-V shows low leakage. A Ni gate was thermally evaporated along with an Al back contact.

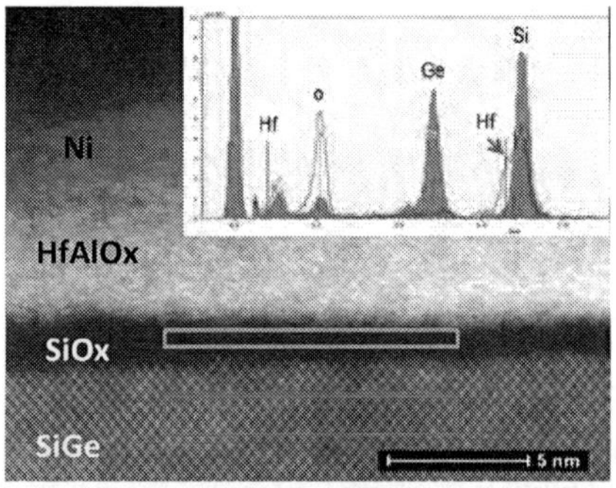

Figure 4: STEM-EDX characterization of Al_2O_3-HfO_2 alloy/ $Si_{0.7}Ge_{03}$(001)/Si(001). Blue box shows a region of interlayer for EDS revealing a Ge-free SiOx interlayer. The orange box is a control region in the SiGe. The EDS spectra of the interlayer is in blue while the spectra on the SiGe control region is in orange. Note the absence of Ge in the interlayer.

TEM-EDS of Full Gate Stack: To study the interface in a full gate stack, an Al_2O_3-HfO_2 alloy was deposited as a nanolaminate using $HfCl_4$, TMA, and H_2O at 300°C. FGA was performed at 300 °C. As shown in Fig 3, a high quality interface was produced, characterized by: (a) a low interfacial trap density (small D_{it} bump in C-V near threshold), (b) low boarder trap density (N_{BT}, small frequency dispersion in accumulation C-V), (c) near zero flat-band voltage (V_{FB} ~ 0 in C-V), and (d) low leakage (flat G-V in accumulation). Cross sectional scanning TEM with energy-dispersive x-ray spectroscopy (STEM-EDS) was performed on the sample. The high-angle annular DF STEM clearly shows an interlayer of low atomic number between the $Al_xHf_yO_z$ and the SiGe (Fig 4). EDS spectra of the sub-1nm interlayer shows that it consists of pure SiO_x and contains no Ge.

SUMMARY

DFT-MD molecular dynamics showed that nearly ideal interfaces between a-HfO_2 high-k oxide and $Si_{0.5}Ge_{0.5}$(001) could be formed with either a-SiO_xN_y or a-SiO_x interlayers; exclusion of GeO_x formation is critical. To follow the process, in-situ XPS was employed to shows $Si_xGe_yO_z$ can be annealed to purely SiO_x. The same process was observed in buried interfaces with angle-resolved XPS and the same process could also be observed in full high-k/SiGe gate oxide stacks with TEM-EDS. This work was sponsored by Lam Research Corporation

REFERENCES

[1] E. Chagarov, K. Sardashti, T. Kaufman-Osborn, S. Madisetti, S. Oktyabrsky, B. Sahu, A Kummel, ACS Appl. Mater. & Inter. **7**, p 26275 (2015).
[2] T. Kaufman-Osborn, E. Chagarov, S. W. Park, B. Sahu, S. Siddiqui, and A. C Kummel, Surf Sci **630**, p. 273 (2014)
[3] S. W. Park, S. W. Park, H. Kim, E. Chagarov, S. Siddiqui, B. Sahu, N. Yoshida, A. Brandt, and A. C. Kummel Surf. Sci, **652**, p 322-333 (2016).
[4] K. Sardashti, K.-T. Hu, K. Tung, S. Madisetti, et al Appl. Phys. Lett. **108**, p. 011604 (2016).
[5] K. Sardashti, K-T Hu, H. Kim, S. Park, et al Appl. Surf. Sci., **366**, p455 (2016)

Interfacial Electronic Characterization of Oxides/Metals on High Mobility Semiconductors Using *in-situ* Synchrotron Radiation Photoemission and The Correlation With The Interfacial Electric Properties

T. W. Pi[1], C. P. Cheng[2], K. Y. Lin[3], Y. H. Lin[3], W. S. Chen[2], Y. T. Cheng[1], L. B. Young[3], H. W. Wan[3], J. Kwo[4], and M. Hong[3]

[1]*National Synchrotron Radiation Research Center, Hsinchu, Taiwan, R.O.C.*
[2]*Department of Electrophysics, National Chiayi University, Chiayi, Taiwan, R.O.C.*
[3]*Graduate Institute of Applied Physics and Department of Physics, National Taiwan Univ., Taipei, Taiwan, R.O.C.*
[4]*Department of Physics, National Tsing Hua University, Hsinchu, Taiwan, R.O.C.*

INTRODUCTION

Compound III-V semiconductors have been under intense research studies for replacing Si as the n-channel material because of their high electron mobility, which enables high-speed low-power III-V metal-oxide-semiconductor field-effect-transistors (MOSFETs) without the need for further aggressive scaling. Atomic-layer-deposited (ALD) trivalent Y_2O_3 on GaAs(001) has given record-low interfacial trap densities (D_{it}), small frequency dispersion in capacitance-voltage (*C-V*), and high temperature thermal stability, critical for a compatible complementary MOS (CMOS) process. [1] Furthermore, Self-aligned inversion-channel InGaAs MOSFETs of 1 μm gate length using the gate dielectrics of ALD tetravalent HfO_2 and trivalent Al_2O_3 have given high extrinsic drain current of 1.5-1.8 mA/μm, drain current of 0.84 mS/μm, I_{ON}/I_{OFF} of $>10^4$, low sub-threshold swing of ~100mV/decade, and field-effect electron mobility of 1100 cm^2/Vs. The devices have achieved very high intrinsic drain current and drain current of 2-2.3 mA/μm and 1.2 mS/μm, respectively. [2-4] It is important to well control the fabrication of and to characterize the high κ dielectrics and metals on (In)GaAs to understand/tailor the interfacial electronic structures in order to achieve low D_{it} and ohmic contacts, vital for high performance devices.

Large frequency dispersion in the *C-V* curves at accumulation has been an important issue for dielectric oxides on III-V compound semiconductors such as (In)GaAs. An ideal set of *C-V*s is the one with identical capacitance without dispersion in the accumulation region. Large frequency dispersion indicates a high interfacial trap density (D_{it}) at the oxide/(In)GaAs interface. The literature unanimously reports that the atomic-layer-deposited (ALD) Al_2O_3 and HfO_2 on a *n*-type GaAs(001) sample shows profoundly greater frequency dispersion than the *p*-type sample in the accumulation region of *C-V* characteristics by over 20%, irrespective of dielectric oxides. [5]

In contrast to the conventional method to obtain the interface information of the metal/semiconductor devices by a fit of a measured I-V curve with thermionic emission model, synchrotron radiation photoemission (SRPES) can keenly show, in an atomic scale, the evolution of the interfacial electronic structure from the initial deposition of atoms to metallic films from sub-nm to nm thick. The Schottky barrier height (SBH) and the interfacial dipole potential have been directly determined without any presumption. An example of an *in situ* SRPES study of the Ag/GaAs(001)-2×4 is shown as follows [6]. To acquire extreme surface-sensitive spectra, the photon energy was tuned to 46 and 120 eV such that the inelastic mean free path (IMFP) for photoelectrons emitted from the valence band, As 3d and Ga 3d core levels was small of 4 Å. according to a statistical approach [7].

AS-GROWN SAMPLES AND OXIDES/METALS COVERED SURFACES

Precise knowledge of the interfacial electronic structure of oxides (metals)/high-mobility semiconductors requires the starting surfaces to be atomically clean and ordered. A clean III-V (Ge) surface is impossible to obtain by conventionally cleaning methods as chemical treatments, annealing, or alternative sputtering and annealing, which results in a random, non-structural surface. Hence, the III-V (Ge) surfaces for interfacial studies must be in ultra-high vacuum (UHV) without exposing them in air; air exposure for one minute has generated a considerable amount of notorious native-oxides.

The As 3d Core-level Spectra: Clean and Oxides/Metals Covered Surfaces

Examples of high-resolution As 3d core-level spectra with an atomically clean GaAs(001)-2×4 surface with a fit are shown in Fig. 1(a). Normally, the 3-fold coordinated surface As atom has more charge than the 4-fold coordinated bulk As, thereby giving rise to distinct surface core-level shifts in a binding energy lower than that of the bulk, as in the case of InGaAs (001)-4x2 (not shown). As shown in Fig. 1, however, anomalously positive shifts appear in higher binding energies than the bulk for the GaAs(001) surfaces, marked as As(1) and As(2) in Fig. 1(a). These surface atoms reveal the electronic states being deficient in charge compared to the bulk, which affects the electric behaviors.

Figure 1(b) displays the As 3d core-level spectrum of ALD 10 cycles of *tri*-methylaluminum (TMA) and water as precursors grown onto GaAs(001)-2×4 surface. The induced As* component suggests the bonding of the precursors with the surface As atoms. However, the As(1) and As(2) components remain detectable, suggesting that partial coverage of the precursors. These un-passivated surface atoms remain in a charge-deficient state, which could trap the electrons at the interface under positive gate bias to manifest a great frequency dispersion. The *p*-type GaAs sample does not suffer the same undesirable trapping because the carriers of the accumulation region are holes.

The above proposal is justified with an experiment that uses adatoms with high Pauling electronegativity as the noble metals. A fit to the Au- and Ag-covered As 3d core-level spectra is shown in Fig. 2. The adsorption of Au (Ag) results in a shift of the As(1) component towards lower binding energy to become induced $As_{Au(Ag)}(1)$ component, suggesting that the corresponding As atoms have drain charge from the contacted noble-metal atoms. This result suggests that the GaAs(001) surface behaves as charge acceptors, or electron trappers for the matter of CMOS.

The behavior of *C-V* characteristics in the inversion region for a GaAs MOS capacitor is very different from that of a Si MOS capacitor, because the carrier generation time of GaAs is much longer than that of Si. In the measurement frequency range of 100 Hz to 1 MHz, the minority-carrier generation rate of GaAs is too slow to respond in the measured capacitance, such that the capacitance remains almost the same from the depletion region to the inversion region. Since the amount of minority carrier itself at the interface is negligible, the un-passivated GaAs surface atoms will not give dispersion for capacitance in the inversion region for either n- or p-GaAs.

The Schottky Barrier Height Formation at the Metal/Semiconductor Interface

Figure 3 shows the valence band spectra near the Fermi level (E_F) and low-energy cutoff obtained with 46 eV photon energy at various Ag coverages. A Fermi edge bulges out at 2.5 Å of thickness, indicating that the metallic film emerges. The ionization potential (IP) of semiconductor and the work function (ϕ_M) of metal can be determined from the formula of (photon energy - spectral width). The IP of pristine *p*-type α2 GaAs(001)-2×4 and the ϕ_M of the 102.5-Å Ag coverages were thus found to be 4.78 and 4.08 eV, respectively.

Figure 4 presents the As 3d and Ga 3d core level spectra obtained with 120 eV photon energy. Decay of the As3d/Ga3d core level intensity shown in an inset of Fig. 2 deviated from an ideal laminar growth character, suggesting an island growth of Ag. Furthermore, the different rate of attenuation for As 3d and Ga 3d states suggests out-diffusion of the As. Out-diffusion causes an induced component As-Ag and Ga-Ag shown on the low binding energy side of the As 3d and Ga 3d spectra, which is attributed to the As(Ga)-Ag bond. A rigid peak shift of 0.08 eV toward higher binding energy is observed at 0.25 Å of thickness and is attributed to the band bending effect (ΔV_{bb}) induced by the Ag adatoms. The fixed energy position above 0.25 Å of coverage suggests that the band bending effect is complete at this coverage. In α2 GaAs(001)-2×4, a total coverage of the As-As dimers in the topmost layer requires 0.27 Å. Herein, we found direct evidence that the band bending effect is stabilized if the surface dimers are passivated. The SBH, which was then determined here using the experimentally deduced values of ΔV_{bb} and VBM (Fig. 3), was found to be 0.38 eV (Fig. 5).

To interpret the band bending direction, we have de-convoluted the As 3d core level spectra to find how Ag atoms are adsorbed on GaAs surface. See Fig. 2. The silver atoms were adsorbed upon the As-As dimers. Note that Ag remains atomic character at this thickness. When the thickness of the Ag film is greater than 2.5 Å, some of the surface As atoms begin to diffuse to the top Ag layer. The surface dipole potential at various Ag coverages shown in Fig. 5(a) has been measured/quantified by measuring the difference between ionization potential of each Ag thickness with that of clean surface. Evidently, the surface dipole has undergone a sign change from pointing outwardly to inwardly. The former, type I, occurs when Ag sits on top of the As-As dimers. For thicker Ag deposition, we find that the dipole potential energy decreases due to the As(-)-Ag(+) surface dipole of type II forms, which can be explained by the As segregation.

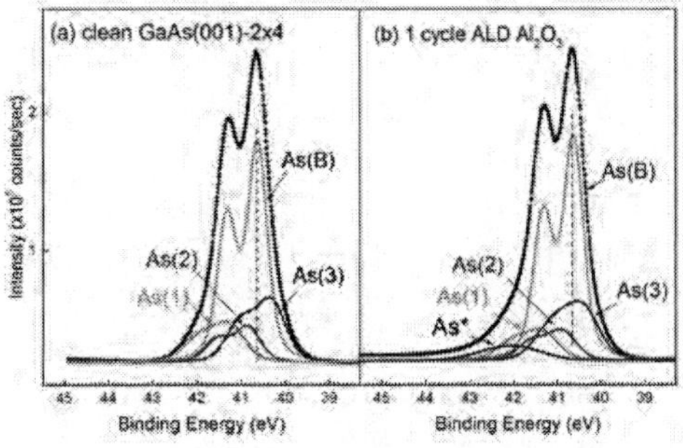

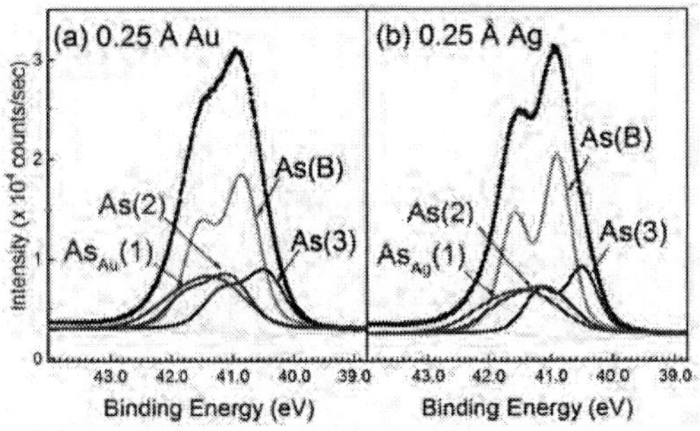

FIGURE 1. (Left) A fit to As 3d core-level spectra. (a) clean GaAs(001)-2×4 and (b) 1 cycle of ALD Al_2O_3 on GaAs(001)-2×4.
FIGURE 2. (Right) A fit to As 3d core-level spectra. (a) 0.25 Å of Au and (b) 0.25 Å of Ag on GaAs(001)-2×4.

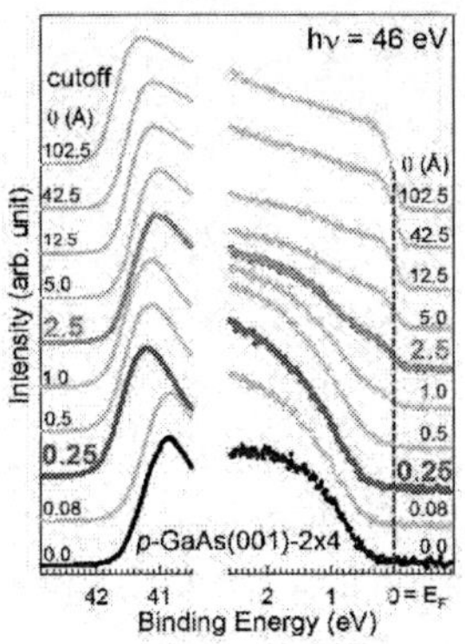

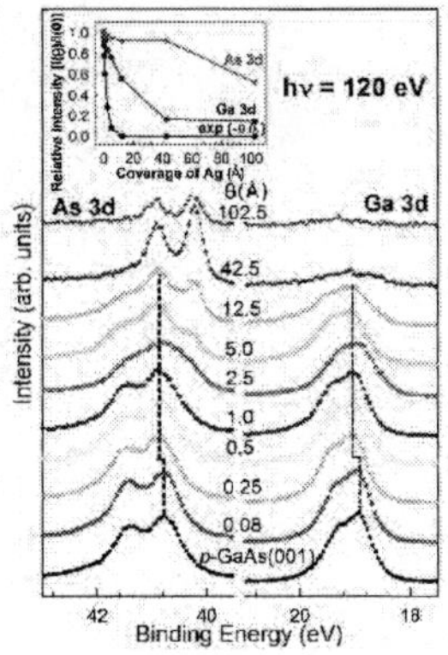

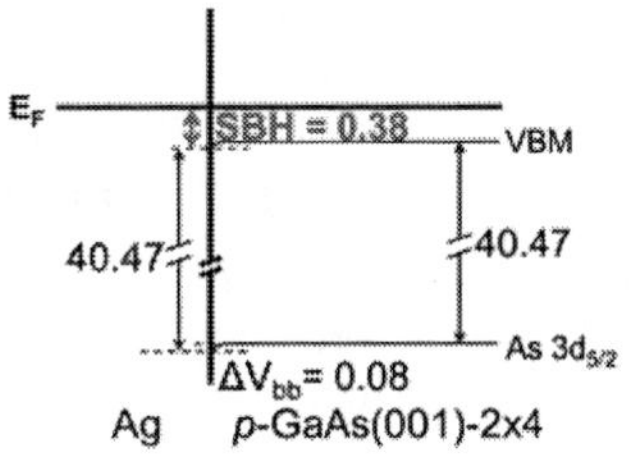

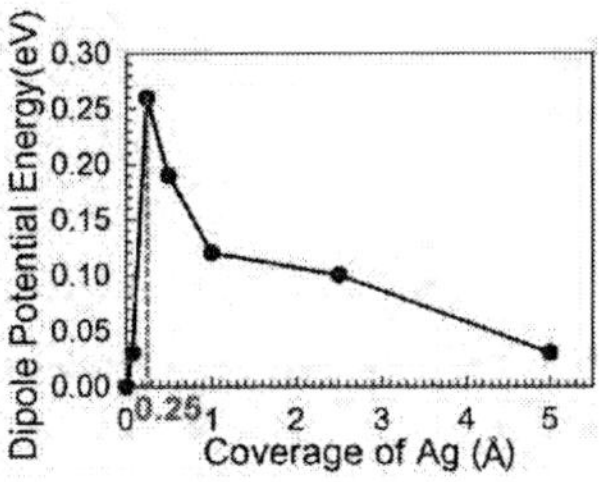

FIGURE 3. (Leftmost) Valence-band spectra in the vicinity of the Fermi level (E_F) and low-energy cutoff recorded with a photon energy of 46 eV for Ag on a *p*-GaAs(001)-2×4 surface.
FIGURE 4. (2nd Left) Evolution of the As 3d and Ga 3d core-level spectra taken with a photon energy of 120 eV. The inset is signal relative intensity as a function of the Ag coverage for the As 3d and Ga 3d cores
FIGURE 5. (2nd Right) The energy-level diagram for Ag on a *p*-GaAs(001)-2×4 surface derived from the corresponding photoemission data.
FIGURE 6. (Rightmost) Surface dipole potential energy with respect to the Ag coverage.

REFERENCES

1. Y. H. Lin, C. K. Cheng, K. H. Chen, C. H. Fu, T. W. Chang, C. H. Hsu, J. Kwo, and M. Hong, *Materials* **8**, 7084 (2015).
2. T. D. Lin(林宗達), W. H. Chang, R. L. Chu, Y. C. Chang, Y. H. Chang, M. Y. Lee, P. F. Hong, Min-Cheng Chen, J. Kwo(郭瑞年), and M. Hong(洪銘輝), *Appl. Phys. Lett.* **103,** 253509 (2013).
3. T. D. Lin (林宗達), Y. H. Chang (張宇行), C. A. Lin (林俊安), M. L. Huang (黃懋霖), W. C. Lee (李威縉), J. Kwo (郭瑞年), and M. Hong (洪銘輝), *Appl. Phys. Lett.* **100**, 172110 (2012).
4. H. W. Wan, K. Y. Lin, M. H. Chen, L. B. Young, Y. H. Lin, Y. C. Chang, T. D. Lin, Y. T. Cheng, W. S. Chen, S. H. Chen, T. W. Pi, J. Kwo, and M. Hong, 47th *IEEE Semiconductor Interface Specialists Conf.* (*SISC*), San Diego, CA. Dec. 8-10, (2016).
5. Y. H. Chang, C. A. Lin, Y. T. Liu, T. H. Chiang, H. Y. Lin, M. L. Huang, T. D. Lin, T. W. Pi, J. Kwo, and M. Hong, *Appl. Phys. Leters.* **101**, 172104 (2012).
6. C. P. Cheng, W. S. Chen, K. Y. Lin, G. J. Wei, Y. T. Cheng, Y. H. Lin, H. W. Wan, T. W. Pi, R. T. Tung, J. Kwo, and M. Hong, *Appl. Surf. Sci.* **393**, 294-298 (2017).
7. M. P. Seah and W. A. Dench, *Surf. Interface Anal.* **1**, 2-11 (1979).

KEYWORDS

GaAs(001)-2x4, ALD HfO_2, Al_2O_3, and Y_2O_3, Schottky barrier height, noble-metal adsorption, self-aligned inversion-channel InGaAs MOSFET

Electron Optics in Graphene Heterostructures with Nanopatterning

Scott Dietrich[1], Shaowen Chen[1,2], Zheng Han[1,7], Mirza M. Elahi[3], K. M. Masum Habib[3], Lei Wang[4], Bo Wen[1,8], Huan Hu[8], Yuanda Gao[5], Takashi Taniguchi[6], Kenji Watanabe[6], James Hone[5], Avik W. Ghosh[3], and Cory Dean[1]

[1]*Department of Physics, Columbia University, New York, NY 10027, USA.*
[2]*Department of Applied Physics and Applied Mathematics, Columbia University, New York, NY 10027, USA.*
[3]*Department of Electrical and Computer Engineering, University of Virginia, Charlottesville, VA 22904, USA.*
[4]*Department of Physics, Cornell University, Ithaca, NY 14853, USA.*
[5]*Department of Mechanical Engineering, Columbia University, New York, NY 10027, USA.*
[6]*National Institute for Materials Science, 1-1 Namiki, Tsukuba 305-0047, Japan.*
[7]*Shenyang National Laboratory for Materials Science, Institute of Metal Research, Chinese Academy of Sciences, Shenyang 110016, P. R. China.*
[8]*IBM Thomas J. Watson Research Center, 1101 Kitchawan Road, Yorktown Heights, NY 10598, USA. USA.*

INTRODUCTION

Ballistic electrons in a uniform two-dimensional electron gas behave in close analogy to light: Electrons follow straight-line trajectories, and their wave nature can manifest in a variety of interference and diffraction effects. This makes it possible to manipulate electrons like photons by using components inspired by geometrical optics. When transmitted across a boundary separating regions of different density, charge carriers undergo refraction, much like light rays crossing a boundary between materials with different optical index. The possibility of exploiting this feature for use in low power transistors has generated continued interest since the discovery of graphene. However, the realization of an electrostatic gate-controlled transistor has not reached the full promises of early expectations for graphene. Here we summarize recent investigations into quantifying transport across p-n junctions as well as several paths toward realizing efficient graphene transistors utilizing electron optics and band structure engineering.

MAGNETIC STEERING ACROSS ULTRA NARROW JUNCTIONS

P-n junctions in graphene provide several interesting opportunities for new transistor architectures. Owing to the Klein tunneling phenomenon an angled gate with can theoretically be designed to reject all non-normally incident electrons providing a novel switching mechanism and without the need for a bandgap. Sharply defined p-n junctions admit the possibility of shaping electron flow via snake state trajectories and Veselago focusing of divergent electron beams. Despite the significant technical advancements in graphene device technology, robust demonstration of these various effects have remained limited. In an effort to more fully understand the electron transmission characteristics across a graphene p-n junction we have fabricated devices and reduced the junction width below 100 nanometers using a local graphite gate and a global silicon gate, and use transverse magnetic focusing to probe the propagation of carriers across the junction (Figure 1A). We demonstrate unambiguous agreement with the predicted Snell's law for electrons, including the observation of both positive and negative refraction [1]. Additionally, we utilize resonant transmission across the p-n junction to provide a direct measurement of the angle-dependent transmission coefficient. Comparing experimental data with simulations reveals the crucial role played by the effective junction width, providing guidance for future device design.

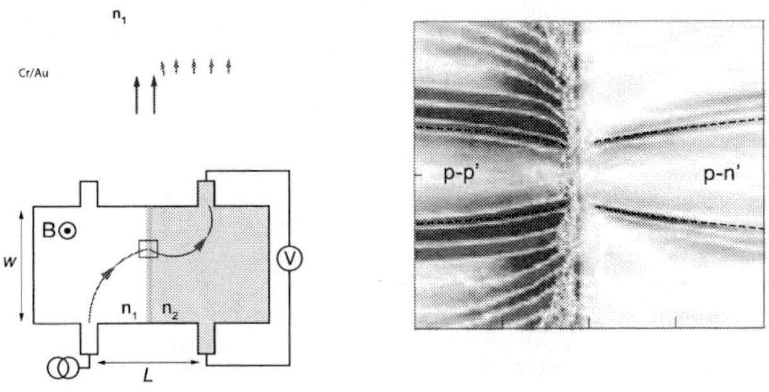

FIGURE 1. **(A)** Split-gate device design showing regions of different densities n_1 and n_2 (top) and the transverse magnetic focusing experimental design. **(B)** Resistance measured as depicted in (A) versus magnetic field and Si gate voltage while the graphite gate fixes the first region as p-type. **(C)** Snell's law demonstrated experimentally for the two cases of refraction. **(D)** Angle-dependent transmission intensity shown for experiment (blue), a simulated graded junction (red), and a theoretical abrupt junction (black).

ELECTRONIC COLLIMATION FROM SUPERLATTICE GATING

Band structure engineering of two-dimensional materials has become attainable in recent years and is now directly in the spotlight with the recent experimental discovery of Hofstadter's butterfly in graphene on hexagonal boron nitride (hBN). In these structures, the lattice mismatch between the graphene and BN substrates gives rise to a moiré-superlattice with wavelength controlled by the relative twist angle between the crystals. Fabricating device structures with features beyond the limitations imposed by the moiré-superlattice potential requires the capability to define artificial superlattices. The task at hand consists of fabricating synthetic lattices that are clean enough to leave transport uncorrupted, are located near enough to the two-dimensional material to create a well-defined structure, and are small enough to affect measureable changes in the band structure of the material. Van der Waals materials overcome the first two points rather easily, since pristine devices can be made using a variety of materials and a few layers of insulating hBN separate superlattices from the studied material. Recently we have met the latter challenge by utilizing a novel e-beam lithography technique.

In this talk we will discuss recent developments in the study of synthetic 1D superlattice potentials. We construct superlattice structures in two different approaches: superlattice building with negative e-beam resist or superlattice etching with positive e-beam resist. The strategy remains the same in either scenario: the superlattice creates a periodic variation in the dielectric medium beneath the graphene layer that modulates the effect of a bottom gate. A top gate controls the overall carrier density. Figure 2 (left) shows a schematic of the device. This approach yields superlattices down to 20nm full pitch as seen in Figure 2 (center).

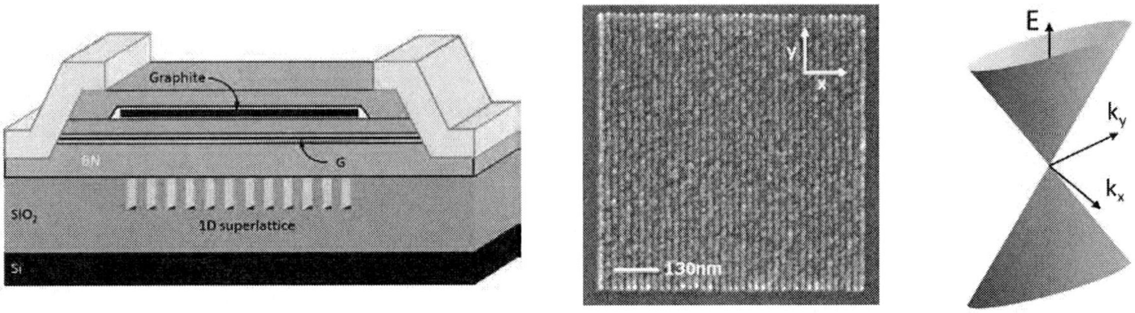

FIGURE 2. **Left**: schematic of electrostatic gating through a 1D superlattice. The top gate modulates the carrier density while the lower tunes the superlattice strength. **Center**: 20nm full-pitch superlattice of thin HSQ on top of SiO_2. **Right**: highly squeezed Dirac cone demonstrating the anisotropic transport resulting from the 1D superlattice.

These superlattices open up the possibility to achieve highly anisotropic transport in graphene heterostructures. As seen in Figure 2 (right), the Dirac cone narrows in the direction perpendicular to the 1D superlattice, resulting in a greatly reduced Fermi velocity while leaving the parallel transport unchanged. Theory also predicts the ability to completely quench the perpendicular Fermi velocity by tuning the lattice parameters [2]. This device architecture could provide electron collimation and restrict boundary scattering that has thus far hindered graphene transistor performance.

REFERENCES

1. S. Chen, Z. Han, M.M. Elahi, K.M. Habib, L. Wang, B. Wen, Y. Gau, T. Taniguchi, K. Watanabe, J. Hone, A.W. Ghosh, C.R. Dean, *Science* **30**, 1522-1525 (2016).
2. C-H. Park, L. Yang, Y-W. Son, M.L. Cohen, and S.G. Louie, *Nature Physics* **4**, 213 - 217 (2008) .

KEYWORDS

Graphene, electron optics, nanofabrication, p-n junctions, transistor

Creating an On/Off Berry Phase Switch in Circular Graphene *p-n* Junctions

Christopher Gutiérrez

Center for Nanoscale Science and Technology, National Institute of Standards and Technology, Gaithersburg, MD 20899, USA
Maryland NanoCenter, University of Maryland, College Park, MD 20742, USA

INTRODUCTION

The geometric phase is an example of anholonomy, where the phase of a quantum state may not return to its original value after its parameters cycle around a closed path; instead the quantum state may acquire a real measurable phase difference, also referred to as a Berry phase. The Berry phase is connected with the geometry of the quantum system, providing a measurable signature of system topology. In this talk, I will present scanning tunneling microscopy/spectroscopy (STM/STS) measurements of graphene quantum dots (GQDs) comprised of circular $p-n$ junctions in graphene/boron nitride heterostructures. I will show how the application of a weak perpendicular magnetic field (B ~100 mT) can act as a topological Berry phase "switch" resulting in sudden and large discrete energy splittings in the GQD resonances, orders of magnitude larger than the orbital or Zeeman shifts. This behavior results from switching on a π Berry phase associated with the topological singularities at the Dirac points in graphene. The electronic states can be switched on and off with small magnetic field changes on the order of 5 mT, producing strong modulation of quantum state energies and opening the door to a variety of novel optoelectronic device applications.

KEYWORDS

graphene, graphene quantum dot, Berry phase

Characterization of 2D Materials: Challenges and Opportunities

Robert M. Wallace

Department of Materials Science and Engineering, The University of Texas at Dallas
800 W. Campbell Road, RL10, Richardson, Texas 75080, USA

INTRODUCTION

The size reduction and economics of integrated circuits, captured since the 1960's in the form of Moore's Law, is under serious challenge. Current industry roadmaps reveal that physical limitations include reaching aspects associated with truly atomic dimensions, and the cost of manufacturing is reaching such values that only 2 or 3 companies can afford leading edge capabilities. To address some of the materials physical limitations, "2D materials" such as graphene, phosphorene, h-BN, and transition metal dichalcogenides have captured the imagination of the electronics research community for advanced applications in nanoelectronics and optoelectronics. This presentation will examine the state-of-the-art of these materials in view of our research on semiconductors, and the challenges and opportunities they present for electronic and optoelectronic applications. [1]

2D MATERIALS OPPORTUNITIES

Recent reviews have articulated the various opportunities presented through the potential utilization of the properties exhibited by 2D materials. [2-7] Among 2D materials beyond graphene, some exhibit semiconductor behavior, such as transition-metal dichalcogenides (TMDs), and present useful bandgap properties for applications even at the single atomic layer level. Examples include "MX_2", where M = Mo, W, Sn, Hf, Zr and X = S, Se and Te [8,9]. Research on TMDC materials has an extensive and rich bibliography dating back 50 years or more. Reviews of the earlier work have been published [10] and was the subject of a conference book series that extend well beyond TMDCs and continued through the 1990s. [1] The case for exploring these 2D TMDC materials for advanced nanoelectronic applications has also been recently summarized. [11] In addition to the potentially useful bandgaps at the monolayer thickness scale mentioned above, the atomically thin layers should enable thorough electric field penetration through the channel, thus enabling superior electrostatic control. Further, with such thin layers, the integration with suitable gate dielectrics can result in a mobility enhancement. From an interface perspective, the ideal TMDC channel material should have a dearth of dangling bonds on the surface/interface, resulting in low interface state densities which are essential for efficient carrier transport. Additionally, the fact that TMDCs incorporate d-orbital electrons may impart an increased functionality to devices not previously exploited in the more conventional semiconductors. Examples include spin-based and superconducting devices. The ideal TMD materials have much appeal, but the reality of significant densities of defects and impurities will surely compromise the intrinsic performance of such device technologies.

2D MATERIALS CHALLENGES

Integration of these materials with semiconductor industrial fabrication processes presents a number of challenges. Several synthesis methods have been employed to study 2D material thin film properties including mechanical/liquid exfoliation, chemical bath deposition, vapor phase deposition, and molecular beam epitaxy (MBE). [1] From a manufacturability and cost perspective, vapor phase (including chemical and atomic layer) deposition are the subject of intense research activity in the electronics industry, while MBE methods facilitate the research of large thin films in advance of precursor development for CVD and ALD.

Among key properties of interest in device applications include the control of impurities and defects as well as their passivation. Evaluations of geological and synthetic TMD materials have been performed using inductively-coupled plasma mass spectrometry (ICPMS). [1, 12,13] Impurity levels exceeding industry roadmap specifications ($>1\times10^{10}/cm^2$) is observed for some elements, drawing attention to growth conditions and precursor purity in the synthesis process. Surface/interface defect densities and classification has also been explored, from a physical (and electrical) characterization perspective, where many TMD material surfaces appear to exhibit densities on the order of $10^{12}/cm^2$ ($10^{12}/cm^2$ eV). [1,12-14] The presence of such defects (e.g. extended vacancies) and impurities can impact contact and mobility behavior [15,16], indicating that, like conventional semiconductor interfaces, passivation treatments will likely be required. [17,18].

In this presentation, we will review these challenges and the progress reported in meeting them in recent years, where the community appears to be poised to make several breakthroughs toward enabling these materials for advanced device concepts.

This work is supported in part by the SWAN Center, a SRC center sponsored by the Nanoelectronics Research Initiative and NIST. It is also supported in part by Center for Low Energy Systems Technology (LEAST), one of six centers supported by the STARnet phase of the Focus Center Research Program (FCRP), a Semiconductor Research Corporation program sponsored by MARCO and DARPA, and the US/Ireland R&D Partnership (UNITE) under the NSF award ECCS-1407765.

REFERENCES

1. S. J. McDonnell and R.M.Wallace, "Critical Review: Atomically-Thin Layered Films for Device Applications based upon 2D TMDC Materials", *Thin Solid Films*, **616**, 482-501 (2016).
2. A.K. Geim, I.V. Grigorieva, "Van der Waals heterostructures," *Nature* **499,** (2013) 419–425.
3. L. Colombo, R.M. Wallace, R.S. Ruoff, "Graphene growth and device integration," *Proc. IEEE* **101,** 1536–1556 (2013)
4. D. Jena, "Tunneling transistors based on graphene and 2-D crystals," *Proc. IEEE* **101**, 1585–1602 (2013).
5. G. Fiori, et al., "Electronics based on two-dimensional materials," *Nat. Nanotechnol.* **9**, 768–779 (2014).
6. M. Xu, T. Liang, M. Shi, H. Chen, "Graphene-like two-dimensional materials," *ACS Chem. Rev.* **113**, 3766–3798 (2013).
7. G.R. Bhimanapati, et al., "Recent advances in two dimensional materials beyond graphene," *ACS Nano* **9,** 11509–11539 (2015).
8. C. Gong, et al., "Band alignment of two-dimensional transition metal dichalcogenides: application in tunnel field effect transistors, *Appl. Phys. Lett.* **103,** 053513 (2013); *Appl. Phys. Lett.* **107**, 139904 (2015).
9. C.Zhang, et al., "Systematic Study of Electronic Structure and Band Alignment of Monolayer Transition Metal Dichalcogenides in Van der Waals Heterostructures," *2D Materials*, **4**, 015026 (2017).
10. J.A.Wilson and A.D. Yoffe, "Transition metal dichalcogenides discussion and interpretation of observed optical, electrical and structural properties," *Adv. Phys.* **18** 193–335 (1969).
11. D. Jena, "Tunneling transistors based on graphene and 2-D crystals," *Proc. IEEE* **101**, 1585–1602 (2013).
12. R. Addou, at al., "Impurities and electronic property variations of natural MoS_2 crystal surfaces," *ACS Nano* **9**, 9124–9133 (2015).
13. R. Addou and R.M.Wallace, "Surface Analysis of WSe_2 Crystals: Spatial and Electronic Variability," *ACS Applied Materials and Interfaces*, **8** (39), 26400-26406 (2016).
14. C. D. Young, wt al., "Evaluation of Few-Layer MoS_2 Transistors with a Top Gate and HfO_2 Dielectric," *ECS Transactions*, 75(5) 153-162 (2016).
15. S. McDonnell, et al., "Defect dominated doping and contact resistance in MoS_2," *ACS Nano* **8**, 2880–2888 (2014).
16. N. Ma and D. Jena, "Charge scattering and mobility in atomically thin semiconductors," *Phys. Rev. X* **4** 011043 (2014).
17. M. Amani, et al., "Near-unity photoluminescence quantum yield in MoS_2," *Science* **350** (6264), 1065-1068 (2015).
18. M. Amani, et. al, "Recombination kinetics and effects of superacid treatment in sulfur and selenium based transition metal dichalcogenides," Nano *Letters*, **16** (4), 2786–2791 (2016).

KEYWORDS

2D materials, dichalcogenides, Van der Waals, scaling, defects, impurities, passivation

High Resolution EBI for Pattern Fidelity Monitor

Fei Wang, Brian Lee, Kevin Liu, Jack Jau

HMI, an ASML Company: 1762 Automation Parkway, San Jose, CA, 95131

INTRODUCTION

As design rules continue to shrink down to sub-10nm, multiple patterning technologies have been intensively implemented in order to transfer fine features from design to wafer. Pattern fidelity is becoming more and more important in determining process window and device yield. Traditionally, CD measurement by CD-SEM was used to characterize pattern fidelity, especially for one-dimensional pattern. However, due to limited throughput, CD-SEM could not meet the increasing requirement on the volume of measurements, both from research and development phase as well as from process monitoring stage. A high speed metrology method is needed for pattern fidelity characterization and monitoring. In this work, we will study the metrology capabilities of an advanced e-beam inspection (EBI) system, which can provide order of magnitude throughput improvement relative to CD-SEM technology. High speed metrology can be used for OPC model calibration and verification in research and development phase; it can also be used for pattern fidelity monitoring during production phase.

METROLOGY CAPABILITY OF ADVANCED EBI

Over the past decade, image resolution of e-beam inspection systems has continuously improved and image quality has become sufficient to perform high precision measurement. In this work, HMI's eP4, a 2 nm resolution e-beam inspection system, was used to evaluate the metrology capabilities and to study feasibility for pattern fidelity measurement.

Study of Measurement Sensitivity

Measurement sensitivity is a critical spec, which defines the minimal CD change a metrology system could robustly measure. In order to meet requirements for sub-10nm technology node, a metrology system needs to capture 0.25nm CD change. A special-designed wafer with known pitch CD change was used to quantify measurement sensitivity in the experiment.

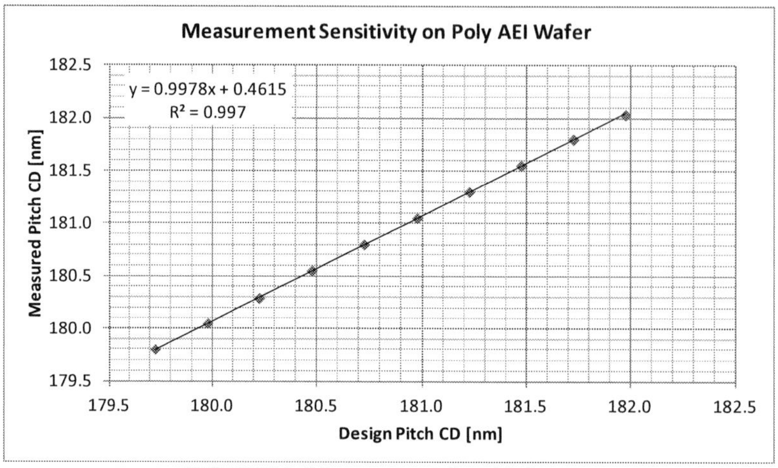

FIGURE 1. Measurement Sensitivity

Study of Contour Extraction on Complex 2D Pattern

To characterize a two-dimensional (2D) pattern, the information carried by simple cutline CD becomes limited. Edge placement (EP) information, which can be measured on contour, provides a solution for pattern fidelity study. In this work, we will study contour extraction from high resolution SEM image scanned by eP4 system. A basic work flow includes: image scan on eP4, align SEM image with design data, contour extraction, and edge placement measurement.

FIGURE 2. Contour Extraction from SEM Image

Cutline CD defined on contour was used to benchmark measurement results from eP4 against CD-SEM. High correlation can be achieved for both horizontal and vertical CD.

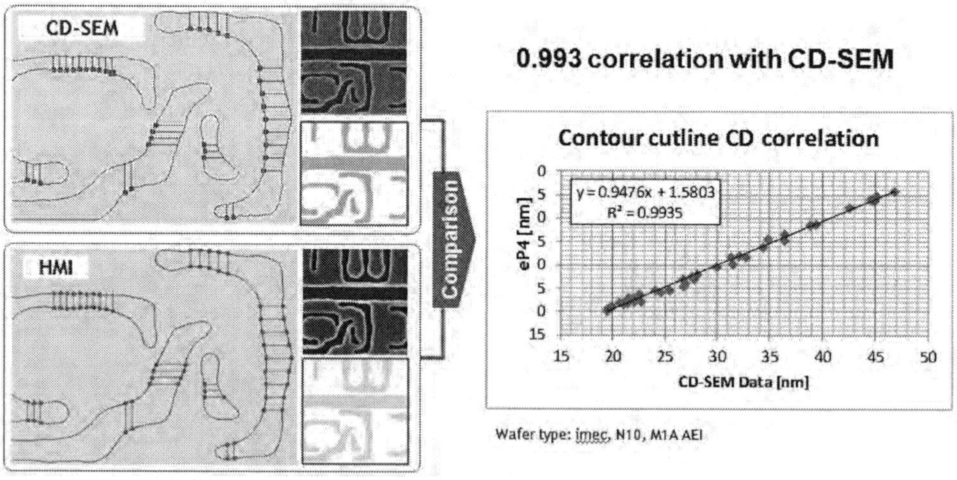

FIGURE 3. Benchmark with CD-SEM of cutline CD measurement on contour

APPLICATIONS

High speed metrology enabled by advanced EBI brings desirable capabilities for applications in R&D and production phase.

OPC Model Calibration

OPC model calibration needs large volume of high quality metrology data to get good model accuracy and shorten iteration cycles. Limited by throughput of metrology system, current OPC model calibration utilized tens of thousands of gauges, which can't provide sufficient pattern coverage and process condition coverage. Moreover, limited repeat measurements on the same pattern type can't efficiently reduce random process variation, which caused OPC mode wrongly compensate for the process variation but not the true mean value. To overcome the difficulties, high speed metrology system will be utilized to provide massive measurement data, in the volume of million gauges.

Pattern Fidelity Monitoring

Pattern fidelity monitoring (PFM) is very important element of ASML's holistic lithography. Defects library, which is observed by Lithography Manufacturability Check (LMC) prediction from OPC or wafer inspection result, provide high potential defects hotspots. With those hotspots and wafer topography and focus information which are observed by scanner and YieldStar, computation engine can predict high risk patterns and their wafer positions. High throughput EBI can measure pattern fidelity of these patterns then give feedback to user to judge pattern health. It's very helpful on yield enhancement.

SUMMARY

Foe leading edge technology, pattern fidelity is becoming an important factor limiting processing window and device yield. A high speed metrology provides capabilities of characterizing and monitoring large volume of critical patterns. New applications can be developed for OPC model calibration in R&D phase and pattern fidelity monitoring in production phase.

KEYWORDS

e-beam, massive measurement, pattern fidelity monitor, die to database, contour

Extending Optical Inspection to the VUV

Keith Wells, Grace Chen, Matthew Derstine, Steve Lange, and David Shortt

KLA-Tencor Corporation, One Technology Drive, Milpitas, CA 95035

INTRODUCTION

Optical wafer inspection is the only technology, over the last two decades, that has proven to be cost effective for wafer-excursion monitoring in a high volume manufacturing (HVM) environment. Solutions have been commercially developed that enable optical inspection wavelengths down to 190 nm and allow inspection rates of approximately 1200 cm^2/hour. These rates in turn allow effective wafer-level signature monitoring at 1–2 wafers per hour that enable design of experiments for process debugging and will eventually enable statistical process control in factories currently developing logic devices at 7 nm and 5 nm design rules. As customers continue to shrink design rules, the question arises as to what optical technologies exist that will enable HVM inspection for 3 nm design rules.

This paper explores both the justification and technologies needed to extend optical inspection to the 3 nm design rule through the implementation of vacuum ultraviolet (VUV) wavelengths, defined for the purposes of this paper as the range 100–200 nm. Delivering inspection capabilities within this range of wavelengths increases resolution of the optical systems as well as creating material contrast that will enable new inspection capabilities for optical inspection. The technology discussed in this paper includes optics, coatings, light sources, sensors and image-formation techniques to enable detection capability at 3 nm. The challenge is not only to enable the core technologies, but to combine them to enable inspection rates that make the technology cost effective for inline monitoring.

VALUE OF VUV WAVELENGTHS

The most obvious reason to extend an optical inspection platform to shorter wavelengths is to increase the resolving capability of the inspector. The resolving power of an optical microscope is proportional to the ratio of the optical wavelength to the effective numerical aperture of the imaging system, or λ/NA. If the inspection problem can be solved by increasing resolving capability, we would expect a resolution benefit of almost 2× over the current state of the art when moving from a mean wavelength of ~200 nm down to a mean wavelength of ~110 nm. This would result in the ability to resolve line and space features with a minimum separation of 125 nm. While this is a major step forward for optical inspection capability, it must be supported by at least 3 other factors: cross-sectional scatter enhancement, material contrast enhancement, and optical-enhancement techniques for pattern (noise) suppression. This section will discuss the first two mechanisms. We will discuss pattern-suppression techniques later in the paper.

Signal production for an isolated particle in wafer inspection can be described by Mie scattering, where the particle's scattering cross section, or dark field (DF) response, changes with wavelength as $1/\lambda^4$. In bright field (BF) imaging the signal depends on resolution and varies as $1/\lambda^2$. Our experience with actual measurements of patterned wafers shows an exponent slightly lower than predicted for both BF and DF. Nevertheless, these effects are nonlinear and will result in increased signal on the order of at least 3× to 12× for isolated defects detected in BF and DF modes, respectively. While resolving power and Mie scattering enhancement are clear justifications for moving to VUV wavelengths, they are not even the most interesting value that VUV wavelengths could provide for patterned wafer inspection.

Calculated values of reflectance and penetration depth for some common semiconductor materials are shown in Figure 1. Reflectance fluctuates widely over the wavelength range 100–1000 nm, although for some of the metals there is a general trend toward lower reflectance at shorter wavelengths. Most notable is the onset of strong absorption "divides" for silicon, silicon nitride, and silicon dioxide at different wavelengths, shown by the penetration depth in the lower panel of Figure 1. The penetration depth is given by $\lambda/4\pi k$, where k is the absorption coefficient. Values of the optical constants n and k were taken for bulk materials from Palik (1). The onset of these absorptive transitions, coupled with broadband inspection, offers the ability to tune the inspection wavelengths to lie on either side of one of the divides. Thus the material can be rendered either transparent (by choosing wavelengths longer than the divide) or opaque (by choosing shorter wavelengths).

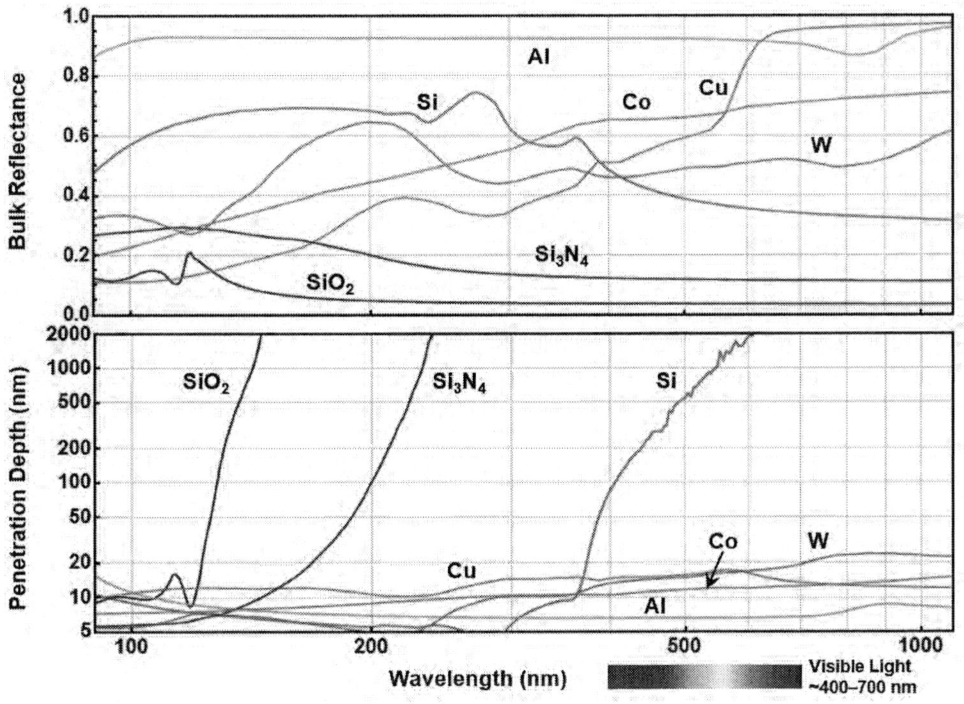

FIGURE 1. Reflectance and penetration depth for some materials common in semiconductor manufacturing. Note logarithmic scale for wavelength and penetration depth.

It has long been observed that the optical properties n and k of the defect and its surrounding structures can make huge changes to the general signal-to-noise ratio (SNR) trend with shrinking wavelength, involving both the defect signal and the associated wafer noise from current-layer and underlying-layer process variations. One major change we see across a broad waveband is the effect on signal and SNR when a material changes from being opaque to transparent. Historically, we first observed this with polysilicon, where the transition occurs at about 400 nm (see the silicon penetration depth in the lower panel of Figure 1). This creates a corresponding need to have inspection wavelengths both above and below for maximizing SNR. Later on, we observed that silicon nitride (commonly called nitride) has an absorption edge at about 230 nm, again creating the need for wavelengths above and below the 230 nm transition. For example, defect detection is required in cases where the defect and the surrounding materials are nitride, as well as cases where the defect is nitride but the surrounding structure is a completely different material. This can only be realized if the inspector contains wavelengths above and below the 230 nm absorption edge.

These material transitions lead us to a more interesting and impactful capability of VUV inspection: the change in material contrast generated by the very shortest wavelengths in the VUV spectrum. In the VUV, silicon dioxide (commonly called oxide) has an absorption edge at about 130 nm, again suggesting the need for wavelengths above and below in order to maximize SNR. Looking at the optical properties of oxide, we can predict the absorbing effect of a layer of that material. Figure 2 shows the penetration depth (light intensity $1/e^2 = 0.135$) as a function of wavelength for oxide over the short-wavelength portion of the VUV range.

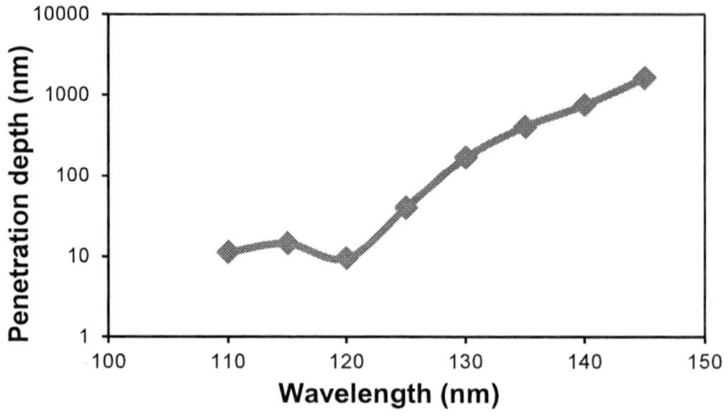

FIGURE 2. Penetration depth of silicon dioxide vs. wavelength. Note logarithmic vertical scale.

The major effect of VUV wavelengths for patterned wafer inspection is the absorption of the light by the top oxide layer. The inspection of the patterned oxide layer on a fully processed advanced logic wafer, at wavelengths longer than the VUV range, is generally limited by strong scattering signals produced by underlying metallic materials and their associated process variations. These sources usually produce much more signal than a defect we wish to detect in the oxide layer. Thus, inspecting with a wavelength where oxide is absorbing can minimize the unwanted signal from the layers below. Oxide layers have a thickness of about 50 nm, and light needs to pass through to the lower layers and back up through the oxide to be detected, so a combined thickness of 100 nm is a general guideline for absorption. From Figure 2 we see that a wavelength less than about 130 nm satisfies this criterion and would greatly remove the underlying noise problem.

To study the oxide effect on real semiconductor structures, we can employ a rigorous coupled wave approximation (RCWA) simulation to solve Maxwell's equations. Figure 3 shows a model for an oxide inspection where lines and contact holes are etched into the top oxide and line-edge roughness (LER) on tungsten metal is a noise source just below.

FIGURE 3. Simulation model of silicon dioxide (gray) over tungsten (yellow).

The result of the RCWA simulation on this example is shown in Figure 4, where signals for a number of BF inspection modes are plotted as a function of wavelength. This type of layer has always been a gap for traditional inspectors because the oxide material does not produce much signal, and at longer wavelengths the oxide is totally transparent, allowing underlying metallic wafer noise to dominate. Typically, SNR < 1 for current inspection wavelengths greater than 190 nm. However, VUV wavelengths below 130 nm produce SNR in the detectable range, assuming threshold of about 2 to ensure detection. Such performance would close this long-term gap.

Today, inspection of oxide defects is routinely performed by scanning electron microscopes (SEM). Penetration of electrons and generation of secondary electrons occurs within a few nanometers of the oxide surface, making this technique sensitive to detecting the presence of single-digit nanometer residual oxide thickness. The invention of voltage contrast (VC) modes on SEMs have helped to dramatically increase inspection rates, but state-of-the-art VC scan rates are still projected to be two orders of magnitude slower than a properly engineered VUV optical inspector. The ability to implement VUV inspection would mean that for the first time the industry could inspect under-etched contacts or oxide bridges at speeds that support full wafer inspection and enable detection of random low-density etch excursions in HVM applications.

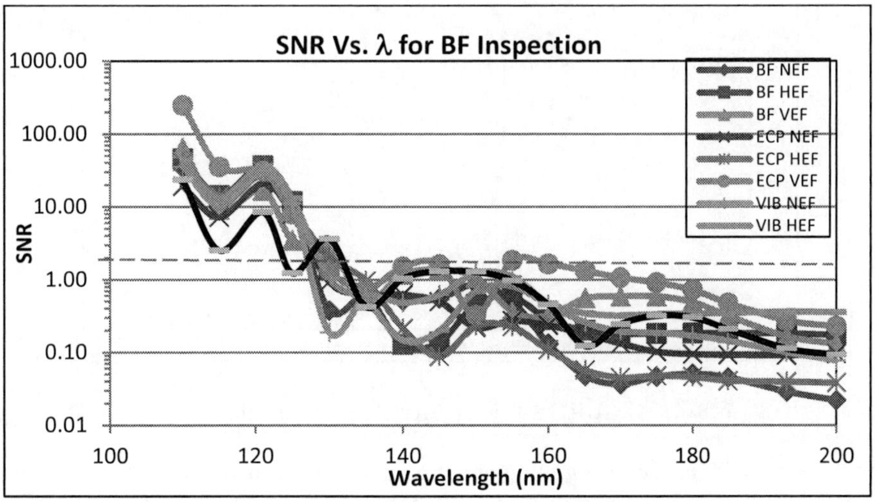

FIGURE 4. Simulation result of SNR for oxide over tungsten. A threshold of SNR = 2 is shown as a red dotted line. Note logarithmic vertical scale.

OPTICAL FEASIBILITY

Few optically refractive materials transmit light at VUV wavelengths. For the few materials available, the residual absorption and large dispersion in the bulk material make a conventional catadioptric broadband objective lens impossible. Consequently, as in the extreme ultraviolet (EUV) wavelength range, the objective and most of the rest of the optics will be reflective. Unlike the EUV steppers, however, inspection systems can have a central obscuration and rotational symmetry. This enables the design of two- and four-mirror configurations that have high resolution and reasonable optical field. Figure 5 shows a Schwarzschild design form for a two-mirror objective. Aberration corrections to achieve high Strehl across a useful field size drive objective mirror elements to be aspheric in form. Surface-figure requirements are also challenging, but forming and polishing techniques developed for EUV technology can be relied upon to produce the required surface figure. The remaining challenges of the objective design are mounting stability, thermal stability, and vibration reduction. The tight optical specifications drive the requirement for active real-time corrections: micro-positioning capability of the lenses relative to one another, as well as active cooling, are expected to be inherent design features with real-time control capability. Vibration is expected to be controlled through the design of the individual objective mounting elements. Existing lithography applications have retired problems of similar scope, and the objective lens itself is not expected to be a major obstacle to the technical feasibility of a VUV inspection system.

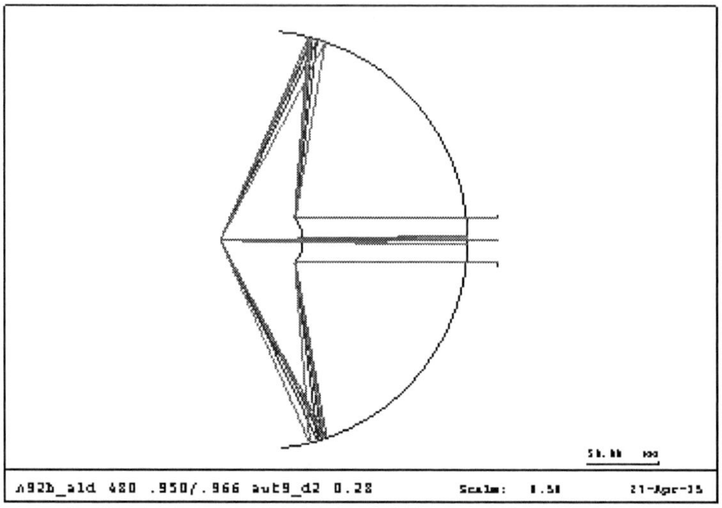

FIGURE 5. Two-mirror objective design.

The real challenge for a VUV inspector is getting enough light on the imaging camera to capture defects with high signal-to-noise while enabling inspection speeds that are practical for HVM monitoring applications. Part of the solution is a bright, powerful light source (see the next section) and the other is high-reflectance optical elements (>85% per element) that are stable over the life of the product. Solving the light-source problem and the coating stability issue are keys to ensuring that sufficient light reaches the detector to enable both design objectives: sensitivity and speed.

High-reflectance mirrors in the VUV are traditionally based on aluminum due to its very high reflectance throughout the VUV (see Figure 1). These mirrors are often enhanced and protected with MgF_2 or other dielectric materials. Our initial experiments with these coatings at high VUV fluences show a significant degradation in reflectance at short wavelengths for a wide range of exposure conditions. Figure 6 shows the comparison between the reflectance of a sample stored for a year and one with about 100 hours of VUV exposure. We have confirmed that this degradation is associated with VUV exposure and is likely related to trace gases in our purge system interacting with the mirror surface. Early data would suggest that our current coatings could be employed to create an inspection tool down to 160nm, but this is not sufficient to enable the opacity of oxide. Work continues to understand these processes and to find ways to improve the purity of the environment and to make the protective dielectric coatings more effective over long times. Creating stable and long-lasting coatings for optical elements represent one of the highest technical risks in developing a practical VUV inspector.

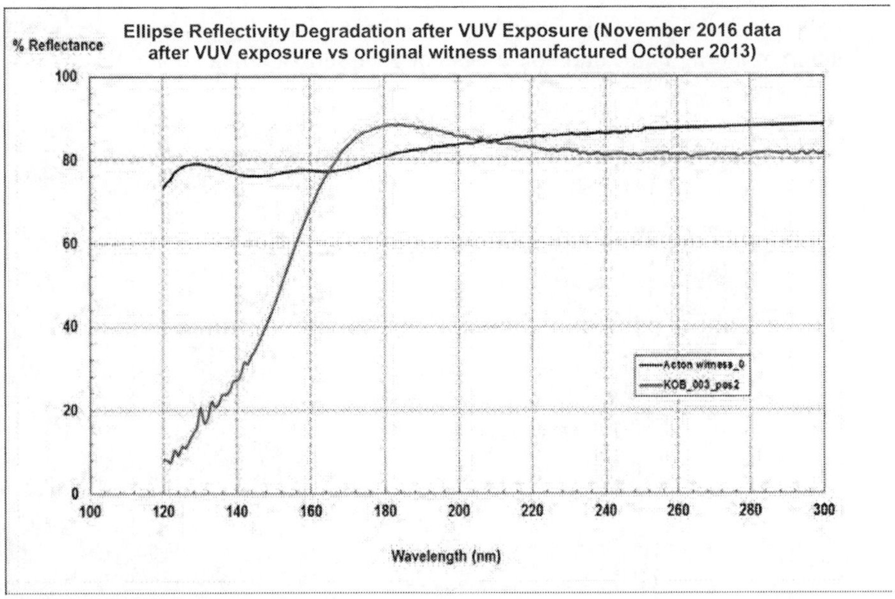

FIGURE 6. VUV degradation of the reflectance of protected aluminum. Unexposed sample (black); exposed sample (red).

For other optical elements, outside of the main objective, there are some materials like MgF_2 and LiF which are transmissive down to about 115 nm. These materials will enable development of components such as homogenizers, prisms and optical windows. Utilization of these are critical to improve the overall system performance and provide options for manipulating the illumination (see later section). Unfortunately, high VUV fluence leads to rapid creation of color centers in these materials. In MgF_2 this leads to increased absorption outside of the VUV band. We expect that color-center creation can be minimized but not eliminated. This is not likely an issue for components that see relatively low power, but does lead to problems for illuminator components, especially for the light-source windows. It is expected that these components will require replacement at regular PM intervals and are another significant risk to the development of a VUV inspector.

LIGHT SOURCE CHALLENGES

KLA-Tencor has successfully used continuous-wave (CW) laser-sustained plasmas (LSP) for inspectors with wavelength ranges from 190–450 nm. The source for VUV inspectors can make use of this same approach but with significant changes in the implementation and configuration.

Fused silica bulbs are used to contain the gas used for plasmas in the 190–450 nm range. However, fused silica is not suited for the VUV because it is opaque below 185 nm. This means that alternative high-pressure containment systems must be used. The major challenge in the architecture of such a system is finding an optical element for an exit window that can withstand the pressure and thermal gradients of the source, while still being able to maintain high transmission. From our current understanding, only a fluoride crystal exit window will contain pressures up to 100 bar and still transmit the VUV light. The major challenge for windows is that temperature gradients in fluoride crystals must be kept below a few degrees Celsius per millimeter to avoid rupture. This is exacerbated by color-center generation in the exit window. Color-center formation results in increased absorption and has the potential for significantly higher heating (and thus temperature gradients) than would otherwise be expected, leading to tighter constraints on the window designs and cooling. We are investigating approaches to minimize these problems.

The plasma containment system is also a major challenge. Current lamp houses use cold-mirror technology to permit the same solid angle around the plasma to be used for pumping the plasma and collecting the light. VUV coating technology seems unlikely to produce a suitable cold mirror, suggesting that the solid angles used for pumping and collection will need to be different. This will drive efficiency lower and push up the requirement to make sure all 4π of the solid angle be used for either pumping or collection. As lamps cannot be used, one option is to combine the pump focusing and collection optics together and then seal this at the inputs and output. Figure 7 shows the configuration and example of a prototype lamp house designed with this implementation.

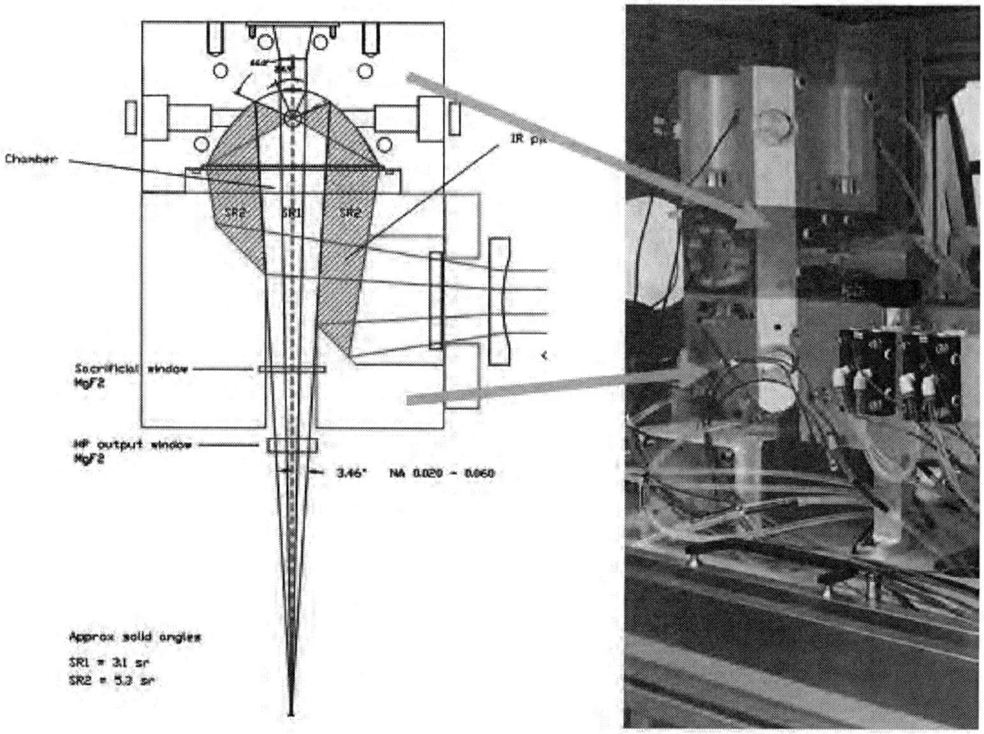

FIGURE 7. Prototype lamp house configuration.

Beyond finding an acceptable container and windows to get the light out, the absolute amount of light needed for a full HVM tool is quite large. Assuming aluminum mirrors, with stable reflectance at ~85%, the HVM requirement drives a requirement of more than 100 W of collectable light in the VUV. This drives radiance of the source to be high, which requires the plasma be hotter and larger than the plasmas used on current inspectors. The plasmas need to be hotter so that we have a higher radiance limit (set by Planck's black body radiation law), and the plasmas need to be larger so that the path length through the plasma approaches the black body condition. At the same time, too hot a plasma would quickly lower the efficiency because of radiation emitted outside of the working wavelengths of the system. We are currently working to optimize the pump/collection configuration as well as the operating conditions. Current results are within a factor of 10 of the requirements, and it is believed improved lamp house designs and optimization of the pump power, gas mixture and pressure will meet the output requirements for HVM inspection speeds.

THE FEASIBILITY OF OPTICAL SIGNAL ENHANCEMENT TECHNIQUES

An optimized VUV optical system with a high Strehl >90%, 0.9 NA objective, operating at 110 nm mean wavelength will have a resolving power no better than 125 nm. At the 3 nm design node, feature and defect sizes will be less than 1/10 of the resolution of such an optical inspector. Therefore, as beneficial as resolution may be, inspection techniques relying solely on raw optical resolution are no longer sufficient for wafer inspection. Several optical signal enhancement techniques have been developed to boost defect signal and reduce noise. Typical techniques are spectral filtering, illumination optimization, DF aperturing (also known as Fourier filtering), and polarization selection. These optical enhancement techniques have extended the sensitivity of optical inspectors such that they can detect defects as small as 1/20 of the optical resolution today, and we expect this trend to continue.

These techniques are vital to keep optical inspection relevant at the 3 nm design node. Figure 8 illustrates the power of suppressing the collection of spatial frequencies from the actual line and space pattern and collecting only the spatial frequencies associated with the defect. The detection of the defect does not depend on the resolving power of the optics—imaging the line and space separation is not required—but rather detection is dependent upon selecting either an illumination and/or collection mode that suppresses the ability of the system to resolve the line and space pattern while still collecting light scattered by the defect.

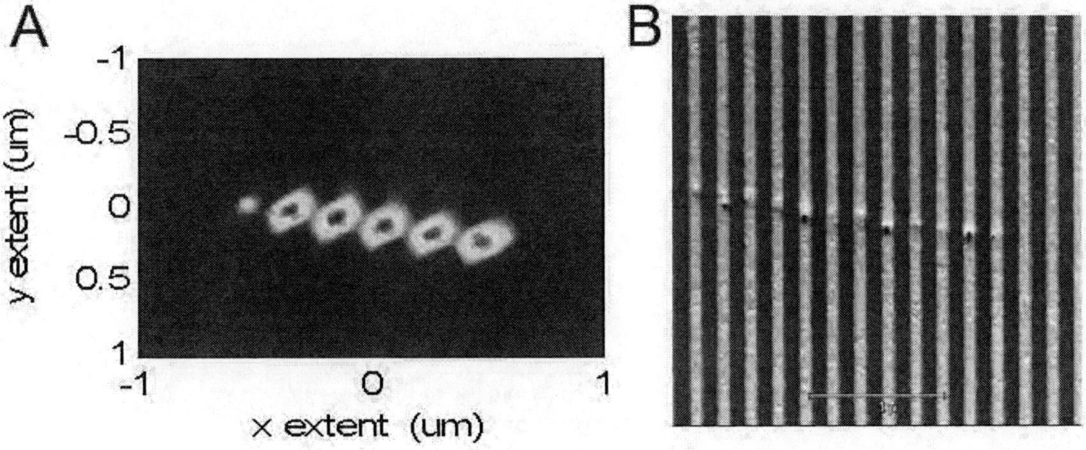

FIGURE 8. Suppression of line and space pattern while maintaining scratch signal detection. (A) Pattern-suppressed optical image of scratch defect; (B) SEM image of pattern and defect.

These techniques provide sensitivity that is dependent upon the pattern and the defect location in the pattern. It is extremely difficult to predict the detection sensitivity resulting from a particular optical and pattern-noise suppression technique unless the exact defect size, orientation in the pattern, pattern geometry and *n* and *k* values of the defect and surrounding pattern are known. Today we see routine capture of defects in the sub-10 nm range with use of 190 nm light employing optical configurations designed to suppress pattern or directly enhance defect signal. The overall objective is to detect the relevant defect types at sufficient capture rate to allow statistical monitoring, rather than to guarantee absolute capture rates. Coupling the resolving power of a VUV objective and the resulting increase in scattering cross-section with appropriate optical enhancement techniques will allow sufficient capture rates of sub-10 nm defects to enable high-speed full-wafer inspection for the 3 nm design rule.

For a VUV inspection platform, implementation of these signal enhancement techniques is challenging. Illumination-angle optimization and DF aperturing require the system to provide well-corrected, accessible pupils in both illumination (for illumination-angle optimization) and imaging paths (for DF aperturing). Additional mirror elements are needed to create physically accessible pupil planes in the illumination and imaging paths. VUV mirror reflectance (~80%) is significantly lower than transmission of high-end refractive lenses at longer wavelengths (~99%). Total system transmission T follows $T = R^n$, where R is the reflectance of a single mirror and n is number of mirror elements in the system. Figure 9 shows the system transmission with reflectance projected for VUV as a function of the number of elements. Adding requirements for well-corrected, accessible pupils is likely to increase the number of mirrors in the system. As it is already difficult to create enough light to meet the throughput requirement, implementing illumination optimization and DF aperturing schemes would impose additional challenges to the light source development to maintain high throughput for the platform. To create a practical HVM system, we estimate the number of reflective surfaces at $R \sim 80\%$ will need to be constrained to 16 or fewer. While challenging, we believe this constraint is not a barrier to the development of the system.

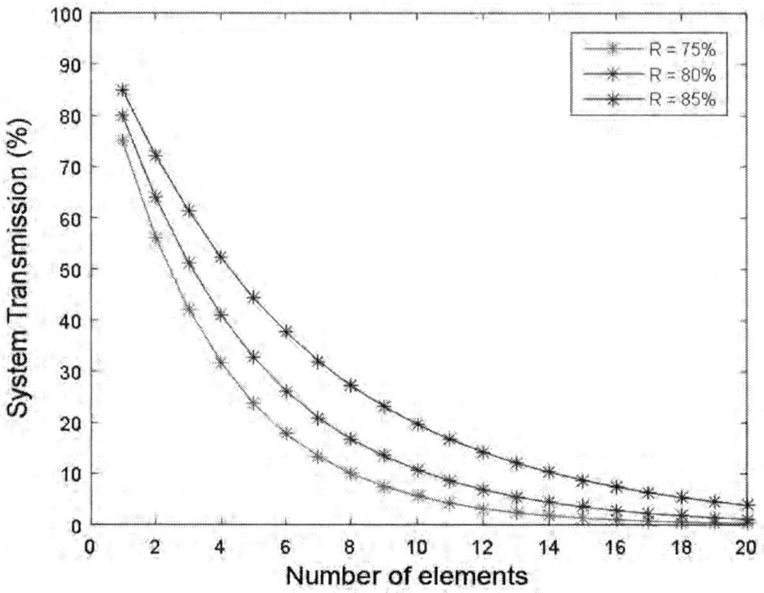

FIGURE 9. Total system transmission as a function of mirror reflectance and the number of mirrors employed in the system.

As stated earlier, spectral filtering is a major requirement for an optimal system. The ability to operate with wavelengths in the range 110–130nm will enable unique oxide defect detection. However, achieving a practical means of wavelength selection is another major challenge. For wavelength greater than 193 nm, optical filtering can be achieved with multi-layer coating designs. The typical spectral filter in the UV and visible range contains 50–100 layers to achieve high transmission and sharp bandpass edges. For a VUV system, the use of multi-layer dielectric coatings is not practical to achieve the bandwidth range, transmission, and lifetime. KLA-Tencor is investigating dispersive materials for spectral separation, and early analysis and modeling show promise.

SENSOR DEVELOPMENT FOR VUV APPLICATIONS

For any DUV or VUV platform, light generation and high transmission efficiency are difficult. The limitations of these design constraints have been historically mitigated by sensor architecture selection and optimization. KLA-Tencor's VUV system will continue to leverage the innovation of sensors employing time delay integration (TDI) which provides higher system data rate (several tens of giga-pixels per second) while at the same time allowing long integration time (up to fractions of a millisecond) via the number of integration stages. KLA intends to develop TDI sensors to further increase the number of integration stages. Increasing the number of stages can increase the risk of sensor-induced blur, present yield challenges due to increased size, and introduce mechanical packaging difficulties. None of these concerns is a limiting factor in the creation of the system.

However, the sensor does face the same lifetime challenges as other low-fluence optical elements in the system. An unprotected sensor exposed to even low-fluence VUV wavelengths is expected to have minimal useful lifetime. To prevent VUV damage on the sensor, a protective coating must be applied, but the coasting itself cannot have a significant impact on the sensor quantum efficiency (QE). Fortunately, a common solution for sensor protection has already been developed for astronomical applications. A thin boron coating can potentially harden the sensor against VUV damage while maintaining acceptable QE. This enables KLA-Tencor to extend its traditional sensor technology to the VUV with lower technical risk.

CONCLUSION

As design rules for semiconductor devices shrink toward the 3 nm node, existing optical inspection wavelengths and technology are not likely to be sufficient to defect defects on advanced structures. VUV wavelengths offer the potential to meet the inspection challenge, as they provide significant value for defect detection not only due to resolution improvement but also through important material-contrast mechanisms. However, significant innovative developments in optics, coatings, light sources, sensors and image-formation techniques will be required to enable a practical VUV inspection system. We have performed initial investigations in these areas and conclude that a HVM inspector operating at VUV wavelengths is difficult but feasible. Looking forward, the challenge is twofold: to enable and refine the core technologies, and to combine them with appropriate systems tradeoffs to achieve a cost-effective solution for inline monitoring.

REFERENCES

1. E. Palik, *Handbook of Optical Constants of Solids*, Academic Press, 1998.

KEYWORDS

Optical inspection, defect detection, semiconductor manufacturing, vacuum ultraviolet, VUV.

Optical Inspection at 10 nm Node and Beyond: Perspectives and Challenges

Mehdi Vaez-Iravani

Applied Materials, 3340 Scott Boulevard, Santa Clara, CA 95054, USA

INTRODUCTION

For decades optical inspection has played a central role in enabling Moore's law. In the past, high resolution (at times referred to as "bright field") inspection was able to provide broad coverage of the defects of interest fundamentally through its resolution. This basic capability, however, has not been sufficient in the recent nodes. Accordingly, the industry has progressively relied on other, mostly imaging-independent attributes to provide acceptable levels of performance in defect detection. Naturally, the impediments in the way of defect detection are rapidly growing as Moore's law marches on.

This talk will discuss the nature of the problem, and a number of potential solutions, in semiconductor defect detection at the sub 10 nm nodes. It will be shown that despite compounding levels of difficulty, optical defect inspection is expected to remain relevant for the foreseeable future. The increasing interplay between optical and ebeam based techniques will be emphasized.

KEYWORDS

Optical inspection; defect detection; pattern attributes

Current and Future Critical Dimension Metrology Perspective for Sub-10nm Process

Mari Nozoe

Hitachi High-Technologies Corp., 24-14, Nishi-shimbashi 1-chome, Minato-ku, Tokyo 105-8717, Japan

1. INTRODUCTION

Semiconductor device performance has been enhanced, bit cost reduced by miniaturization following Moore's Low. But in recent years, to obtain even better performance, the technology is expanding into more diversified and complex areas. EUV lithography, new materials, multiple patterning, 3D structures, new memories, system-in-package approach are being employed aggressively. These movements result in an increase in process steps and complexity of structures, which in turn decrease tolerance to fabrication and layer-to-layer alignment. Because of these changes in needs, the metrology becomes more important than ever, in order to grasp more comprehensive and detailed information.

Although novel metrology has many different approaches, this presentation will focus on the critical dimension (CD) metrology.

2. CD METROLOGY OVERVIEW

Metrology applied to high volume manufacturing (HVM) lines should be non-destructive, automated, high-speed, and stable. From this point of view, high-precision 2D measurement technology such as CD-SEM and high-speed scatterometry such as Optical CD (OCD) are mainly used depending on applications. Characteristics of these technologies are summarized in Table 1. While CD-SEM, which is based on high resolution imaging, can measure any local pattern accurately, it has difficulties when it comes to capturing height profiles or covering large areas. OCD on the other hand, can measure large areas quickly and obtain height profile factor. However, it cannot collect information from one specific pattern and its measurement is based on calculations from spectral models, making it an indirect measurement that takes time for optimizing the model.

TABLE 1. Summary of major metrology technology characteristics

Metrology method	Pros	Cons
CD-SEM (image-based)	➢ Measure any complex arbitrary feature ➢ Direct measurement from image (no modeling) ➢ Automated, stable, precise	➢ Low throughput for large area coverage ➢ Height measurement
Optical CD (model-based)	➢ High throughput (for global monitoring) ➢ High sensitivity, CD/ 3D profile measurement ➢ Automated, stable, precise	➢ Average measurement only (unavailable for complex pattern) ➢ time for build model (recipe setup) (reference needs)
CD-AFM (image-based)	➢ Measure 3D profile of arbitrary feature	➢ Measurable pattern is limited ➢ Low throughput
X-ray CD (model-based)	➢ High sensitivity, CD/ 3D profile measurement	➢ Average measurement only (unavailable for complex pattern) ➢ Low throughput
Cross section TEM/ STEM (image-based)	➢ High resolution, CD/ 3D profile measurement	➢ Destructive, low throughput

In the area of offline analysis, to realize "High-resolution and high-precision measurement which complement CD-SEM" and "Direct image-based profile measurement which complement OCD", CD-AFM profile measurement, Cross-section SEM/TEM profile measurement, and X-ray high-precision measurement have been studied. Currently, these techniques are limited to lab-analysis or reference applications, because of lack of automatization and low throughput. [1]

3. CHANGES IN MEASUREMENT NEEDS AND TECHNOLOGY GAPS

CD-SEM has been popular as a standard equipment for HVM in-line CD measurement. Its performance such as resolution, precision, throughput has been enhanced as device feature size shrinks. However, new needs and technology gaps have emerged to keep up with manufacturing process changes.

(1) Global Trend Monitoring/ Mass Measurement

One of these new needs is "mass measurement" of more than 10 billion features to control variations of their shapes which have significant impacts on yield. As seen in Fig.1, global variation control as well as statistical local CD uniformity (CDU) is indispensable for controlling edge placement errors (EPE). Especially in multi-patterning process such as SAQP (self align quadruple-patterning), final CDU is determined by the result of repeated processes of litho, etching and deposition, thus key parameters for process control are increasing.[2] To monitor global variations of EPE, high-throughput and yet high-precision mass measurements are needed. Significant errors found in CD metrology are also detected as defects in inspection systems which have higher throughput compared to conventional CD-SEM. This indicates that inspection system has the potential for mass measurement, if it has precise metrology functions. Also, because EPE measurement consists of 2D-measurement, overlay measurement and CD uniformity including line edge roughness measurement, the boundary between 2D-measurement and overlay is becoming fuzzy.

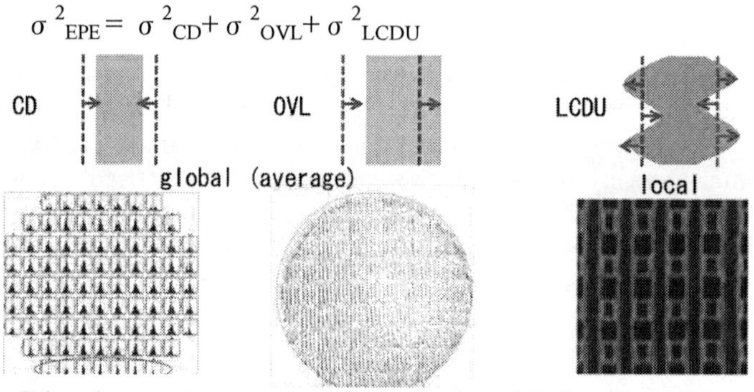

FIGURE 1. Edge placement error measurement requires global and local CD measurement

(2) Diversity of Metrology Needs due to Device Structure

The second need is the support for various metrology items due to the diversity of device structures. The necessary technology differs between 5~7nm logic device and 3D-NAND which has 10um-high stacked layers. For 5~7nm logic, CD+OVL(overlay)+LCDU(local CD uniformity)+LER(line edge roughness)/ LWR(line width roughness)+D2DB(die to database) combined measurement algorithm is required to support EPE control mentioned above, and EUV lithography. On the other hand, measurement of bottom dimensions of deep hole or trenches is strongly desired for 3D-NAND memories, because they consist of nearly 100 stacked layers and have holes and trenches with 10um in depth. This diversity makes it difficult to support these different measurement items by a single tool.

(3) Technology Gaps

The third need is the technology gaps from requirements. 3D-profile measurement of arbitrary local pattern has always been required. In pursuing this need, a tilted image capturing by CD-SEM, enhancement of modeling of OCD, extra-fine tip for AFM are being pursued - but are still inadequate. Thus, several approaches such as injection of CD-SEM data to OCD data are investigated.[3]

Another gap is a resolution of image-based metrology. The pace of CD-SEM resolution improvement is slower than the speed of device feature size shrinkage. On the other hand, CD-SEM has kept its performance of stability, precision, and tool to tool matching of 0.1nm order, as the standard ruler for HVM measurement tool.

4. FUTURE DIRECTION

At sub-10nm process, the pattern fidelity measurement will be more important to control and optimize critical process for the yield enhancement of each device (see Fig.2).[4] Because the number of critical check points becomes numerous and the shape of each feature becomes complex, image-based metrology such as CD-SEM or CD-AFM remains important. Therefore, the enhancement of throughput for numerous points and resolution for single nm pattern without damage continues to be required.

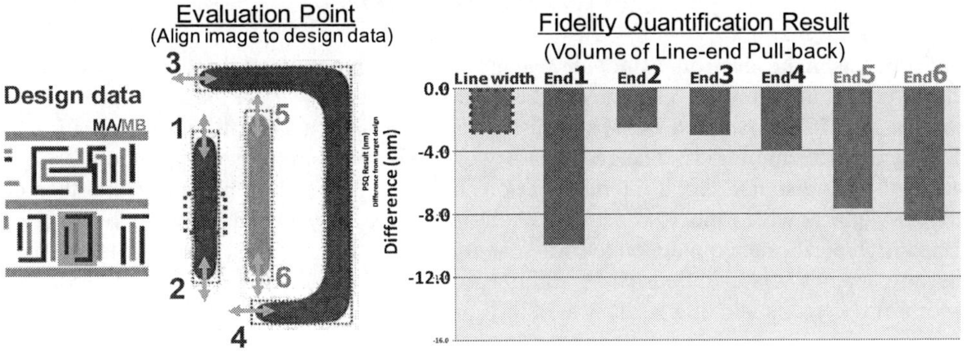

. fidelity qualification of LELE process

For high aspect ratio contact hole and trench such as 3D-NAND (>5um depth), one of the challenges is a high voltage CD-SEM to obtain signals from underlayer features by detecting back scattering electrons.[5] Another approach is the combination of FIB tilted milling and CD-SEM measurement, which will be necessary to measure precisely.

Needless to say, the ideal solution is a high-throughput, high-sensitivity "Super tool" that can cover large area and yet measure any specific local pattern. Such a tool will be able to measure devices globally and locally in the same measurement. But current solutions cannot cover all these needs by single tool. Therefore, it is important to establish an holistic approach metrology to achieve this ultimate goal.

REFERENCES

1. B. Bundy, *HVM Metrology Challenges towards the 5nm Node*, Proc of SPIE Vol.9778
2. H. Fukuda and Y. Momonoi, *Electron beam-based metrology and inspection, overcoming the limitation of scaling, variability, and productivity*, The 24th Lithography Workshop
3. A. Vaid, et al, *3D Technology & Advanced Patterning*, Semicon Korea 2015, Semi Technology Simposium
4. D. Fuchimoto, et al, *Contour-Based Metrology for Complex 2D Shaped Patterns Printed by Multiple Patterning Process*, Proc of SPIE 9050-30
5. O. Inoue, S. Halder, et al, SEM based Overlay measurement between Resist and buried patterns, Proc of SPIE 9778-47

KEYWORDS

critical dimension, metrology, CD-SEM, OCD, CD-AFM, X-ray, EPE, hybrid metrology

Metrology And Characterization Challenges For Complex 2.5D And 3D Packaging

W. R. (Bill) Bottoms PhD

Chairman, 3MTS
3101 Alexis Drive,
Palo Alto, Ca. 94304
Bill_bottoms@3mts.com

INTRODUCTION

The pace of progress in the electronics industry can no longer be maintained through Moore's Law scaling. Today performance, power requirements, size, latency and cost at the system level are all dominated by the packaging and interconnect rather than IC characteristics. The industry has addressed this reality with innovation in packaging and system level integration technologies. The result is a world where electronics components are being moved ever closer together with long distance signal interconnections being addressed by photonic rather than electrical signals. The problem was well known in the 1980s but the solutions were very expensive and were only used in systems which demanded high performance independent of cost and energy requirement such as mainframe computers. The technology used to support this included IBM's multilayer co-fired ceramic packages and dense thin film packages from DEC, Motorola-Siemens and others illustrated in figure 1 below.

FIGURE 1. MCMs typical of the 1980's

These solutions were very expensive with investment over 10 years of more than $1 billion for IBM to commercialize their approach[1]. These solutions did not impact high volume products since the performance was less important for those markets than cost and power requirements. The expanding size and complexity of FPGAs begin to exceed the wafer fab capability of a single chip in the early 2000s. The introduction of the smart phone in June 2007 began changing this for consumer products as system level integration in a smaller form factor was required. Packaging innovation became the enabler for higher performance, lower power requirements, decreased latency and lower cost for consumer products and, high performance systems. These new architectures range from 2.5D packaging to complex 3D System in Package (SiP) products. New materials, new production processes and requirements for 3D heterogeneous integration in these new package architectures presented new challenges for metrology and characterization.

METROLOGY FOR 2.5D PACKAGING

The complexity of electronics packaging expanded with the complexity of the integrated circuits. The increase in frequency and physical density moved the limitation in system performance from the individual ICs to the assembly and packaging of those components into sub-systems and systems. Multi-die packaging was essential to meeting the market requirements. The MCM was not a new idea for system solutions. They were introduced in the early 1980s for specialized applications. When interconnect density and bandwidth requirements exceeded the capability of conventional package substrates the ITRS forecast that silicon would be used as a package substrate by 2010. Xilinx began production of the first commercial 2.5D product in 2011 delivering substantial advantages in performance. The work to develop this 2.5D solution began in late 2003 and the initial product deliveries were more than 8 years later. The technical problems were resolved but cost remained a limitation when production began. Characterization of the processes was required to resolve remaining yield issues and this is still a work in progress.

The metrology and characterization capabilities developed for IC production can meet the technical requirements for the silicon interposer but there were cost issues primarily associated with the through silicon vias (TSVs) that limited the widespread use of this technology. Probing of TSVs results in yield loss, adding a probe pad adds capacitance and increased area required for the connection. There are a variety of alternatives for probing the TSVs that attempt to address these problems that will be discussed. However, today there is no proven low cost solution. TSVs also have defects that might not be detected in a conventional probe test such as partial voids in the TSV. They may pass probe tests and still represent a reliability problem when the product is in use. The characterization to test for such defects involve more measurements such as complex capacitance measurements that require more expensive test equipment and require significant additional time.

The initial 2.5D technology used for a package interposer between the IC and the package substrate was fabricated on a silicon wafer using the materials and processes developed for IC production. Today similar structures are being explored using glass, organic and ceramic materials. Each has their own advantages and disadvantages and they may offer cost and other advantages which in some cases will require new metrology and characterization techniques.

CHALLENGES FOR 3D PACKAGING

The evolution of packaging became a revolution as 3D integration, heterogeneous integration of diverse materials, incorporation of new circuit type such as photonics and plasmonics and passive components were combined with the issues identified for 2.5D above. The list of components that must be incorporated into a SiP over the next 15 years include:

- Monolithic integrated photonic ICs (photonics, electronics and plasmonics)
- Other optical components that are not integrated in the SiPh-ICs.
- Si based logic and memory ICs
- MEMS devices
- Sensors
- GaN power controller circuits
- RF circuits
- Compound semiconductor lasers
- Optical interconnects to and from the outside world
- Electrical interconnects to and from the outside world
- Passive components
- New devices and new materials that will be developed over next 15 years

The initial learning cycles for early 2.5D and 3D integration were approximately a decade each. The learning curve for integrating these known solutions will be more rapid and the evolution will become revolution as new materials, new devices and new architectures arrive in parallel. This evolution in packaging that is rapidly becoming a revolution that is already underway is presented in figure 2 below[2].

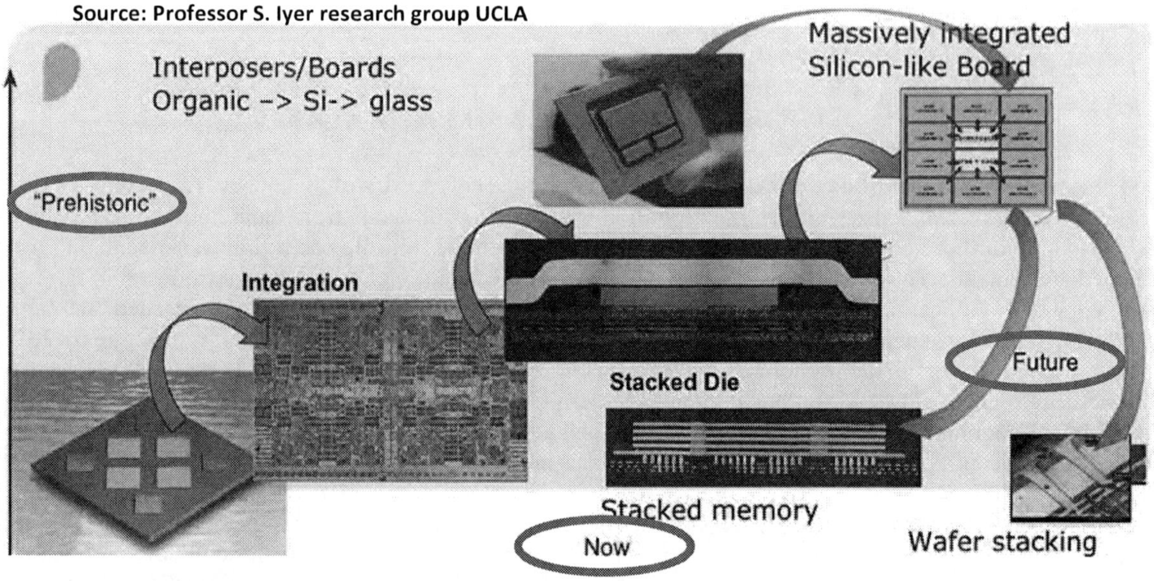

FIGURE 2. Evolution of System Level Heterogeneous Integration in a package

The introduction of the third dimension does not resolve any of the issues associated with 2.5D but adds many new challenges. They include:

- Access for characterization and testing of components post assembly.
- Characterization of mechanical, electrical and thermal properties of the thin layers used in 3D integration that do not exhibit bulk properties.
- Characterization of mechanical, electrical, chemical and thermal properties of interfaces between layers. These interfaces typically dominate the behavior characteristics of 3D structures.
- Characterization and metrology for stress and strain in 3D stacks due to heterogeneous integration of dissimilar materials with different mechanical properties.
- Characterization of stress and strain associated with thermal cycling of materials with different coefficient of thermal expansion.
- Characterization of properties for new composite materials and their processing characteristics.

The challenges for metrology and characterization are not limited to the static issues listed above. We are in an era where transistors wear out during the life of the products that incorporate them. No complex 3D-SiP product will succeed if a single point failure can bring the system. The solutions must incorporate the hardware and software for continuous test while running and intelligent redundancy support dynamic self-repair. These Challenges and potential solutions will be discussed.

REFERENCES

1. *Multichip Modules*, Integrated Circuit Engineering Corporation, 1997, pp. 12-1 through 12-76.
2. Subramanian Iyer, *Chips Program web site*, UCLA

KEYWORDS

Characterization, Packaging, 3D Packaging, Heterogeneous integration, Test

Novel Metrology Solution for Advanced Packaging Based on Multi-energy X-ray Microscopy and Tomography

Ehrenfried Zschech[1], Wenbing Yun[2]

[1]*Fraunhofer Institute for Ceramic Technologies and Systems, Maria-Reiche-Str. 2, 01109 Dresden, Germany*
[2]*Sigray Inc., Concord/CA, USA*

INTRODUCTION

Advanced packaging including 3D through-silicon-via (TSV) integration technologies and the resulting 3D-stacked products challenge materials and process characterization. For 3D TSV stacking of wafers or dies, die-to-die interconnections like micro solder bumps (e. g. Sn-rich solders) and Cu pillars are used [1]. The control of the micro-bump quality of 3D stacked products, e. g. high-bandwith memories (HBM), is a particular issue. The geometrical shape of microbumps and their chemical composition (including formed intermetallic phases) have to be monitored, and defects like pores and microcracks (also in relation to formed intermetallic phases) have to be detected with high resolution (for micro bumps usually better than 700 nm). In addition, the requirements from industry, that metrology for advanced packaging should be nondestructive and with high throughput (short time-to-data), have to be considered. We demonstrated the capabilities for nondestructive imaging of multi-die stacks with Cu TSVs and SnAg micro solder bumps using sub-micron XCT (currently, best resolution about 700 nm) and nano XCT (currently, best resolution about 50 nm) [1,2]. Sub-micron XCT of a multi-die stack provided information about major filling defects in TSVs, mismatches in relative positioning and variability in the shape of micro bumps, micron-size pores, and the distribution of intermetallic phases in solder connections. Nano XCT studies (using Cu-Kα radiation, 8.05 keV) demonstrated clearly, that voids in Cu TSV with a size of about 100 nm can be visualized.

In this paper, the potential of lab-based sub-micron XCT and nano XCT for process monitoring and failure analysis of 3D IC stacks as well as a novel solution based on multi-energy X-ray microscopy and tomography are described.

EXPERIMENTAL STUDY OF MICRO BUMPS IN HBM STACKS

A commercial HBM stack was studied using micro XCT, which provided a resolution of about 1 μm (Figure 1). The arrangement of the bumps is clearly visible, however, the resolution does not allow to provide details about the shape of the micro bumps and defects. Details like the geometrical shape of the solder connection (micro bump) and defects like pores and micro-cracks can be detected based using nano-XCT (see Figure 2).

FIGURE 1. Two virtual 2D images (plane view and cross-section view) of a HBM stack, based on a 3D data set from micro X-ray tomography. Nondestructive imaging.

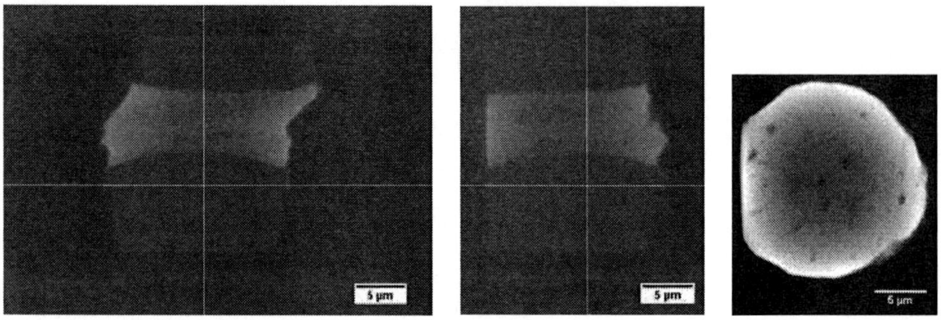

FIGURE 2. Three virtual 2D images (two perpendicular cross-section views and one planar view) through a solder connection (micro-bump of a HBM stack), based on a 3D data set from nano X-ray tomography. Imaging of a small extracted sample.

NOVEL SOLUTION: IMPROVED EXPERIMENTAL SETUP AND COMPONENTS

High-flux X-ray sources provide potential advantages for both micro XCT and nano XCT (reduction of measurement time and/or reduction of source size with the potential to reduce the resolution to about 300 nm to 500 nm). However, in this paper we explain nano XCT in more detail. The design of an improved laboratory X-ray microscope includes

- a high-flux X-ray source with multiple targets, including the option of photon energies > 10 keV (FAAST™ and MAAST™ concepts), combined with an efficient X-ray mirror optics
- an X-ray lense that can be operated also at photon energies > 10 keV (MLL concept).

Figure 4 shows a scheme of a nano XCT setup for high-throughput nondestructive studies of solder connections..

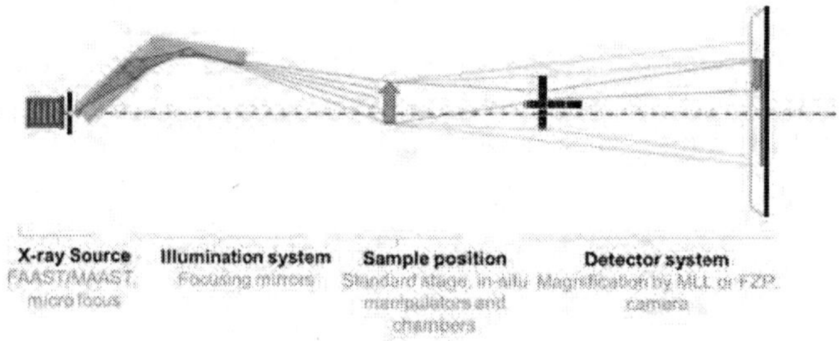

FIGURE 4. Scheme of a nano XCT setup for advanced packaging metrology, using novel X-ray sources and optics.

A new type of a super bright laboratory source, applying the so-called Fine Anode Array Source Technology (FAAST™) with a microstructured anode that incorporates diamond with its outstanding material properties is presented. Figure 5 shows a scheme of the anode, i. e. a metal target embedded into a microstructured diamond substrate, and the resulting improved temperature distribution within the anode. The X-ray generating material may be any one of the commonly used metals, such as Cu, Mo, W, Cr, or many materials not suited for anode in conventional X-ray sources. This design offers major benefits: (1) incorporating the outstanding thermal properties of diamond to increase average thermal conductivity of the anode, (2) utilizing the dependence of electron energy deposition rate on mass density to create a favorable energy deposition profile, (3) creating large thermal gradients between the Cu elements and the surrounding diamond to produce enhanced thermal dissipation, and (4) highly efficient accumulation of X-rays along the X-ray beam axis. The Multiple Anode Array Source Technology (MAAST™) allows to accumulate photons generated from multiple anode array sources. The illumination optics consists of a mirror optics, e. g. paraboloids.

The use of conventional Fresnel Zone Plates (FZPs), fabricated using electron beam lithography and subsequent etching, as focusing lenses is limited to photon energies of about 10 keV since the focusing efficiency is too low for higher energies. An new type of X-ray lenses, so-called Multilayer Laue Lenses (MLL), consist of two perpendicularly assembled components which are cut out of a multi-layer stack. The thicknesses of those layers, manufactured by thin film deposition, e. g. magneton sputtering, follow the zone plate law, and layers with a thickness down to 1 nm can be produced with high quality (Figure 6 left) [4,5]. This design offers major benefits: (1) extension of the use range of X-ray lenses to higher photon energies (> 10 keV), (2) improvement of resolution to 10 nm and higher. The focusing efficiency of MLLs can be increased for wedged multilayer stacks. The degree of tilting can be controlled by diffraction patterns (Figure 6 right).

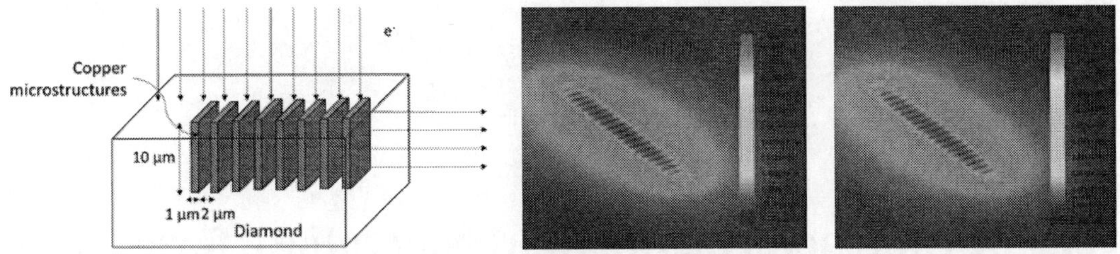

FIGURE 5. Scheme of the microstructured target of the Sigray Fine Array Anode Source Technology (FAAST™) source (left) as well as thermal modeling of the Sigray FAAST™ (middle) and of a conventional Cu X-ray target (right)

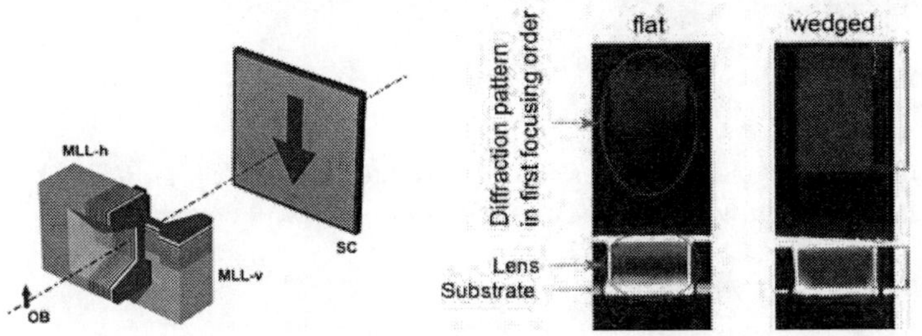

FIGURE 6. Crossed MLL after FIB preparation and assembly acting as a lens to image the object (OB) to the screen (SC) (left) [5] and diffraction patterns of flat and wedged multilayers (right).

CONCLUSION

Lab-based sub-micro XCT and nano XCT offer intrinsic advantages for 3D imaging and metrology for advanced semiconductor packages. There is a strong need to develop a high throughput XCT tool operating at X-ray energies >10 keV with a resolution between about 200 nm and 700 nm (known as the "resolution gap"). It removes two major limitations of the current nano XCT tools: requiring sample preparation (typically about 50 µm size if 8.05 keV photons are used) and low imaging throughput. It offers significantly higher resolution than the current sub-micron XCT tools. A novel solution for advanced packaging metrology that uses the novel X-ray sources with high flux and the option of multi-energy photons and novel X-ray optics with high efficiency even at photon enegries > 10 keV, is presented.

ACKNOWLEDGEMENT

The authors thank Jürgen Gluch, Kristina Kutukova, Peter Krüger, Rüdiger Rosenkranz and Christoph Sander (all with Fraunhofer IKTS Dresden, Germany) for their support during sample preparation and experiments. Fruitful discussions with Martin Gall (Fraunhofer IKTS Dresden, Germany), Jürgen Wolf (Fraunhofer IZM-ASSID Dresden, Germany), Norman Huber (Huber Diffraktionstechnik Rimstein, Germany) as well as Reiner Dietsch and Sven Niese (both with AXO Dresden, Germany) are acknowledged.

REFERENCES

[1] E. Zschech, S. Niese, M. Gall, M. Löffler, M. J. Wolf, „3D IC Stack Characterization using Multi-Scale X-Ray Tomography", Proc. 20th PanPacific Microelectronics Symposium, Kolao/HI 2015

[2] E. Zschech, M. Löffler, J. Gluch, M. J. Wolf, "Micro and Nano X-ray Tomography of 3D IC Stacks", Invited talk at MRS Spring 2016, Phoenix/AZ 2016

[3] W. Yun, J. Kirz, S. J. Lewis, A. Lyon, D. Reynolds, S. Chen, I. Spink, B. Stripe, "A New Type of Super Bright Laboratory X-ray Source", Compact EUV & X-ray Light Sources Meeting, Long Beach/CA 2016, OSA Tech. Dig., EM4A.2. (2016)

[4] S. Niese, "Lab-based in-situ X-ray microscopy – Methodical Developments and Applications in Materials Science and Microelectronics", PhD thesis, BTU Cottbus (2015)

[5] S. Niese, P. Krueger, A. Kubec, S. Braun, J. Patommel, C. Schroer, A. Leson, E. Zschech, "Full-field X-ray Microscopy with Crossed Partial Multilayer Laue Lenses", Optics Express 22, 20008-20013 (2014)

KEYWORDS

X-ray tomography, X-ray microscopy, advanced packaging, micro bump, high-bandwith memory

Future Needs of Characterization and Metrology for Silicon Qubits in Quantum Computing

Neil M. Zimmerman, Rick Silver, Xiqiao Wang

Quantum Measurement Division, National Institute of Standards and Technology, MD 20899, USA; +1-301-975-5887; neilz@mailaps.org; ftp://ftp.nist.gov/pub/physics/neilz/papers.html

INTRODUCTION

In this paper, we will attempt to lay out some of the interesting future needs of characterization and metrology for Si qubits (and to a limited extent, generic qubits) for quantum computing. We will attempt to avoid duplicative discussion, by putting these needs in the context of those for both CMOS and beyond CMOS (BC). We consider two prototypical architectures for Si-based qubits: 1) Single atom devices, where the computing elements are electrons bound on individual dopants encased in crystalline Si; 2) Quantum dot devices, where the electrons are induced below gates, similar to MOSFETs. In general, there are both new and more demanding measurement challenges in both structural/analytical and in electrical characterization. Unlike classical computers, the state of qubits in a quantum computer would not be protected by the energy barrier that suppresses bit flipping; an example of a new challenge which arises for this reason is a variety of quantum metrologies such as quantum tomography [Giovanetti]. An example of a more demanding challenge is the ability to locate individual atoms below the surface of the three-dimensional structure [Usman].

CONTEXT

Introduction to Qubits

Quantum computers are one of the candidates for beyond CMOS devices, for instance, in the "Big Data" section of the 2015 ITRS Emerging Research Materials chapter. Quantum computers hold the promise of the next exponential increase in computing speed, and thus the potential to solve problems such as the "traveling salesman problem" or factorising large numbers (crucial for encryption) [Nielsen and Chuang]; in the field of Computer Science, these types of problems are considered fundamentally intractable with any classical computer. In addition, simpler quantum information elements offer fundamentally secure communications, free from undetected eavesdropping by the laws of physics.

As a notional prototype of qubits, we can consider two atoms floating in space. This pair would satisfy the basic criteria necessary for quantum processing [DiVincenzo]:

A scalable physical system with well characterised qubits.
The ability to initialise the state of the qubits.
Long coherence times.
A “universal” set of quantum gates.
A qubit-specific measurement capability

Estimates of the number of qubits necessary for a general purpose quantum computer range up into the hundreds of millions. The physical implementations of possible qubits fall into two general classes: i) Collections of atoms or ions in vacuum systems or photons, all of which have very long coherence times, but will be challenging to scale to many integrated qubits; ii) Solid-state (superconducting or semiconducting) "artificial atoms", for which there is a straightforward scalability, but for which long coherence times are challenging. In this paper, we concentrate on semiconducting qubits, and thus much of the discussion will be centered on achieving and measuring long coherence times. In particular, with gate or switching times of order 1 ns, we require coherence times of at least about 100 ms to achieve error-free quantum computing with surface code quantum error correction [Devitt].

Two Prototypical Architectures

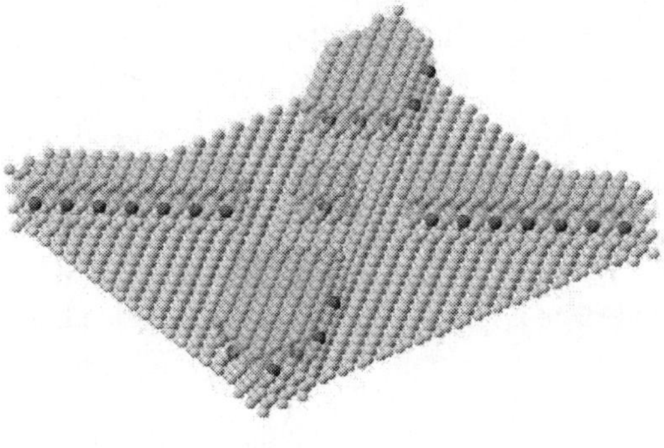

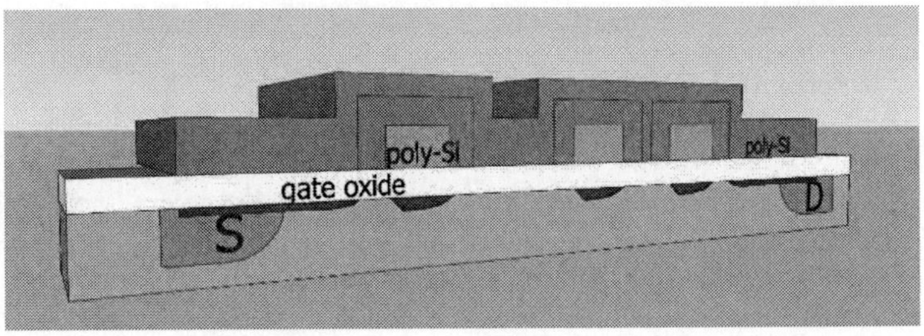

FIGURE 1. Two prototypical devices: Left: single-atom device with electron localized to P+ dopants (color: red, B/W: dark dots); dopants are surrounded by crystalline Si. Right: quantum dot device with two-dimensional electron gas (2DEG) (color: red, B/W: black) at interface between Si and oxide.

Figure 1 shows sketches of the two prototypes we consider. In both cases, the crucial computing element is one (or a small number) electron localized due to the electrostatic field from either the dopant or the gate above. The computing degree of freedom is generally the spin of the electron. As mentioned above, the fact that the electrons are surrounded by solids (crystalline Si, dopants, impurities, isotopes, phonons, etc.) means that achieving long coherence times is crucial.

A SMATTERING OF CHARACTERIZATION AND METROLOGY NEEDS for Si QUBITS BEYOND CMOS AND BEYOND BC

General Issues of Decoherence

As mentioned above, we will consider both new and more demanding challenges beyond those discussed for various classical BC candidates (see the ITRS 2015 Metrology chapter). First of all, we must acknowledge that most of these techniques are either destructive or time-consuming, and thus not well-suited for in-line characterization. The general metrology need for all possible qubit candidates is that associated with demonstrating sufficiently long coherence times, and more specifically sufficiently high initialization, coherent manipulation, and measurement fidelities for both one- and two-qubit gates. In Si, workers have demonstrated single-qubit gate manipulation fidelities as high as 99.99% using randomized benchmarking [Muhonen]. This type of sophisticated fidelity measurement requires amongst other things spin resonance microwave techniques with complicated pulse sequences. This represents at the least a more demanding application of previous techniques; extending this to thousands or millions of qubits in a single quantum computer represents a daunting task. One possible avenue (not yet demonstrated) to achieve this is to use faster, simpler, less expensive electrical characterization (C-V, G-V, charge pumping, etc.), and correlate these with coherence times; we would then routinely use the simpler measurements to characterize candidate materials, geometries, architectures. We hope to demonstrate this correlation as we progress.

Structural/Analytical Characterization

Isotopic Enrichment: One of the major sources of decoherence for spin qubits in Si and GaAs is the nuclear spin of some of the isotopes, because those nuclear spins in the nearby environment generate fluctuating local magnetic fields which perturb the electron spin. For this reason, recent advances in Si qubits with long coherence times have been in devices with isotopic enrichment to avoid the ^{29}Si isotope[Muhonen]. With a new growth technique, workers have achieved thin films with ^{29}Si concentrations below one ppm [Dwyer]; clearly, routinely achieving such materials will push established techniques such as SIMS, PL and ESR well beyond their typical uses. Approximate sensitivity need: 1 – 100 ppm.

Dopant Monolayer Growth: For the single atom prototype discussed above, we desire to grow a single monolayer of dopants, and in particular to maintain as small as possible the vertical thickness of the monolayer after thermal treatments during overgrowth of crystalline Si. This will put very challenging constraints on SIMS, atom probe, etc. Approximate sensitivity need: 1 – 100 ppm.

Si Overgrowth Thickness/Crystallinity/Purity: For the single atom prototype discussed above, in addition to standard challenges in characterizing these parameters, there arises a new one: As exemplified in Figure 2, for good crystalline material, the interface (the location of the monolayer in the crystal) should be perfect. This perfection yields a challenge: How do we even determine in a microscope where the interface is? Approximate sensitivity need: 1 nm.

Location of Single Atoms: For the single atom prototype discussed above, it is becoming apparent that the location (in the lattice and with respect to each other) of each individual dopant strongly determines the energetics and coherent properties. Recently, workers have begun to demonstrate how to determine the lattice location of dopants within 5 nm of the surface using STM [Usman]. Another issue is that such characterization inherently samples on very small volumes, and may not be representative (2015 ITRS Emerging Research Materials chapter). Approximate sensitivity need: 0.15 nm.

Strain-Induced Quantum Dots: For the quantum dot prototype discussed above, it turns out that extra quantum dots (isolated regions of the 2DEG) are generated by the elastic stress from the gate geometry [Thorbeck]. While electron diffraction measurement techniques and finite element simulations for nanoscale strain measurements are well-established, for CMOS devices, these techniques have been used solely for enhancing mobility. Extending these to the quantum dot regime will at least require refocusing these techniques on different issues. Approximate sensitivity needs: 0.01 % strain and 0.01 meV conduction band modulation.

FIGURE 2: TEM micrograph of (bottom to top) Si, monolayer, overgrown Si; the crystallinity is clear, but where exactly is the interface?

Electrical Characterization

Dopant Metrology: In devices such as single atom transistors which depend on conduction through a single dopant, simply identifying the chemical species of the dopant can be a challenging measurement; a technique has been established by measuring energetics in the single electron regime. Extending this time-consuming and complicated measurement to many quantum dots will require new metrology techniques. Approximate sensitivity need: species of single atom in solid.

Defect-Induced Instability: Because these devices are inherently operating as single-electron devices, they are exquisitely sensitive to the motion of individual charge defects, such as interface traps. This topic has had great attention for a long time in CMOS due to effects such as threshold voltage shifts; these quantum devices are more sensitive, and often have gross time-dependent instabilities due to the charge offset drift [Stewart], which can greatly affect integration. These measurements are very time-consuming, and it is unclear how to extend them to multiple devices in an effective way. Approximate sensitivity need: charge change 0.001 e.

Defect Identification and Characterization: For qubits, the defect can also quantum mechanically alter the single atom or quantum dot device via energy exchange. These interactions have been identified as the primary cause of decoherence in superconducting qubits; in semiconducting qubits, the field has just started to investigate this issue. One proposed technique has been to use electron spin resonance (ESR) to probe these defects [Tenorio]. Approximate sensitivity need: 1 neV.

We note that many of the techniques we have discussed require cryogenic temperatures, which will pose an additional challenge to these characterization and metrology measurements.

Acknowledgments: Over the years, we have learned a great deal from my NIST colleagues, including Garnett Bryant, Kevin Dwyer, Hyun Soo Kim, Roy Murray, Josh Pomeroy, Ryan Stein, and Michael Stewart, Jr,; we also received very valuable comments from James Clarke (Intel) and Alan Seabaugh (UND).

REFERENCES

1. Vittorio Giovannetti, Seth Lloyd and Lorenzo Maccone, *Nature Photonics* **5**, 222 – 229 (2011), “Advances in quantum metrology”
2. M. Usman et al, *Nature Nano* **11**, 763 – 769 (2016), “Spatial metrology of dopants in silicon with exact lattice site precision”
3. Michael Nielsen and Isaac Chuang, *Quantum Computation and Quantum Information*, Cambridge: Cambridge, 2000, pp. 120-160.
4. D. P. DiVincenzo, *FORTSCHRITTE DER PHYSIK-PROGRESS OF PHYSICS* **48,**:771-783 (2000) “The physical implementation of quantum computation”
5. Simon J Devitt, William J Munroe and Kae Nemoto, *Rep. Prog. Phys.* **76,** 076001-1 to -35 (2013), "Quantum error correction for beginners"
6. Juha T. Muhonen et al, *Nature Nano* **9**, 986 - 991 (2014), "Storing quantum information for 30 seconds in a nanoelectronic device"
7. Christopher Granade et al, arXiv:1610.00336v1 (2016), “QInfer: Statistical inference software for quantum
8. Applications”
9. K J Dwyer, J M Pomeroy, D S Simons, K L Steffens and J W Lau, *J. Phys. D: Appl. Phys.* **47** (2014) 345105, “Enriching 28Si beyond 99.9998% for semiconductor quantum computing”
10. Ted Thorbeck and Neil M. Zimmerman, *AIP ADVANCES* **5**, 087107 (2015) "Formation of strain-induced quantum dots in gated semiconductor nanostructures"
11. M. D. Stewart, Jr. and Neil M. Zimmerman, *Appl. Sci.* **6**, 187 (2016), "Stability of Single Electron Devices: Charge Offset Drift"
12. J. O. Tenorio-Pearl et al, *Nature Materials* DOI: 10.1038/NMAT4754 (2016), “Observation and coherent control of interface-induced electronic resonances in a field-effect transistor”

KEYWORDS

Characterization and metrology; Si qubits; experimental

In-Line Sheet Resistance Measurements of Nanometer-Wide Semiconducting Fins

J. Bogdanowicz[a,1], S. Folkersma [a,b], A. Schulze[a], A. Moussa[a], C. Merckling[a], B. Kunert[a], W. Guo[a], Y. Mols[a], D. H. Petersen[c], M.-L. Witthøft[c], O. Hansen[c], H. H. Henrichsen[d], P. F. Nielsen[d] and W. Vandervorst[a,b]

[a] *IMEC, Kapeldreef 75, B-3000 Leuven, Belgium*
[b] *Instituut voor Kern- en Stralingsfysika, KU Leuven, Celestijnenlaan 200D, B-3001 Leuven, Belgium*
[c] *Department of Micro- and Nanotechnology, Technical University of Denmark, DTU Nanotech Building 345 East, DK-2800 Kgs. Lyngby, Denmark*
[d] *CAPRES A/S, Scion-DTU, Building 373, DK-2800 Kgs. Lyngby, Denmark*
[1] *email address: janusz.bogdanowicz@imec.be*

INTRODUCTION

Since the advent of non-planar metal-oxide-semiconductor (MOS) transistors, embodied by the three-dimensional fin field-effect transistors (finFETs), there has been a rising need for fast in-line metrology solutions for the electrical characterization of nanometer-wide trenches filled with semiconducting materials. Indeed, on such narrow structures, conventional four-point probe fails, mostly due to the challenging alignment of the four millimeter-sized and -spaced probes to the nanoscale trenches. The electrical characterization of these state-of-the-art devices must therefore be performed on fins connected by metal contacts in Kelvin resistor or transmission line structures [1]. However, this is done at metal-1, i.e. at the expense of extra processing cost and time. As a consequence of the absence of a metrology to access the electrical properties of semiconducting fins immediately after their growth, little is known or understood of the impact on these properties of the increased surface/volume ratio via e.g. sidewall roughness, dimension-dependent epitaxial growth or dimension-dependent diffusion/activation effects [2] etc.

In this paper, we demonstrate the capabilities of the micro four-point probe technique, as implemented in the fully automated microHALL®-A300 tool of CAPRES A/S, for the electrical resistance measurement of narrow fins without the need for metal contact pads. This four-point probe technique where the probes and probe pin distances are downscaled to a few micrometers has been successfully applied to a myriad of blanket materials, ranging from multilayered metal stacks [3] to semiconductors [4] and two-dimensional materials [5]. Advantageously, micro-probes allow a smaller sampling volume and a minimized sensitivity to junction leakage in blanket samples [6] and, as we show in this paper, they are also much easier to align to nanoscale trenches. We start this paper with an introduction to the technique. We then experimentally extract the sheet resistance of implanted and annealed Si fins as well as of epitaxially grown III-V fins. Importantly, this work demonstrates experimentally that the technique can determine the sheet resistance of conductive material confined to widths down to 20 nm.

FUNDAMENTALS OF MICRO FOUR-POINT PROBE ON NARROW FINS

Four-point probe measurements are widely used for the accurate determination of the sheet resistance R_s of metal and semiconducting layers [3,4]. In such measurements, a current is injected into the investigated specimen via two electrodes while the induced voltage drop is measured between another two electrodes [1,4] (Fig. 1, left). As opposed to a two-point probe measurement, this indeed allows to eliminate any sensitivity to contact resistance [1]. The use of micro four-point probes (μ4PP), where the probe size and probe pin distance d are downscaled to micrometers, furthermore allows to reduce the sampling volume and the impact of possible junction leakage [6]. As shown in Fig. 1(left), different probe configurations A and B can be used, whereby independent resistances R_A and R_B are measured [7]. In configuration A, the resistance $R_A=V_{23}/I_{14}$ is measured by injecting a current I_{14} between the two outer pins (i.e. 1 and 4) and measuring the voltage drop V_{23} induced between the two inner pins (i.e. 2 and 3). In configuration B, the resistance $R_B=V_{24}/I_{13}$ is measured by running a current I_{13} between pins 1 and 3 while measuring the voltage drop V_{24} between pins 2 and 4. For the sake of completeness, note that the $R_{A'}$ and $R_{B'}$ resistances in the conjugate A' and B' configurations, as obtained by reversing the role of the electrodes (current injection vs voltage measurement) [7], are also used in this paper but are not considered in this short introduction to the technique.

To understand the fundamentals of the transition of µ4PP measurements from blanket samples to nanometer-wide fins, we have to evaluate the impact of the width W of a conductive line on the potential distribution induced by two point injectors on this line. Using Ref. [7] which analytically solves this problem by the method of images, Fig. 1(right) shows that two regimes can be distinguished. First of all, if the electrodes are placed in the center of a wide line ($W/d>>1$), the injected current is not confined by the line edges and therefore flows circularly in a two-dimensional (2D) fashion between the injectors. As a result of this 2D current flow, it can be demonstrated that $R_A/R_B{\sim}1.26$ [7]. This is the common case of blanket sample measurements, where R_A and R_B are combined in a so-called dual-configuration measurement to access the sheet resistance [4]. Most applications of the µ4PP technique belong to this 2D regime. Second, in narrow conductive lines ($W<d$), the current is geometrically confined by the line edges and therefore flows in a straight one-dimensional (1D) path and only between the two injectors. In this case, it can be shown that $R_A=R_B=R_s{\times}d/W$ (Fig. 1, right), i.e. the resistance of the portion of the line included between the two inner probes is measured in both A and B configurations. This paper focuses on this 1D regime.

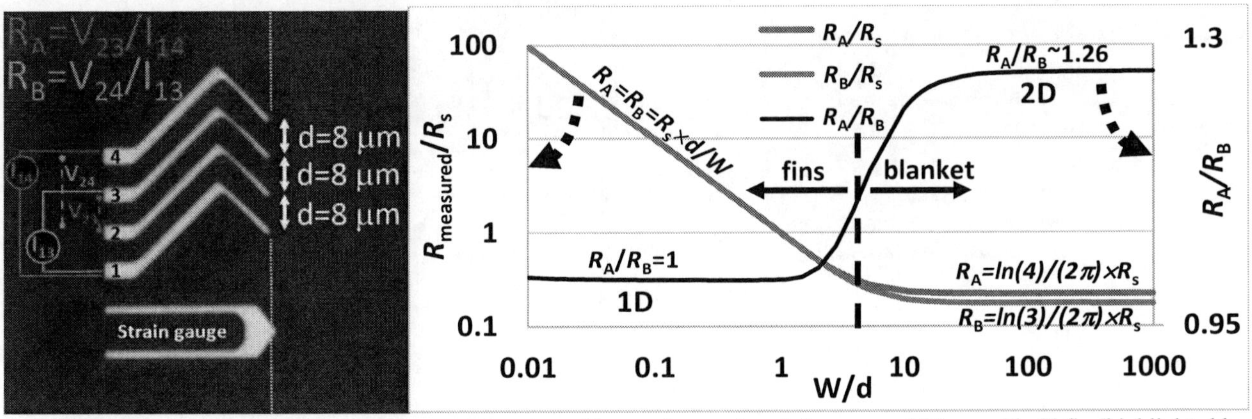

FIGURE 1. (left) Snapshot of a micro four-point probe, with a probe spacing d=8 µm, on a 300 nm wide Si fin (highlighted by the dotted line). In configuration A, the measured resistance $R_A=V_{23}/I_{14}$ is obtained by measuring the voltage drop induced between the inner pins 2 and 3 by a current injected between the outer pins 1 and 4. In configuration B, $R_B=V_{24}/I_{13}$ is obtained by measuring the voltage drop induced between pins 2 and 4 by the current injected between pins 1 and 3. (right) Dependence of the measured R_A and R_B resistances upon line width W, as derived from Ref. [7]. Two regimes can be distinguished. First, in wide lines ($W/d>>1$) i.e. blanket samples, $R_A/R_B{\sim}1.26$ (right vertical axis) and the measured resistances R_A and R_B (left vertical axis) are images of the sheet resistance and independent from width. Second, in narrow lines ($W/d<1$), $R_A/R_B=1$ and the line resistance $R_s{\times}d/W$ is measured in both A and B configurations.

EXPERIMENTAL

The capabilities of the µ4PP technique to measure the resistance of narrow fins were evaluated using the fully automated microHALL®-A300 tool on implanted and annealed Si fins of different widths ranging from 500 nm down to 20 nm. The Si fins were implanted with B ($3{\times}10^{15}$ cm^{-2}, 5 kV) and subsequently annealed. On the same wafer, three different annealing conditions were used in three different regions. One region was only heated from the chuck (450 °C, ~20') while the other two regions were laser annealed with 7 laser scans respectively at 1150 °C and 1250 °C. The electrical isolation from the substrate was ensured by the implantation of a deep n-well. Epitaxially grown III-V fins were also measured (not shown in this abstract). For all samples, the µ4PP measurements were run by scanning the four probes across the fin series with a step size of ~150 nm and using a 1 µA injection current. An accepted measurement was characterized by, first, $R_A/R_B=1$ and, second, a null phase shift between the injected current modulated at 13 Hz and the measured modulated voltage drop. Indeed, when all four probes are aligned with a conductive fin, 1D current flow is expected ($W/d<<1$ even on the 500 nm wide fins) and the electrical contact to the conductive fin should ensure a relatively short RC time constant [8].

As can be observed in Fig. 2(left), the resistances of all fins of the three different regions were captured, including the narrowest 20-nm wide features. As expected, the measured resistance $R_{measured}=(R_A+R_B+R_A'+R_B')/4$ proves to scale inversely with fin width and the impact of an increased thermal budget is observed as a decrease in the measured resistance. Looking more quantitatively at the values of these resistances, as also highlighted in Fig. 2(left), they approximately follow the lines given by $R_s^{pad}{\times}d/W$, where R_s^{pad} is the sheet resistance measured in the center of an 80×80 μm^2 pad having received the same implant and annealing conditions as the narrow fins. This interestingly means that the electrical properties of the material inside the narrow fins are roughly the same as in the pad. To evaluate more precisely the impact of dimensions on the sheet resistance, R_s was extracted from the data of Fig. 2 (left) using the reverse equation $R_s = R_{\text{measured}} \times W/d$. As Fig. 2(right) shows, the sheet resistance of the narrow fins actually monotonically increases with decreasing width. This indicates that the material inside the narrow fins is actually more resistive than in wider features. This dependence of resistivity upon fin dimensions decreases as the thermal budget is increased and vanishes at the highest thermal budget (red symbols). Such dimension-dependent resistivity in narrow features has been observed previously on similar implanted fins with the Scanning Spreading Resistance Microscopy technique [2]. For completeness, note that measurements on epitaxially grown fins showed an opposite decrease in sheet resistance as their width decreased (not shown in this abstract).

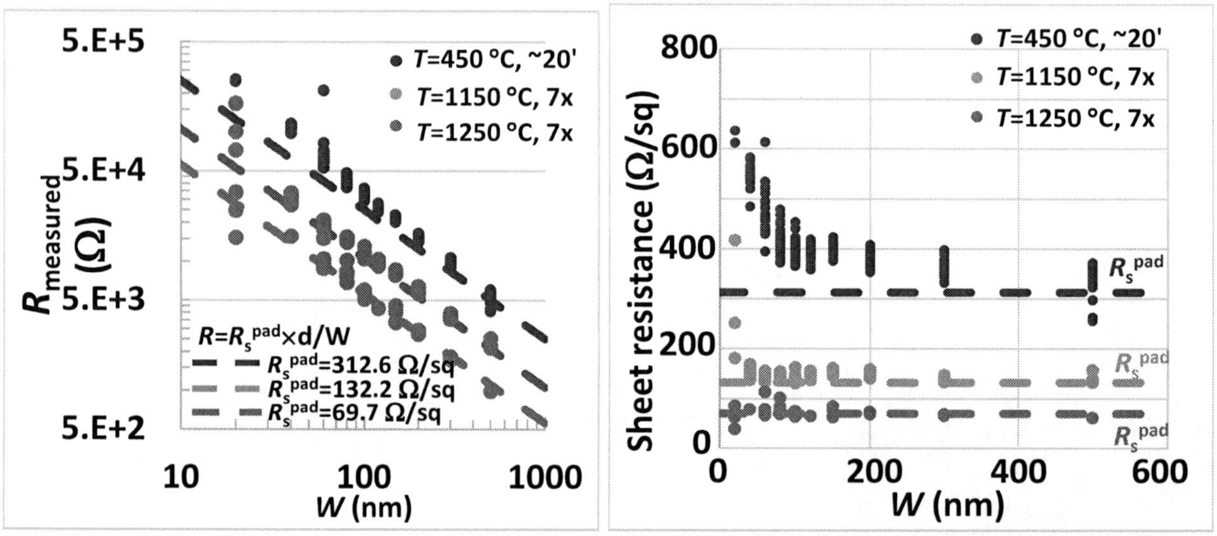

FIGURE 2. (left) Measured resistance $R_{measured}=(R_A+R_B+R_A'+R_B')/4$ as a function of fin width W on B-implanted (3×10^{15} cm^{-2}, 5 kV) Si fins annealed under different conditions, i.e. 450 °C for ~20' (purple symbols), 7 laser scans at 1150 °C (blue symbols) and 7 laser scans at 1250 °C (red symbols). The dashed lines correspond to the resistance using the equation $R=R_s^{pad} \times d/W$, i.e. assuming simple geometrical confinement of the sheet resistances R_s^{pad} measured in wide pads.
(right) Sheet resistance obtained from the raw data of the left figure using the reverse equation $R_s=R_{measured} \times W/d$. The sheet resistance is the same in the widest fins as in the pads (dashed lines) but it increases as width decreases, except at the highest thermal budget (red symbols).

CONCLUSION

This paper demonstrates the capabilities of the micro four-point probe technique as implemented in the microHALL®-A300 tool of CAPRES A/S to measure the resistance of nanometer-wide conductive fins. The technique can measure the sheet resistance of fins with widths as narrow as 20 nm, which allows evaluating the resistivity of material as a function of their (confined) dimensions. As we show, implanted and annealed Si fins are characterized by a resistivity increase as width decreases. This variation, however, is observed to vanish as the thermal budget is increased. The ability to probe the electrical properties of nm-wide semiconducting lines excitingly opens the way to the study of the impact of sidewall roughness, of dimension-dependent epitaxial growth or dopant diffusion/activation, which has so far been hampered by the absence of such metrology solution.

REFERENCES

1. D. K. Schroder, *Semiconductor material and device characterization*, 3rd ed., Chap I, Wiley-Interscience (2006)
2. A. Schulze et al., Appl. Phys. Lett. 102, 052108 (2013)
3. D.C. Worledge et al., Appl. Phys. Lett. 83, 84 (2003) ; D. Kjaer, et al., Meas. Sci. Technol. 26, 045602 (2015).
4. D. H. Petersen et al., J. Vac. Sci. Technol. B 28, C1C27 (2010) ; T. Clarysse, et al., Mater. Sci. and Eng. B 154, 24 (2008)
5. J.D. Buron, et al., Nano Lett. 12, 5074 (2012) ; J.D. Buron, et al., Nano Lett. 14, 6348 (2014).
6. R. Lin et al., AIP Conf. Proceedings 1496, 175-178 (2012)
7. D. H. Petersen et al, J. Appl. Phys. 104, 013710 (2008)
8. P. Horowitz and W. Hill, *The Art of Electronics* 3rd ed., Cambridge University Press (2015)

KEYWORDS

Electrical characterization, sheet resistance, finFET, confined volumes, in-line measurements

Reflective Small Angle Electron Scattering to Characterize Nanostructures on Opaque Substrate

Wen-Li Wu[1], Lawrence H. Friedman[2], Wei-En Fu[3] and Yunsan Chien[3]

[1]*Materials Science and Engineering Division, Materials Measurement Laboratory National Institute of Standards and Technology, Gaithersburg, MD 20899, USA*

[2]*Materials Measurement Science Division, Materials Measurement Laboratory National Institute of Standards and Technology, Gaithersburg, MD 20899, USA*

[3]*Center of Measurement Science Industrial Technology Research Institute, Hsinchu, Taiwan*

INTRODUCTION

Features sizes in integrated circuits (ICs) are often at the 10 nm scale and are ever shrinking. ICs appearing in today's computers and hand held devices are perhaps the most prominent examples. These smaller feature sizes demand equivalent advances in fast and accurate dimensional metrology for both development and manufacturing. Techniques in use and continuing to be developed include X-ray based techniques [1,2], optical scattering and of course the r electron and AFM microscopy techniques. Each of these techniques have their advantages and limitations. To complement these methods, we have explored the use of small angle electron beam scattering measurements in a reflection mode (RSAES) to characterize the dimensions and the shape of nanostructures on flat and opaque substrates. [3] The feasibility of RSAES for dimensional metrology is demonstrated using both experimental and theoretical evidence.

X-ray based techniques [1,2] have been initiated by one of the authors as a viable dimensional and shape metrology where the X-ray wavelength used is about a tenth of a nanometer. However, the major drawback of the X-ray based methods is the lack of high flux X-ray source which can be readily collimated to a spot size of a few tens of micrometers or below; a typical target size in many nanotechnology applications. Optical scattering techniques have already encountered severe difficulties in terms of resolution while the X-ray techniques is far too slow unless a much large target size is available and this is often not the case especially in IC fabrication. Microscopy techniques require a great deal of time and often invasive sample preparation.

RSAES might fill the metrological gap left by these other methods. RSAES leverages existing TEM and STEM technologies to probe structures with nm size beams using wavelengths of a few picometers, thereby achieving high dimensional precision with a small footprint on measured nanostructures. Additionally, the beam flux limitations of X-ray methods are overcome by scattering cross-sections between electron and matter that are 10^4 times stronger than for X-rays.

RSAES EXPERIMENTS

RSAES differs from Reflection High-Energy Electron Diffraction (RHEED). [4] in its angular resolution and angular region of interest. RHEED reveals atomic arrangements and nanoscale morphology as a secondary inference from Laue spot intensity. Laue patterns range over a few degrees. RSAES uses electron energies similar to RHEED (< 20 keV), but with smaller incidence and reflecting angles that are less than 1°. To achieve these small angles and fine resolutions requires electron optical elements (Fig. 1a). The necessary configuration cannot be fully realized in existing TEMs but two proof-of-concept experiments have been performed with a commercial TEM.

In the first experiment, electrons were diffracted off a carbon film with a square pattern grooves with a periodicity of 463 nm. The resulting diffraction pattern (Fig. 1b) demonstrated that high resolution diffraction can be achieved with the instrument used.

The second experiment demonstrated the application of RSAES to 3D structures. In this experiment, RSAES was performed in a Cu line grating on Si substrate. The lines had pitch = 200 nm and height = 50 nm. The sample was places in a modified sample holder such that the sample surface tangent can be aligned at a preset angle of a few degrees or less with respect to the incident electron beam. The condenser lens was under focused through the sample with the objective lens off. By adjusting the intermediate and the projection lenses, a resulting diffraction pattern is given in Fig. 1c. Three key phenomena were observed. First, as hoped there were clearly observed electron reflections with strong intensity (Fig. 1c). Second, these reflections showed striping perpendicular to the plane of incidence, indicating that the Cu-line grating had diffracted electrons. Furthermore, intensity oscillations along the diffraction stripes resulted from interference between reflections off the Cu lines and reflections off the lower substrate surfaces between the Cu line. Based on these diffraction observations in the test geometry, simulations were developed and performed for the final geometry that would be of more practical use for measuring nanostructures.

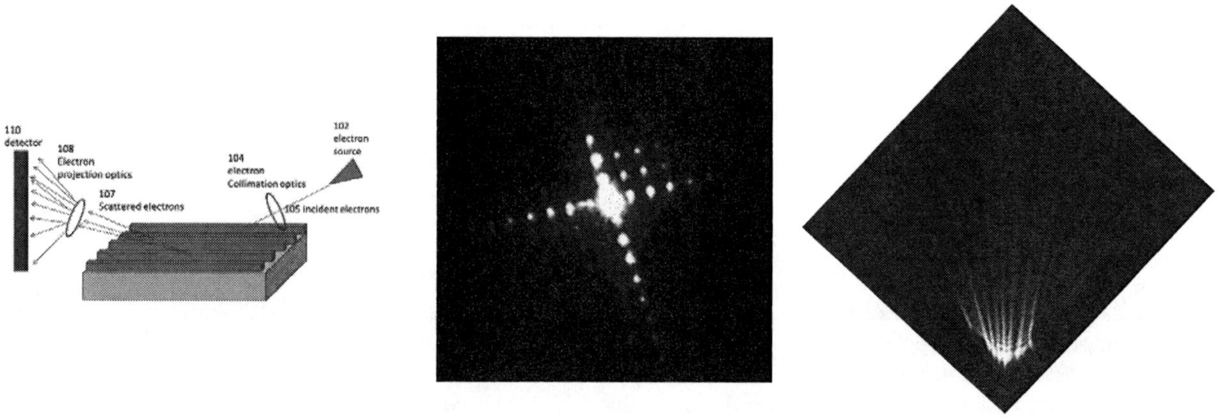

(a) (b) (c)

FIGURE 1. (a) Experimental setup for envisioned RSAES apparatus. (b) RSAES pattern from carbon film grating. (c) RSAES pattern from nanoscale Cu line grating.

RSAES SIMULATIONS

RSAES simulations were performed for simple scattering geometries (Figs. 2b and 2d). Although simpler than the envisioned RSAES apparatus these geometries can be effectively realized using the appropriate electron optics. These simulated experiments answered three important questions regarding RSAES:

1. Can interesting information be obtained using incident and reflection angles less than 1°? i.e. are diffraction patterns sensitive to pitch, height, width and shape for the types of structures likely to be of technological interest?
2. Can strong signals be achieved for structures with dimensions less than 10 nm?
3. Can a multiple-scattering (i.e. fully dynamic) electron scattering simulation be developed and implemented to aid in measurement analysis and design of RSAES apparatuses and experiments?

To simulate the strong electron-material interactions, it was necessary to solve the time-independent Schrödinger's equation (TISE) numerically. For ideal gratings, it is only necessary to simulate a single representative cell in the transverse-vertical plane. (Fig 3a). Two of the more interesting cases result from simulating the reflections of convergent spherical wave focused through the sample (Fig. 3b) or the result of using a vertical slit aperture for the incident beam (Fig. 3c). The fan-like structure in Fig. 3b results from the pitch of the simulated grating. The underlying intensity undulations reveal more information about the grating geometry although they are somewhat difficult to interpret by inspection. Using the vertical slit aperture (Fig. 3c) generates a pattern that can be more intuitively understood. Distances between vertical stripes are inversely proportional to the grating pitch, while vertical intensity oscillations along the stripes are inversely related to the grating height. The simulated reflections are sensitive to other geometric features such as line-width and shape but in less obvious ways. Finally, it is worth noting that the reflections are very strong with peak intensities greater than 10% of the incident beam intensity and usually much higher.

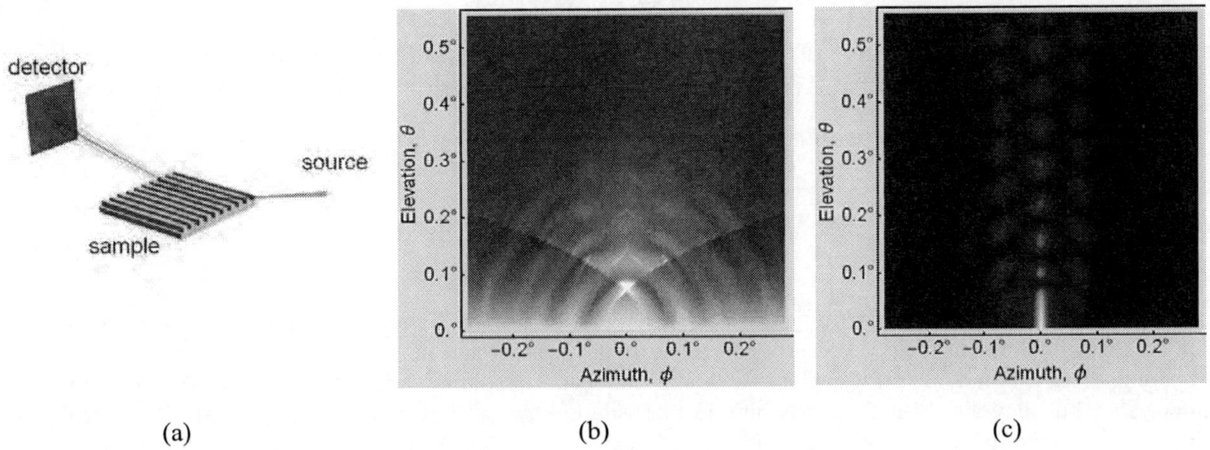

FIGURE 2. (a) schematic of simulated RSAES geometry. (b) simulated spherical wave RSAES pattern from line grating. (c) simulated vertical slit-aperture RSAES pattern from line grating.

REFERENCES

1. T. Hu, R. L. Jones, W.L. Wu, E.K. Lin, Q.H. Lin, D. Keane, S. Weigand and J. Quintana, *J. Appl. Phys.* 96, (2004) pp. 1983-1987.
2. J. Wernecke, M. Krumrey, A. Hoell, R. J. Kline, H.K. Liu and W.L. Wu, *J. Applied Crystallography* 47(6) (2014) pp. 1912-20.
3. W.-L. Wu, TW, Y.-S. Chien, TW, W.-E. Fu, TW, Y.-S. Chen, TW, H.-C. Ho, and TW, U.S. Patent No. 9390888 (12 July 2016).
4. Ichimiya A & Cohen P I (2004). Reflection High Energy Electron Diffraction. Cambridge University Press: Cambridge, UK. pp. 1,13,16,98,130,161. ISBN 0-521-45373-9; Horio Y; Hashimoto Y & Ichimaya A (1996). Appl. Surf. Sci. 100: 292–6. ; Braun W (1999). Applied RHEED: Reflection High-Energy Electron Diffraction During Crystal Growth. Springer-Verlag: Berlin. pp. 14–17, 25, 75. ISBN 3-540-65199-3.

KEYWORDS

diffraction, electron microscopy, dimensional metrology, simulation

Individual Device Analysis Using Hybrid TEM-Scalpel SSRM Metrology

U. Celano[1], P. Favia[1], C. Drijbooms[1], O. Dixon-Luinenburg[1], O. Richard[1], H. Bender[1], E. Vancoille[1], K. Paredis[1], R. Loo[1], A. Schulze[1], A. Hikavyy[1], L. Witters[1], J. Mitard[1], N. Collaert[1], N. Horiguchi[1], W.Vandervorst[1,2]

[1]imec, Leuven, Belgium; [2]KU Leuven, Belgium; email: umberto.celano@imec.be

INTRODUCTION

New architectures of nanoelectronic devices such as FinFETs, TFETs and gate-all-around pose huge challenges for the metrology due to their 3D nature, geometrical confinement and heterogeneous materials integration. Carrier profiling, electrical and chemical analysis with sub-nm spatial resolution within the same device are of paramount importance but they are difficult to combine in one, single physical characterization method. Thus usually only a limited set of materials properties can be probed from the same selected device. While in process development still multiple devices can be used in combination with several characterization methods to obtain all relevant information, access to complementary analysis methods is incompatible with a site-specific device characterization and failure analysis. Whereas in general high-resolution TEM provides the best spatial resolution for local structural analysis, and enables compositional information by the usage of energy-dispersive X-ray spectroscopy, it lacks sensitivity to variations of dopant concentration and resistivity fluctuations. Vice versa, scanning probe microcopy offers a flexible set of techniques for nm-resolved electrical (resistivity) measurements, dopant profiling, matrix compositional variations and/or defectivity. Although inherently a 2D-method, recent work using Scalpel SPM has expanded its application towards 3D-profiling.[1] In this work, we combine the strengths of both techniques into a novel method that uses Scalpel SPM and TEM on the same device *for the nm-resolved 3D electrical and chemical analysis of Si- and Ge-based structures compatible with sub-20 nm technology nodes.*

EXPERIMENTAL

The 3D resistivity profiling is achieved by Scalpel SPM which is carried out using a scanning spreading resistance microscopy (SSRM) sensor equipped with in-house fabricated conductive diamond tips. The TEM analysis is carried out in a FEI Titan, electron beam energy 300 keV. During the consecutive cycles of SPM-TEM analysis, the sample remains fixed to a TEM Cu-grid modified with focused ion beam (FIB) to facilitate both measurements.

RESULTS AND DISCUSSIONS

Scalpel SPM on a fully-integrated FinFET structure: The extension of the 2D concept of SSRM towards 3D analysis with Scalpel SPM creates the opportunity to analyze fully-integrated devices despite their complex geometry and co-integration of multiple materials.[1], [2], [3] Scalpel SPM is based on a slice-and-view approach whereby we add the depth profiling capability to standard SSRM by a tip-induced material removal method. As an example the analysis of state-of-the-art input/output pFinFET transistors is presented in Fig. 1. Once collected, the set of 2D SSRM profiles (Fig. 1b) are interpolated and compiled into a 3D tomogram (Fig. 1c). This allows the observation of the same device along all the directions (Fig. 1d). The 3D tomogram (resistance map Fig. 1c) emphasizes the spatial distribution of the five gates (of which two dummies), the raised source/drain regions, the shallow trench isolation (STI) and the metal contacts. As the entire volume is based on 60 SSRM slices with a depth interval of $\approx$ 0.5 nm/scan, high spatial resolution analysis becomes available and fine details can be extracted from the tomogram. For instance, in Fig. 2, the cross-sectional view of the fin in the channel region (fin width ≈ 8 nm, fin height ≈ 30 nm) is shown. Note that the resistance ranges over many orders of magnitude (5-7) therefore the color-scale can be filtered to highlight the features in the relevant range for the fin channel observation i.e. $10e^{6}$-$10e^{9}$ Ω. A detailed view of one FinFET along the y axis is presented in Fig. 3a-b. This allows to measure the electrical channel length 11 ± 1 nm, and the low resistive source drain regions that extend for ≈ 70 nm from the gate level into the bulk-fin (Fig. 3b). Finally, the selection of the highly resistive or –highly conductive regions clearly show (Fig. 3c) structural details of the FinFET such as the presence of the spacer around the gate (red color), the metal plugs and silicides in the source/drain regions (blue color). Despite the flexibility of Scalpel SPM to obtain high-resolution 3D electrical information, concurrent chemical composition and the sub-nm structural analysis are not yet available limiting a complete characterization.

Complementary TEM and Scalpel SPM: To overcome this limitation and provide nm-resolved 3D electrical and chemical analysis, a method is reported here which combines TEM and Scalpel SPM on the same device. First, FIB is used to prepare a standard TEM lamella of the devices of interest. This lamella will be positioned on a customized TEM sample grid. The TEM-grid (Fig. 4) has a small slit milled by FIB in its central pin as in the inset of Fig. 4. The TEM lamella is positioned in the middle of the slit and anchored by the depositing of Pt on its edges. This design, reported in Fig. 5, allows the same area to be studied by TEM and Scalpel SPM, as it represents a good trade-off between the needs of TEM i.e. electron transparency, and of Scalpel SPM i.e. a mechanically stable and flat surface.

Analysis of strained Germanium pFinFETs: We demonstrate our method to the source and drain (S/D) regions of state-of-the-art pFinFETs using strained-Germanium (Fig. 6a).[4] The sGe channel is placed on top of SiGe strain relaxed buffer (SRB) whereby boron-doped $Si_{0.25}Ge_{0.75}$ is used for the raised S/D regions.[5] The table in Fig. 6a reports the integration steps of the sample, while more information on its electrical performance can be found elsewhere.[4] Thanks to the intrinsic higher hole mobility compared to Si and the presence of a quantum barrier between ground plane and channel, this structure has shown improved performance over the Si-channel FinFET developed at imec.[4] However, in such a technology the crucial process steps such as the sGe pre-epi clean prior to the SiGe D/S growth done at low temperature and the in-situ Phosphorous and Boron doping of the SiGe/Ge fin, require a combined analysis providing nm-precise structural, electrical and chemical information. The latter is crucial to observe and mitigate dopant segregation in the sGe channel, to identify defects-assisted leakage current in the SRB inducing parasitic under-fin paths, compositional variations and other issues all leading to a loss of electrostatic control of the channel. For this reason we have mounted a TEM lamella on our custom sample holder as in Fig. 5a-b. The area of interest includes ≈ 20 devices distributed across ≈ 3 µm. Fig. 5c shows the AFM morphology of the lamella and a 2D profile indicating the FIB-thinned area where the analysis will be carried out. Structural, chemical and electrical information are now accessible for any device within the 3 µm analysis area. A direct comparison between TEM and 2D SSRM is immediately available as seen in Fig. 5b (SSRM inset). Performing our slice-and-view technique, the same area can be investigated by TEM (projection technique) and Scalpel SPM (tomographic technique) repeatedly. Fig. 6b shows this combination; note the presence of a miss-processed device (likely introduced by the TEM preparation), that helps to locate the same area between the two techniques. Fig. 7 shows a magnified view with one single fin probed by multiple techniques namely SSRM, TEM and EDX. The use of FFT-SSRM [6] is essential to obtain the quantified n-type carriers profile in our device as shown in Fig. 8. In addition, we used our method also to study other sources of local resistivity variations. A clear example is the analysis of the resistivity profile within a raised S/D whereby the pre-epi clean treatment which is used on the sGe layer before the B-doped SiGe growth, has a clear impact on the current resistance within that region. Indeed, Fig. 9 clearly shows the formation of a thin interface between the two layers extending for 2-3 nm between the sGe channel materials and the SiGe raised S/D region. This interface is consistently observed in all the measured devices. The direct correlation of this area by TEM and Scalpel SPM, reveals that the interface shows a reduced conductivity. Interestingly, the EDX map does not present within its sensitivity the presence of oxygen at this interface, likely indicating that instead of GeO_x the layer forms due to the presence of a residual roughness and strain on the sGe when the epi-SiGe is regrowth. Obviously this interfacial resistance degrades the final series resistance in the raised S/D area and can easily explain the reduced drive current in the tested devices due to an increased access resistance. Finally, Fig. 10 shows a site-specific analysis, where a single structure having a defective SRB layer is measured. The EDX Si spectra of the ($Si_{0.25}Ge_{0.75}$) SRB shows a high Ge concentration on the left side of the SRB. It is worth noting that the 3D conductive profile of the same device relates the Ge-rich area with a highly conductive region. Scalpel SPM allows the observation of the conductive path and its extension in the third dimension (Fig. 11). The latter relates the defects in the SRB to an enhanced conductivity and represents a threat for the off-state leakage of the transistor, if the area extends under the sGe-channel forming under-fin paths impacting on the performance as in Fig. 12.

In summary, the combination of Scalpel SPM and TEM for the characterization of confined volumes is presented for the first time. After describing in details our method, we demonstrated the application on different ultra-scaled pFinFETs heterogeneous structures, for which we have provided 3D structural, electrical and chemical information with nm-precision.

REFERENCES

[1] U. Celano, et al., "Conductive-AFM tomography for 3D filament observation in resistive switching devices," IEDM Tech. Dig., pp. 21.6.1–21.6.4, Dec. 2013.

[2] J. Mody, et al., "3D-Carrier Profiling in FinFETs using Scanning Spreading Resistance Microscopy," IEDM Tech. Dig., pp. 119–122, 2011.

[3] P. Eyben, T. et al., "Scalpel soft retrace scanning spreading resistance microscopy for 3D-carrier profiling in sub-10nm WFIN FinFET," Tech. Dig. - Int. Electron Devices Meet. IEDM, vol. 2016, no. 32, pp. 14.1.1–14.1.4, 2016.

[4] J. Mitard, et al., "A 2nd Generation of 14/16nm-node compatible strained-Ge pFINFET with improved performance with respect to advanced Si-channel FinFETs," in 2016 Symposium on VLSI Technology (VLSI-Technology), 2016, vol. 1–2.

[5] L. Witters, et al., "Strained Germanium quantum well pMOS FinFETs fabricated on in situ phosphorus-doped SiGe strain relaxed buffer layers using a replacement Fin process," in 2013 IEEE International Electron Devices Meeting, 2013, pp. 20.4.1–20.4.4.

[6] A. Schulze, et al., "Outwitting the series resistance in scanning spreading resistance microscopy," Ultramicroscopy, vol. 161, pp. 59–65, 2016.

KEYWORDS

Scalpel SPM, 3D metrology, Hybrid metrology, SSRM

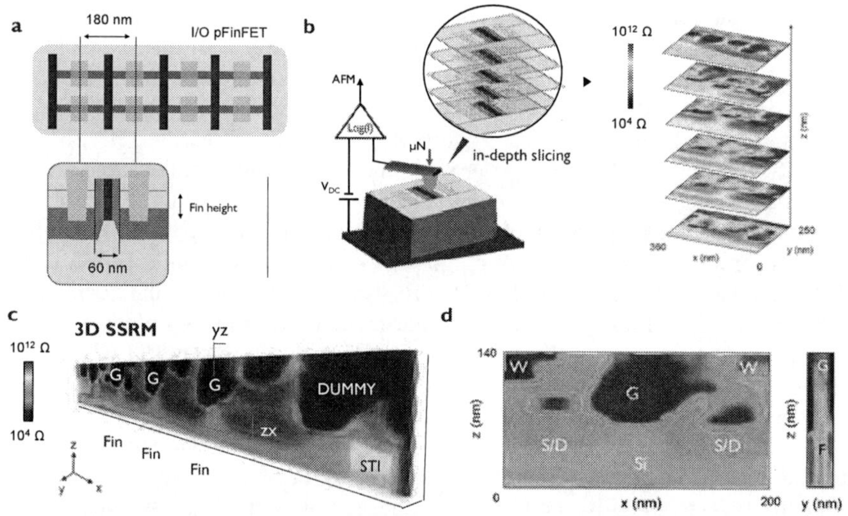

FIGURE 1. (a) Plain view and schematic cross-section of the series of fully-fabricated I/O pFinFET devices measured with 3D SSRM. (b) Schematic of the 3D SSRM implementation enabled by Scalpel SPM. Several 2D SSRM profiles of the structure under test are collected, aligned and interpolated to obtain (c) a 3D tomogram of the sample's volume. (d) 2D extracted profiles in the regions indicated in the 3D tomogram, note the xz-cut is along the fin with the gate and the contacts partially shielding the channel and the yz-cut is a plane across the fin under the gate.

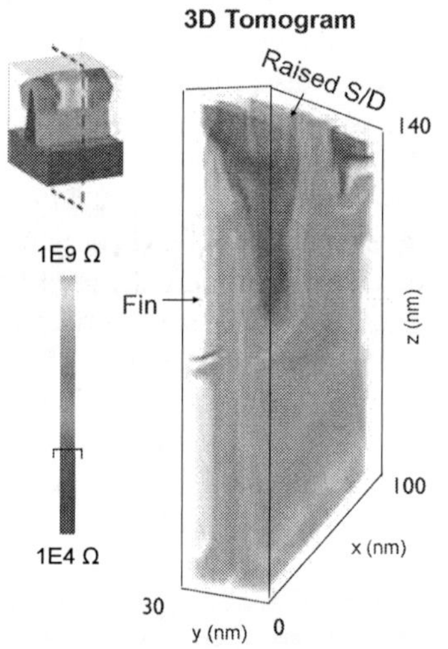

FIGURE 2. A single transistor is extracted from the 3D tomogram and observed in section in the middle of the electrical channel region. Note, the contribution of the contacts and the gate is reduced to enhance contrast in the area of interest.

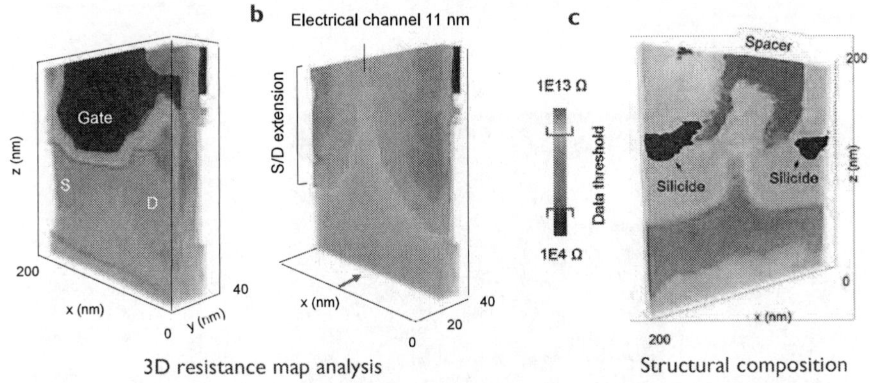

FIGURE 3. 3D-reconstructured volume of a single device visible at different stages along the y axis. (a) Out of the channel where the gate contact is visible and (b) under the gate where the spatial distribution of the S/D implants and the channel region are clearly visible. In addition, the other parts of the device are potentially accessible, for example in (c), filtering respectively the highest and lowest conductive features, the contact silicide and the spacer can be emphasized.

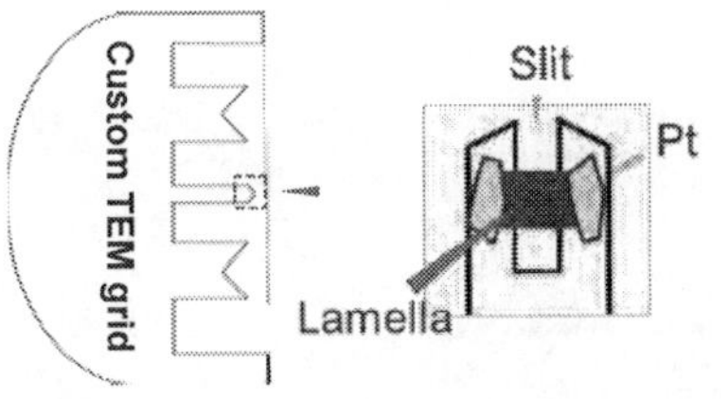

FIGURE 4. In-house fabricated customized TEM sample holder. A standard Cu-grid TEM holder is modified by FIB to create a small slit (≈ 2-3 μm) in the center of the middle pin, where the lamella will be placed and anchored by Pt deposited on the edges of the lamella.

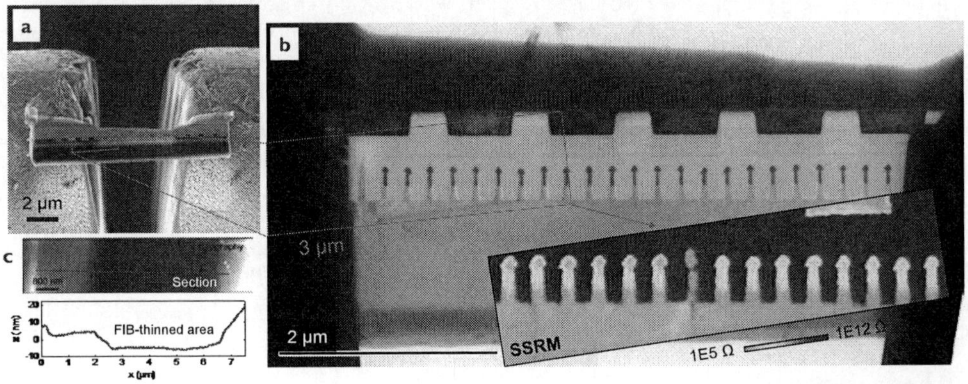

FIGURE 5. (a) SEM image of the lamella as mounted on the custom TEM grid. (b) SSRM and TEM can be alternatively used on the area of interest (red box). The SSRM image of a series of structures is combined with the high resolution TEM image. (c) AFM is used to image the surface of the sample. Note in the AFM profile the step associated to the FIB-thinned region in the middle part of the lamella.

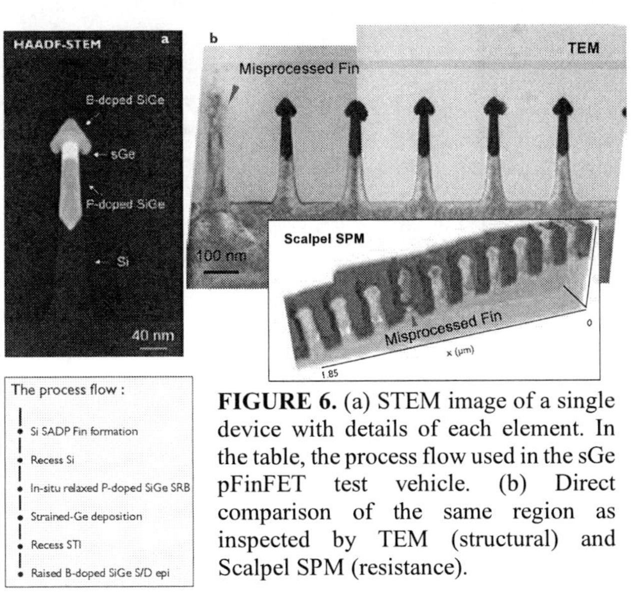

FIGURE 6. (a) STEM image of a single device with details of each element. In the table, the process flow used in the sGe pFinFET test vehicle. (b) Direct comparison of the same region as inspected by TEM (structural) and Scalpel SPM (resistance).

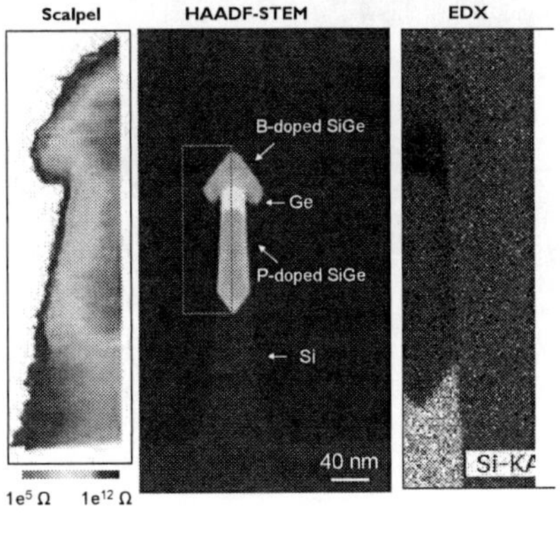

FIGURE 7. Site-specific structural, chemical and electrical information. The same device can be studied by Scalpel SPM and TEM. In this case, the chemical sensitivity of EDX is combined to the carrier profile capability of SSRM and the structural analysis of TEM in the red rectangle

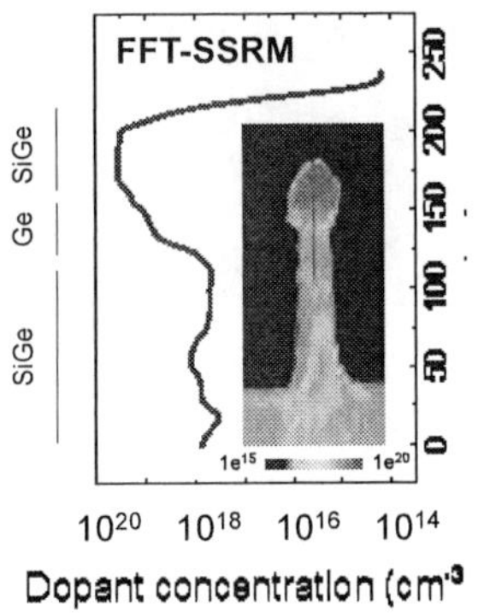

FIGURE 8. Due to the parasitic terms introduced by the small geometries (very high series resistance), FFT-SSRM is used for the direct quantification of the carriers in the SiGe-based raised S/D regions.

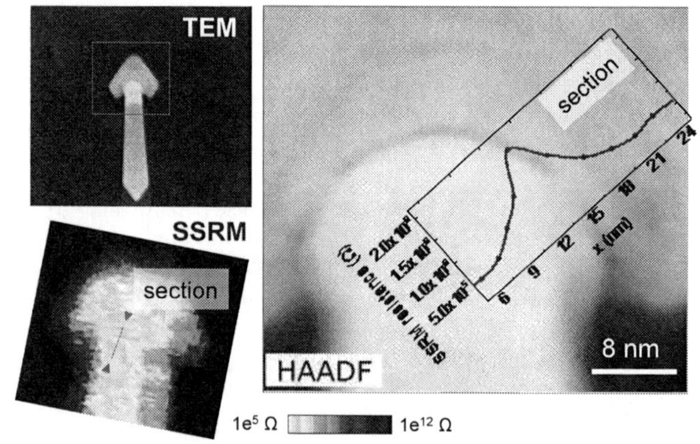

FIGURE 9. The pre-epi clean treatment applied before the growth of the doped-SiGe S/D, introduces a thin interfacial layer. Visible in both the TEM and the SSRM analysis, likely due to the residual strain or roughness between the sGe and SiGe. The conduction properties of this thin interface are visible in the SSRM profile where the interface shows a reduced conductivity. Clearly this interfacial layer increases the extension resistance of this transistor.

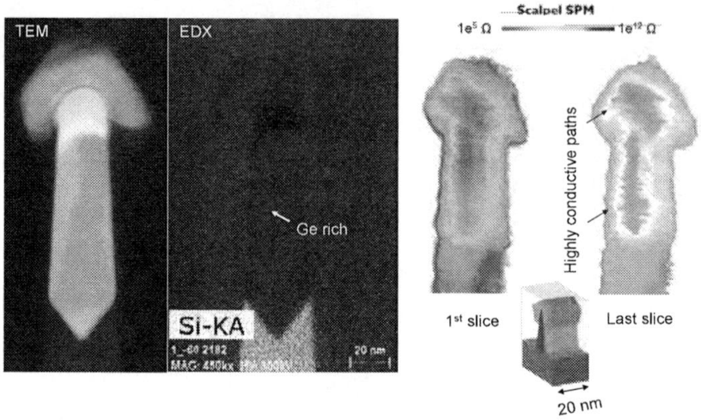

FIGURE 10. TEM, EDX and SSRM comparison for the same fin: the chemical analysis clearly reveals the presence of an extended defect in the (RSB) SiGe layer, where a Ge-rich area appears by the chemical EDX composition. A direct comparison with 3D SSRM of the same structure reveals an enhanced conductivity extending in the area across the 3D tomogram. Thus representing a potential threat for the off-state leakage of the transistor for the formation of parasitic paths under the finFET channel.

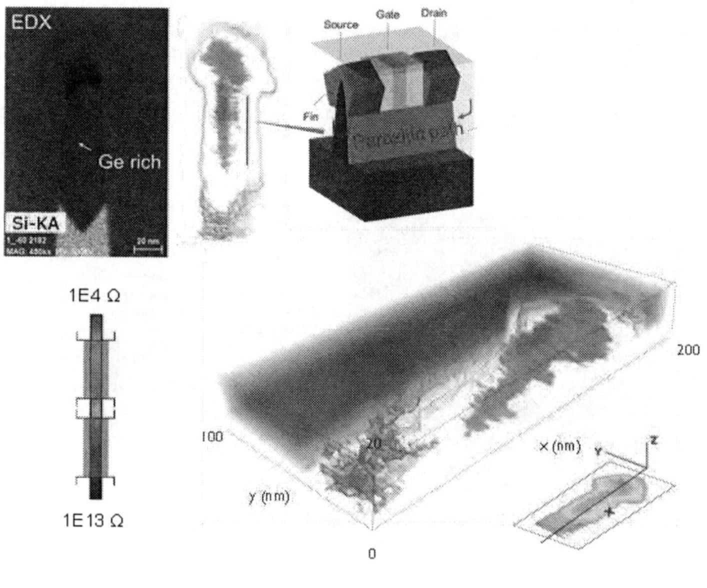

FIGURE 11. The mid-range and high resistive features are suppressed in the 3D tomogram (resistance map) according to the color-scale, to highlight the extension and 3D evolution of the defective area inside the SRB (3D cut as in the bottom inset). The top-right schematic describes a potential parasitic path introduced by such kind of defective SRB. A similar effect has been often ascribed to the threading dislocations density in the P-doped SiGe.

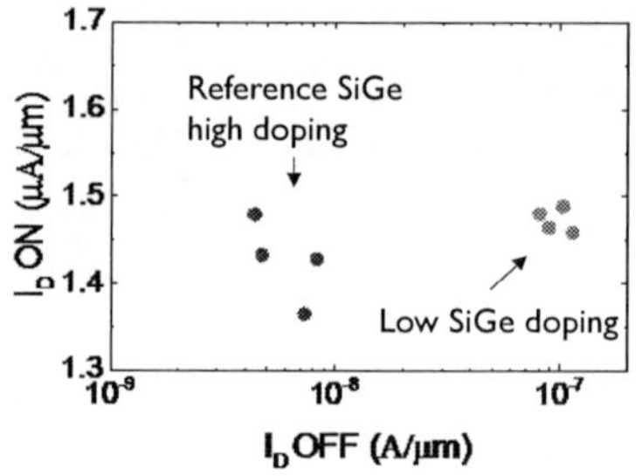

FIGURE 12. Impact of the SiGe SRB doping level (high vs low). In the low doping case, the formation of parasitic paths under the fin similar to the one presented here can induce a shift in the off-state leakage of the transistor.

Super-Resolution Single Molecule Imaging Microscopy On Plasmonic Grating Platform

Aaron Wood[3], Syed Barizuddin[2], Keshab Gangopadhyay[2], Sheila Grant[3], Shubhra Gangopadhyay[1,2*]

[1]*National Science Foundation*
[2]*Dept of Electrical and Computer Engineering,* [3]*Dept of Biological Engineering, University of Missouri-Columbia,*
349 Naka Hall
Columbia, MO 65211 USA
**Corresponding author*

INTRODUCTION

Super resolution imaging has significantly improved over the last 10 years. However, most of the system requirements include bulky, expensive, and complicated optical setups which limits its use as a day-to-day research tool for many research labs albeit their need. This study presents, an inexpensive, plasmonic grating-based platform that was fabricated using HDDVDs and Bluray discs to enable super resolution imaging of single fluorescent molecules in a simple epi-fluorescence microscope. Plasmonic gratings use the surface plasmon resonance (SPR) generated through angular illumination to excite nearby fluorescence molecules. In diffraction-limited imaging, single molecules (SM) of a fluorescent dye are free to re-emit a photon which is captured by the microscope's objective in the far-field. However, if a SM in located within 100 nm of the plasmonic grating, the energy of the photon is non-radiatively transferred to the grating and a photon is emitted at a specific angle via a phenomenon referred to as Surface Plasmon Coupled Emission (SPCE.) The photon emitted via SPCE can carry near-field information into the far-field and, if a photon is captured by a microscope, the resulting image will have sub-diffraction limit resolution.

MATERIALS AND METHODS

The improvement in SM image resolution were tested using nanoscale gratings obtained from a microcontact printing process detailed in previous publications [1,2]. The gratings used were HDDVD gratings (Λ = 400 nm, H = 55 nm) coated with a 100 nm thin silver film with a 10 nm thin alumina (Al2O3) capping layer. This process yields similar results as other grating fabrication methods, such as E-beam lithography, but is much less expensive and less time consuming. Single molecules of rhodamine 6G (R6G) were suspended in PMSSQ thin film (33 nm) that was spin-coated on top of the silver grating using a 1% wt. PMSSQ solution in pure ethanol with 10^{-14} M R6G dye. The gratings were imaged using a BX51WI Olympus epifluorescence microscope equipped with a 100x oil immersion objective, ORCAFlash2.8 CMOS camera, and Lambda XL light source.

RESULTS AND DISCUSSION

A representative bright-field and fluorescence image of the R6G dye film has been provided in Fig. 1. Upon further analysis of the fluorescence image, two populations of SM emissions were observed. The first population exhibited airy disc emission patterns (Fig. 2(a).) This would indicate that the SM image is diffraction limited. However, the second population exhibited a "split" emission pattern where two lobes of the emission pattern were generated by a single molecule. The presence of a single molecule for each split emission was confirmed by blinking spectroscopy where both lobes disappeared simultaneously and in a single step photobleaching event. Upon defocusing of the second population with the microscope objective, angular emission pattern was observed.

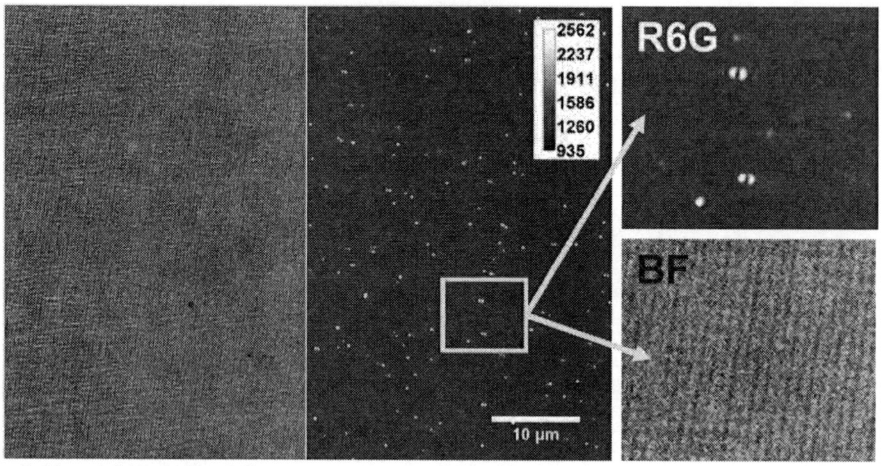

FIGURE 1. Bright field and false-colored fluorescence image (100xOI objective, 10 sec. exposure) of a plasmonic grating with a spin-casted 33 nm thin fluorescent PMSSQ film (10-14 M R6G) with expansions of the same location.

The angular spread of the emission pattern was measured using the spread of the lobes with increasing focal height (Fig. 2(a).) It was found that the angular emission pattern (green line) matched the emission angle range predicted by the SPR dispersion (gray line) for the emission intensity and wavelength range emitted by R6G dye (Fig. 2(b)) but with variations in intensity between the two lobes of the split emissions. In theory, an excited dye molecule located in the middle of a grating groove has an equal probability of non-radiatively transferring its energy to either of the grating ridge parallel to the groove. However, closer proximity to either ridge will result in a higher probability of energy transfer according to the surface energy transfer (SET) model [3]. As the placement of the dye molecule is random within the pits of the grating structure, the molecular position may be located closer to one side of the groove than the other. It may be possible to locate the precise XYZ-location of a molecule within the grating groove based on the relative intensity of the two lobes and focal plane location, but this requires a secondary localization precision method beyond the scope of our current measurement capabilities.

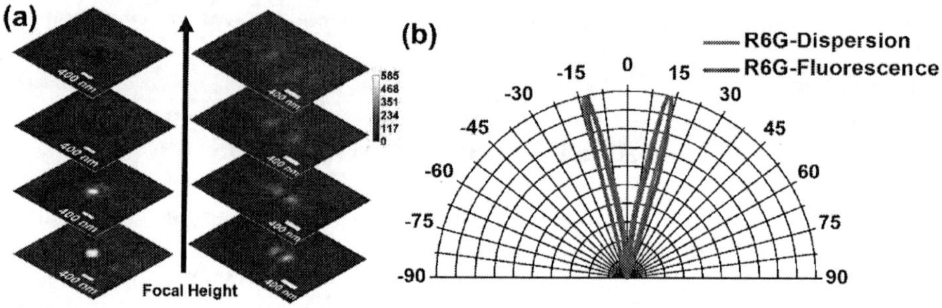

FIGURE 2. (a) Stacked fluorescence images taken at increasing focal plane height above the plasmonic grating. (Left) Angled SPCE and (Right) Airy disc pattern from the isotropic emission of a SM spreading with increasing focal height. Scale bar: 400 nm. Image exposure: 10 sec. (b) The normalized emission angle range for a SM of R6G plotted against the emission range predicted by the SPR dispersion and dye emission properties.

The intensity profiles of two representative SM for each population were also compared to determine the image resolution. Based on the intensity profile for the diffraction-limited population, the FWHM was found to be ~327 nm. Given that the wavelength of most of the photons collected in these images are between 542 and 600 nm and that an airy disc pattern with distinct 0 and 1st-order diffraction modes can be seen, the first population is certainly diffraction-limited. However, the lobes of the split emission have much better resolution with a FWHM of ~233 nm and ~217 nm for each lobe, respectively. Additionally, the spacing between the two lobes was smaller yet at ~164 nm. The diffraction limit can be roughly estimated to be half of the imaging wavelength, ~272 nm in this case, which is much larger than the FWHM and spacing of the split emission pattern. If the split emission is due to SPCE, this confirms that sub-diffraction limit information is being transmitted into the far-field.

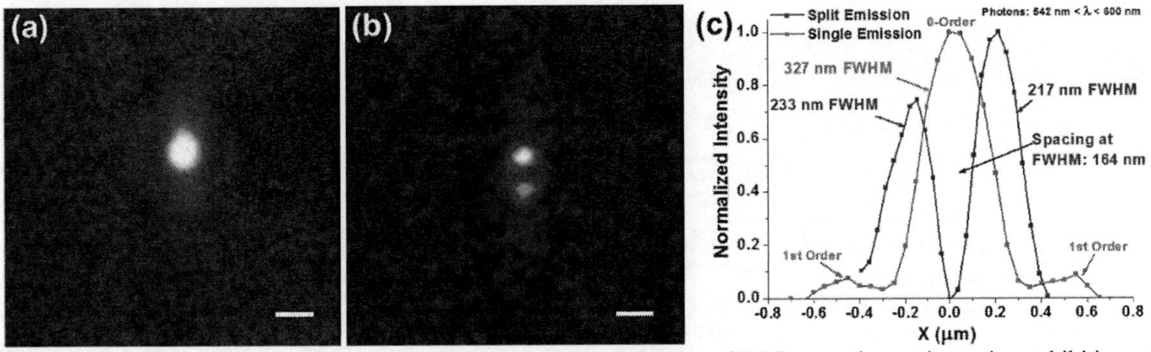

FIGURE 3. Background-subtracted Fluorescence images of single molecules of R6G on a plasmonic grating exhibiting a (a) diffraction-limited airy disc and (b) split emission due to SPCE. 100x Objective (NA 1.49), 10 sec. exposure, Scale bar: 400 nm. (c) Cross-section of the SM in (a) and (b) with calculated FWHM and separation distance of the split emission lobes.

CONCLUSION

In summary, we have demonstrated that it is possible to image SM SPCE from plasmonic gratings using a simple epifluorescence microscope. From these images, we can extract the angular emission profiles for individual dye molecules as well as obtain super resolution images of SM. This imaging technique can be easily applied to other SM research applications to greatly improve fluorophore emission intensity and image resolution.

REFERENCES

1. B. Chen, A. Wood, A. Pathak, J. Mathai, S. Bok, H. Zheng, S. Hamm, S. Basuray, S. Grant, K. Gangopadhyay, P. V. Cornish, and S. Gangopadhyay, *Nanoscale.* **8**, 12189-12201 (2016).
2. A. J. Wood, B. Chen, S. Pathan, S. Bok, C. J. Mathai, K. Gangopadhyay, S. A. Grant, and S. Gangopadhyay, *RSC Advances.* **5** (96), 78534-78544 (2015).
3. C. J. Breshike, R. A. Riskowski, and G. F. Strouse, *The Journal of Physical Chemistry C.* **117** (45), 23942-23949 (2013).

KEYWORDS

Surface Plasmon Resonance, Plasmonic Grating, Super-resolution Imaging.

Direct Observation of Alloyed Contact Formation in Nanowire Cross-section

Renjie Chen,[1] Katherine L. Jungjohann,[2] William M. Mook,[2] John Nogan,[2] and Shadi A. Dayeh[1,3,4]

[1] Department of Electrical and Computer Engineering, University of California San Diego, La Jolla, California 92093, USA. [2] Center for Integrated Nanotechnologies, Sandia National Laboratories, Albuquerque, New Mexico 87185, USA. [3] Materials Science and Engineering Program, University of California San Diego, La Jolla, California 92093, USA. [4] Department of NanoEngineering, University of California San Diego, La Jolla, California 92093, USA

Corresponding email: sdayeh@ucsd.edu

INTRODUCTION

Alloyed and compound contacts between metal and semiconductor transistor channels enable self-aligned gate processes which played a significant role in transistor scaling.[1] Alloyed contacts, formed by thermal anneal of metal-semiconductor junction, are prescribed for lithography-free self-aligned gate design.[2] The advent of semiconductor nanowires evoked reevaluation of the thermodynamics, kinetics, and phases in nanoscale metal-semiconductor reactions, and unveiled large differences between nanoscale reactions and their bulk counterparts.[3] However, existing knowledge concern with reactions in the horizontal nanowire channels, leaving the early stage of reaction in their cross-sections largely unknown. From a materials perspective, the early stage of the reaction can determine the final phases and crystalline orientations when the reaction extends into the horizontal segment of nanowire channels (forming an end-bond-contact geometry), and be largely affected by local defects or interfacial properties. From an engineering perspective, the early stage reaction facilitates a side-contact geometry in nanowire devices that is more desired for self-aligned technology than end-bond-contact geometry. These issues demand a detailed understanding of the early stage reaction in nanowire cross-sections.

INTEGRATION OF SPECIMEN ON IN-SITU TEM HEATING PLATFORM

In this work, we focused on the narrow band-gap high electron mobility III-V semiconductor, InGaAs, motivated by its potential in sub-10 nm metal-insulator-semiconductor field-effect transistors (MISFETs).[4] The Ni-InGaAs solid-state reaction in the cross-section of InGaAs nanowire was observed under transmission electron microscope (TEM) with in-situ heating. Firstly, we fabricated horizontal InGaAs nanowire channels on insulator on top of Si substrate through a novel wafer bonding approach,[5-6] followed by EBL patterning and dry-etch. Secondly, Ni was deposited on top of the nanowire channels, and then focused-ion-beam (FIB) milling was utilized to cut the nanowire from cross-section. Finally, the FIB milled specimen was transferred on top of a commercial thermal heating chip for in-situ TEM observations. The detailed FIB milling and transferring steps are shown in Figure 1.

IN-SITU HEATING TEM OBSERVATIONS

During the in-situ heating TEM experiments, the specimen was heated at 180 °C and the solid-state reactions between Ni and $In_{0.53}Ga_{0.47}As$ were recorded by real-time videos, as shown in Figure 2. At similar temperatures, a crystalline $Ni_2In_{0.53}Ga_{0.47}As$ phase usually grew on planar $In_{0.53}Ga_{0.47}As$ surface, while the nanowire cross-section experienced a solid-state amorphization step and formed $Ni_xIn_{0.53}Ga_{0.47}As$ ($x<2$). During this amorphization process, nickelide reacted in a layer-by-layer manner with ledge moving on {111} facets and along <112> directions. Nickelide ledge movement events were recorded with time-sequenced high-resolution TEM images which showed that the phase transformation occurred through the evolution of stepped edges on non-{111} facets and their consequent elimination to result in an equilateral rhombus like region bounded by {111} planes. The interfacial property between Ni and $In_{0.53}Ga_{0.47}As$ nanowire was found to significantly influence the reaction kinetics and was captured by a model that we developed specifically for the cross-sectional geometry of nanowire channels. Finally, the amorphous $Ni_xIn_{0.53}Ga_{0.47}As$ (x<2) phase regrow into a single crystalline $Ni_2In_{0.53}Ga_{0.47}As$ phase at temperatures above 375 °C by additional incorporation of Ni adatoms from the contact reservoir. Overall, the results presented here provide a general guide for the development of crystalline self-aligned contacts in nanoscale channels and can be generalized to radial reactions of metallic and other alloys into nanowire cross-sections.

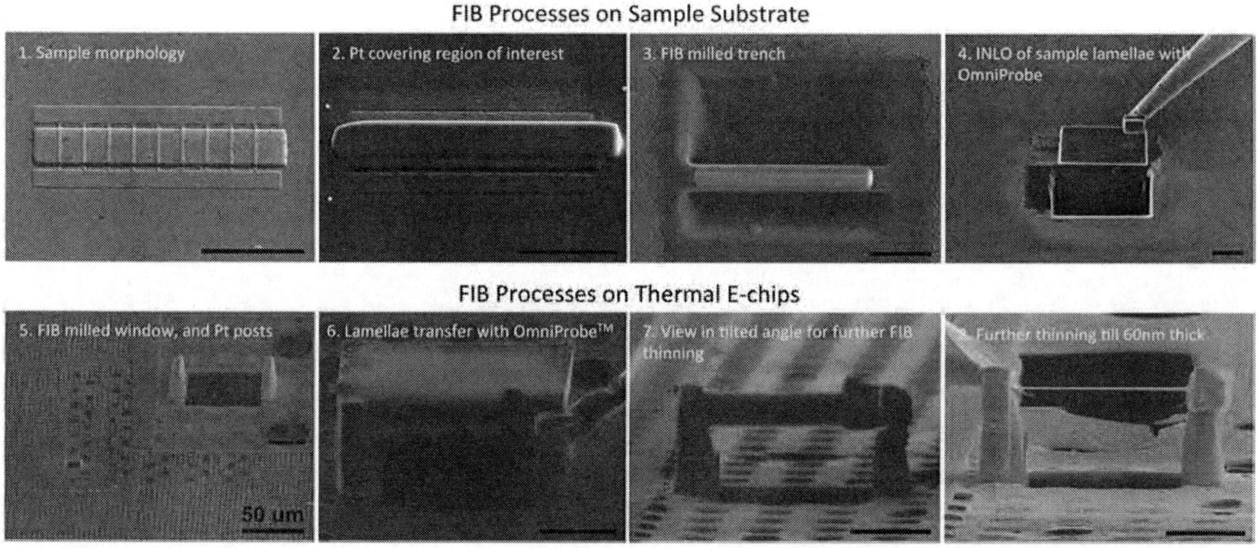

FIGURE 1. SEM images show the fabrication sequences to transfer the FIB cut lamellae from the home substrate onto the TEM membrane window of a TEM compatible thermal chip. All the scale bars are 5μm, except the one labeled separately in step 5.

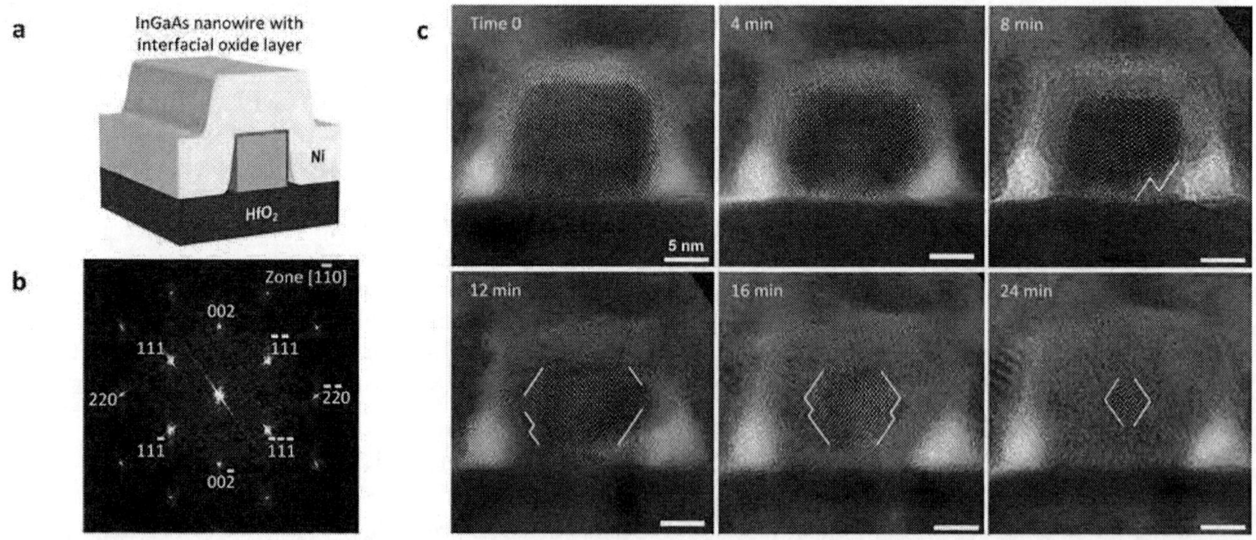

FIGURE 2. (a) schematic of Ni contacting with a InGaAs nanowire cross-section that sits on the HfO_2 dielectric layer. (b) fast Fourier transform (FFT) image of the corresponding nanowire cross-section, showing the diffraction pattern along the zone axis

(projecting direction in TEM). (c) HRTEM sequences extracted from recoded video during the in-situ heating experiment at 180 °C.

REFERENCES

1. F. Léonard and A. A. Talin, *Nature Nanotechnology* **6**, 773-783 (2011).
2. R. Requist, *et al. Nature nanotechnology* **11**, 499-508 (2016).
3. W. Tang, B.-M. Nguyen, R. Chen and S. A. Dayeh, *Semiconductor Science and Technology* **29**, 054004 (2014)
4. J. A. del Alamo, *Nature* **479**, 317-323 (2011).
5. X. Dai, B.-M. Nguyen, Y. Hwang, C. Soci and S. A. Dayeh, *Advanced Functional Materials* **24**, 4420-4426 (2014).
6. R. Chen and S. A. Dayeh, *Nano letters* **15**, 3770-3779 (2015).

KEYWORDS

In-situ TEM, nanowire, alloyed contact, cross-section

Contribution of Luminescence Techniques for the Characterization of Materials and Devices at the Nanoscale

C. Licitra[1], N. Rochat[1], N. Gambacorti[1], S. David[2], G. R. Muthinti[3], F. Olivier[1], A. Faujour[4] and A. Chabli[4]

[1]Univ. Grenoble Alpes, F-38000 Grenoble, France
CEA, LETI, MINATEC Campus, F-38054 Grenoble, France.

[2]Univ. Grenoble Alpes, LTM, F-38000 Grenoble, France
CNRS, LTM, F-38000 Grenoble, France.

[3]IBM Research, 257 Fuller Road, Albany, New York 12203, USA

[4]Univ. Grenoble Alpes, INES, F-73375 Le Bourget du Lac, France
[2]CEA, LITEN, Department of Solar Technologies, F-73375 Le Bourget du Lac, France

INTRODUCTION

Luminescence of semiconductors has been widely used for the study of their fundamental properties such as their optical properties, electronic processes, and crystalline quality including defects identification. Since the semiconductor industry faces major challenges to continue increasing performance and functionality of their devices, new materials are currently introduced in research clean-rooms and fabs. Whereas luminescence techniques have not been particularly popular in fabs due to the poor luminescence of silicon at room temperature, they could gain a higher interest for the characterization of alternative materials and devices. In this paper will present different applications where the characterization of luminescence is improving our knowledge on material and device processing. Applications ranging from III-V monolithic integration on silicon, gate all around CMOS, GaN-based emissive microdisplays and silicon engineering for solar applications will be discussed.

EXPERIMENTAL DETAILS

This work was performed using two luminescence systems. Micro-photoluminescence (μPL) spectra and maps were measured using a Horiba LabRAM tool equipped with several excitation lasers (325, 514 and 633 nm). Cathodoluminescence (CL) spectral maps were recorded using an Attolight system which couples a customized scanning electron microscope (3 to 10 keV) with a light microscope embedded within the electron objective lens for optimal optical collection. For both tools, a helium-cooled cryostat, a CCD detector and an InGaAs array allow the optical response of the sample to be measured from 300 to 1600 nm and from 5 K to room temperature. Depending on the desired excitation energy and spatial resolution, either μPL, CL or both can be used to probe the sample.

RESULTS

III-V monolithic integration on silicon

III-V semiconductors are recognized as an opportunity to improve the micro- and nanoelectronic integrated circuit performances, thanks to high carrier mobility and band gap engineering. To develop this new type of devices, the III-V alloys have to be co-integrated with silicon either by post-bonding of the III-V part on the silicon substrate or by direct epitaxy of III-V materials on silicon (monolithic integration). In the latter case, differences of lattice mismatch, polarity, and thermal expansion coefficient give rise to structural defects such as anti-phase boundaries (APBs), stacking faults (SFs), and dislocations which can affect the device performances. Dedicated growth strategies can be used to trap the defects such as the aspect ratio trapping (ART) which consists in a selective epitaxy inside trenches. In this context we have developed a spatially resolved and co-located characterization methodology using CL and Scanning Transmission Electron Microscopy (STEM) to understand the link between the structural defects and the electronic properties of the InGaAs/GaAs quantum wells formed by ART on Si [1]. An example of combined top-down CL followed by focused ion beam (FIB) preparation and cross-sectional STEM characterization is illustrated below on InGaAs quantum well fins (QWF). We can observe the structural defects crossing the InGaAs QW (d1, d2, d3, and d4) and the corresponding luminescence response over the same area. The luminescence intensity and peak position have been found to vary between the defects indicating that it depends not only on the crystal quality but also on the strain distribution as confirmed by complementary precession electron diffraction measurements. Luminescence is therefore an interesting and complementary tool to monitor the ART process improvements.

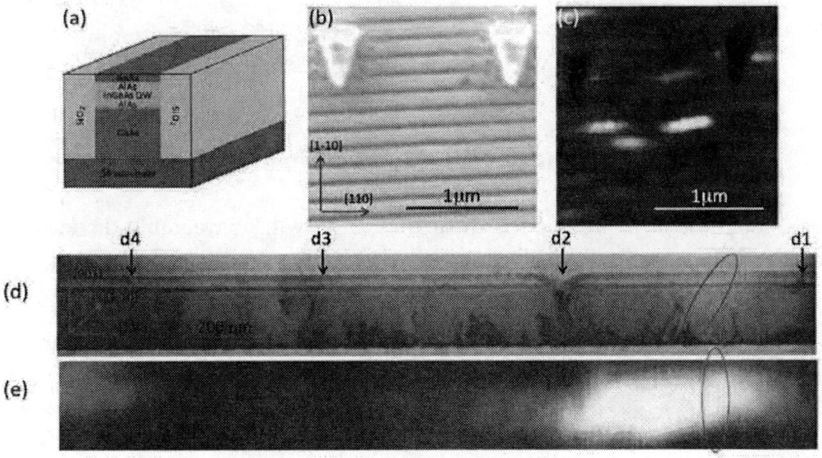

FIGURE 1. (a) Schematic drawing of the III-V QWFs grown on silicon through SiO2 trenches. (b) Top-view SEM image of III-V QWFs. Large light gray and narrow dark gray lines correspond to III-V materials and SiO2 walls, respectively. Two triangular platinum marks are seen on the top of the SEM image. (c) Corresponding panchromatic cathodoluminescence mapping showing high spatial variations of optical emission intensity. The two platinum marks appear in dark. The straight III-V QWFs appear wavy due to a drift during electronic beam scanning. (d) Cross section STEM image of an extracted QWF. Lamella crystallographic orientations correspond to [001] and [110] directions. Structural defects affecting QW are marked with 4 arrows. (e) Top-view CL intensity of the same QWF. Both CL and STEM images are spatially correlated.

Gate all around CMOS devices

Gate-all-around (GAA) nanowire (NW) devices are seen as the ultimate device from an electrostatic scaling point of view compared to single and double gate planar, FinFET, and trigate structures. One way for making GAA devices involves the use of alternate layers of Si and SiGe. After etching a multilayer fin-like structure, the SiGe is selectively removed to form silicon NW channels [2]. During the process, photoluminescence can be used to monitor the properties of Si/SiGe materials. As the stacked Si and SiGe materials are thin (in the order of 10 nm), the incident wavelength can be chosen to probe all the layers at once. At low temperature, the luminescence response will present different contributions: phonon replica of the silicon exciton at around 1.1 eV (FE^{TO}), SiGe exciton and its phonon replicas (X^{NP}, X^{TA} and X^{TO}) between 0.95 and 1.15 eV and possibly other defect-related peaks such as the D-lines and C- or G-lines. Figure 2 shows the general spectrum of Si/SiGe stacked multilayers. The SiGe exciton (and phonon replicas) peak position varies as a function of concentration and strain allowing possible strain or Ge diffusion monitoring. The silicon signature consists of excitons (free and bound exciton FE and BE), possible electron hole droplet (EHD) and sometimes defect lines can be observed after reactive ion etching (here the G-line at 0.97 eV).

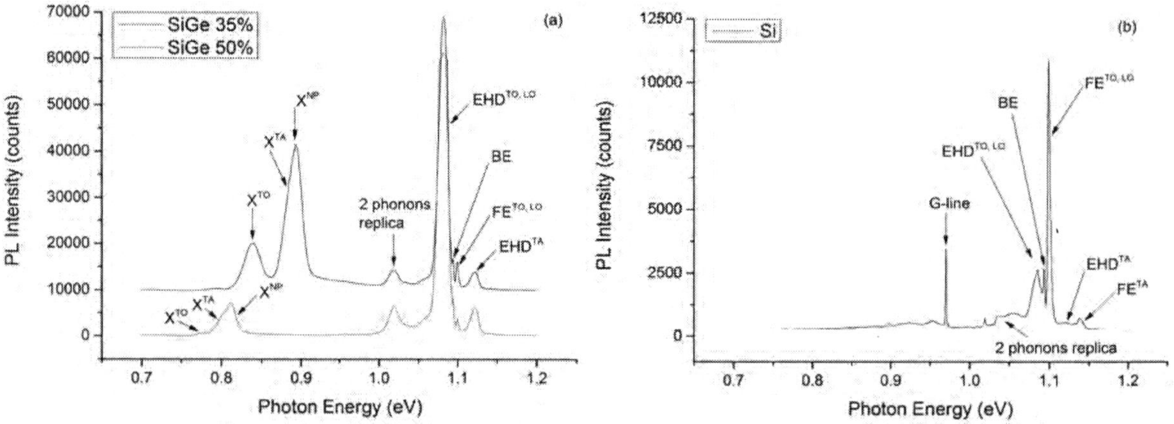

FIGURE 2. (a) PL spectra at 5K of Si/SiGe multilayers with 2 different germanium content. (b) Observation of the radiation-induced G-line after silicon etching (PL at 15K).

GaN-based emissive microdisplays

High performance microdisplays are needed to fulfill the growth of wearable devices. Emissive microdisplays such as organic LEDs (OLEDs) are particularly attractive for these applications because of their compactness and low consumption. However for some applications high-brightness displays are needed. For that purpose GaN LED arrays can be processed on sapphire substrates and subsequently hybridized on silicon CMOS active matrix [3]. The structure consists of sapphire substrates with 440 nm or 525 nm LED Multi Quantum Well (MQW) InGaN/GaN epitaxial layers grown by metal-organic chemical vapor deposition (MOCVD) for blue or green emission, respectively. GaN array process consists of: (i) patterning the pixels by etching the GaN+MQW stack using plasma process, (ii) depositing and patterning the P-contact metal pads on the pixels and (iii) depositing and patterning the N-type contact next to the pixel. The etching process has been characterized by means of cathodoluminescence in order to map the optical response within a single pixel (Figure 3). Indeed each contribution detected in the luminescence signal can be mapped to show the emission zones. We can therefore observe luminescence coming from the GaN epitaxial layer and its dislocations (Near Band Edge: NBE), the MQW emission and homogeneity and the defects (Yellow Band: YB). We can also observe signal from the edges of the structure which may be associated with the etching process.

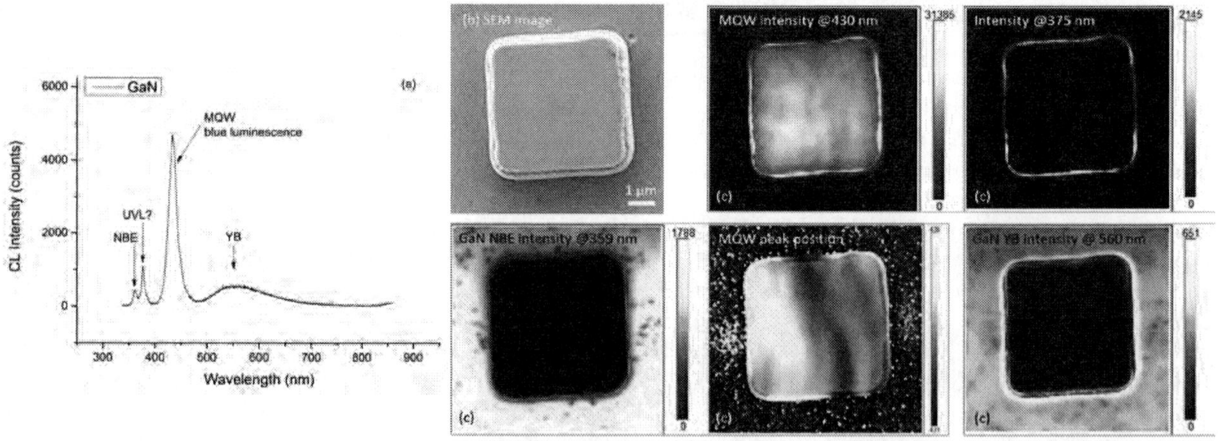

FIGURE 3. (a) Room temperature cathodoluminescence spectra showing all the emission lines. (b) SEM image of a single pixel. (c) Spectral maps of the different luminescence contributions.

Silicon engineering for photovoltaic applications

Monolike silicon [4] is a promising material for Si-based cells, even for advanced cell processes, such as heterojunction, provided that a proper gettering step is applied. This means that the background dislocations present in monolike Si with a typical density of 10^4 cm^{-2} are not efficiency-limiting. However, the local presence of more distorted defects that act as minority-carriers killers is the main performance-limiting factor in the Monolike silicon [5]. 2D extended defects, as sub-grain boundaries (SBGs) domains and grain boundaries, are among these crystalline defects that can be highly active for recombination, especially when decorated with metallic impurities and that can therefore affect the cell performances. The relation between the structures of different types of 2D defects have been previously characterized by Electron Backscatter Diffraction (EBSD) and synchrotron based X-ray topography, and their electrical activities have been analyzed [6]. A further characterization of these extended defects using spectroscopic μPL has been performed at liquid helium temperature. The spectra measured are composed of the well-known D-lines emission peaks, related to point defects around dislocations and to dislocations themselves [7], added to the bound (BE) and the free excitons (FE) radiative recombination peaks. An example of the μPL intensity maps obtained for these different emission peaks is presented on Figure 4 including a new peak at 0.85 eV not yet reported in the published works. These maps show a clear correlation between the μPL intensity distribution and the position of the SBGs determined by EBSD maps. Moreover, D1 (~0.82 eV) and D2 (~0.87 eV) lines are active on different SBGs, depending on the level of the related disorientation while they are usually assumed to have rather the same behavior. The D3 (0.93 eV) and D4 (1.00 eV) lines show the same distribution with some extra features not related with the SBGs positions.

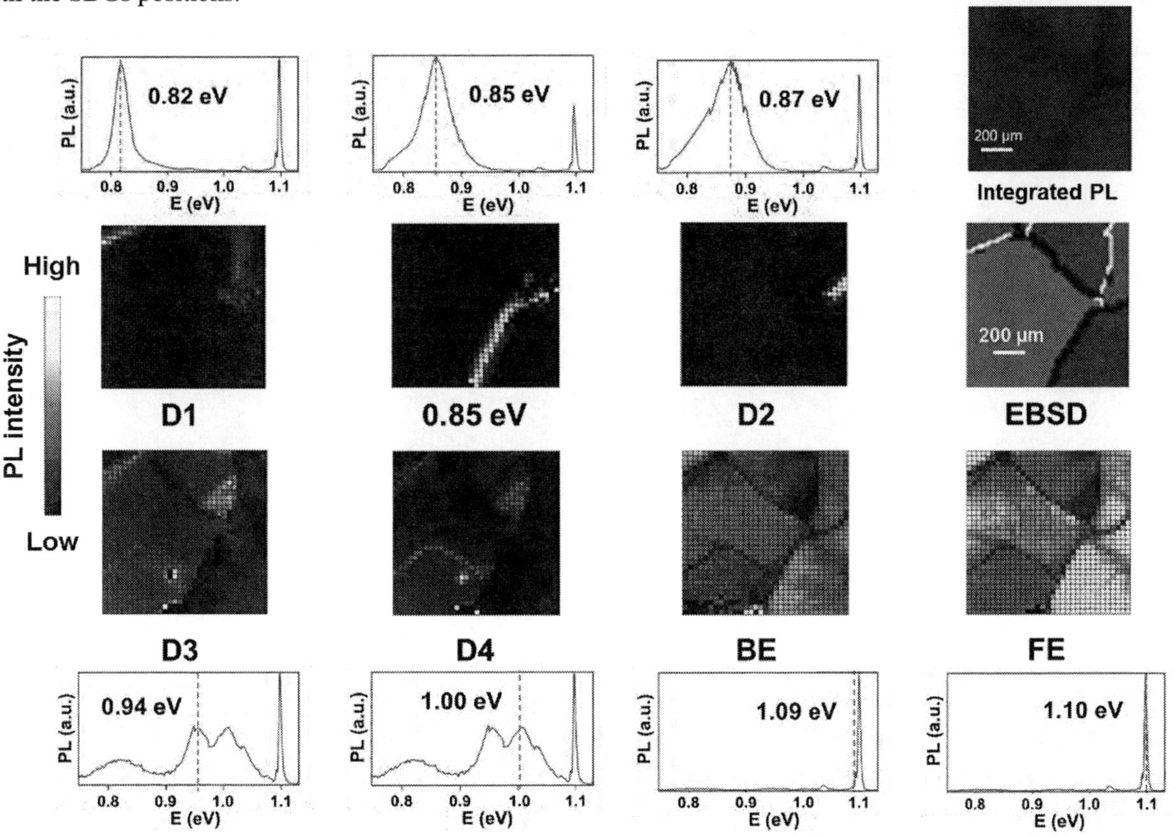

FIGURE 4. μPL intensity maps near sub-grain boundaries in "monolike" silicon showing the distribution of the well-known D-lines, the new line at 0.85 eV, the BE emission and the FE emission, on the same sample area at liquid helium temperature. Integrated PL map at room temperature and EBSD map are given on the same sample area. EBSD: white lines for low angle (<2 deg) disorientation, black lines for high angle (>2 deg) disorientation.

CONCLUSION

Photoluminescence and cathodoluminescence are very powerful tools for many nanoelectronics and nanotechnology applications. Since commercial tools are becoming available, they could become rapidly an in-line characterization technique. Room temperature measurements can be easily performed on wafers (up to 300 mm diameter) whereas low-temperature still require to work on small coupons. However in both cases the luminescence measurements could improve our knowledge on material and device in order to either monitor a process or in a more fundamental scheme of material characterization.

ACKNOWLEDGMENTS

This work has been partially supported by the LabEx Minos ANR-10-LABX-55-01, the National Research Agency (ANR) through the French "Recherche Technologique de Base" Program, the RENATECH programs, MOSINAS project and the Carnot French National Program within the frame of the SiDeRA Carnot project conducted at the "Énergie du Futur" Carnot Institute. This work was also partly supported by the IBM-LETI JDP program. The experiments were performed on the Nanocharacterisation platform (PFNC) at MINATEC.

REFERENCES

1. S. David et al., *APL Mater.* 4 (5), 056102 (2016).
2. I. Lauer et al., Symposium on VLSI Technology, Kyoto, 2015, pp. T140-T141.
3. F. Templier et al., *SID Symposium Digest of Technical Papers* 47 (1), 1013 (2016).
4. F. Jay et al., *Sol. Energy Mater. Sol. Cells,* 1–6 (2014).
5. Y. Zhang et al., *Sol. Energy Mater. Sol. Cells* 132, 1–5 (2015).
6. V. A. Oliveira et al., *Energy Procedia*, 92 755 – 763 (2016).
7. M. Tajima, *IEEE J. Photovoltaics*, 4 1452–1458 (2014).

KEYWORDS

III-V on Si for Advanced CMOS, Channel Engineering, Defects, Thin-Films, Spectroscopic Properties, Display, Solar Cells

Seeing The Invisible: Metrology For Extended Defects in Beyond-Silicon Semiconductor Device Structures

Andreas Schulze[1,*], Anna Prokhodtseva[2], Tomas Vystavel[2], David Gachet[3], Jean Berney[3], Roger Loo[1], Wilfried Vandervorst[1,4] and Matty Caymax[1]

(1) imec, Kapeldreef 75, 3001 Leuven, Belgium
(2) Mat. & Struct. Analysis, Thermo Fisher Scientific, Vlastimila Pecha 12, Brno 62700, Czech Republic
(3) Attolight AG, EPFL Innovation Park Building D, 1015 Lausanne, Switzerland
(4) KU Leuven, Dept. of Physics and Astronomy, Celestijnenlaan 200D, 3001 Leuven, Belgium
(*) Andreas.Schulze@imec.be

INTRODUCTION

The integration of novel semiconductor materials such as Germanium (Ge) and III-V compounds (InP, InGaAs, InAlAs) appears indispensable for CMOS scaling beyond the 7nm technology node [1]. This is mainly due to the fact that charge carriers inside these materials exhibit significantly lower effective masses and hence offer enhanced mobility and injection velocity values compared to silicon (Si). This facilitates the design of transistors operating at lower supply voltage thereby leading to a reduced overall power consumption. Moreover, direct bandgap materials are promising building blocks for tunnel-FETs as well as photonics devices such as lasers and modulators. To date, Ge and III-V compound materials can be grown epitaxially on Si substrates using various different process and integration schemes. However, the large differences in lattice constant and material characteristics between the aforementioned high mobility semiconductors and Si wafers typically lead to extended crystalline defects such as dislocations and stacking faults. Such extended defects can cause a degradation of the material properties (i.e. carrier mobility), significantly increased leakage currents as well as secondary effects such as dopant/impurity segregation. All of the latter lead to a deterioration of the final device performance and reliability. Hence, assessing the crystalline quality of these materials is of utmost importance in order to provide adequate feedback to process engineers and to help them to assess the properties of their materials and devices.

In this conference contribution we will introduce a novel classification of defect metrology approaches where we distinguish between large probe and small probe techniques. In the first case, a large probing beam (e.g. photons, x-rays, ions) is used to analyze a rather large area of ~50 micrometer or larger at once. In case of a small-probe technique, on the other hand, the measurement is performed by scanning a much smaller beam across the area under test. We will present an in-depth analysis of various metrology concepts (e.g. X-ray diffraction, ion and electron channeling, cathodoluminescence, time-resolved photoluminescence), thereby focusing on their respective underlying physical mechanisms as well as their strengths and limitations.

LARGE PROBE TECHNIQUES

Large probe techniques are typically indirect measurement approaches which are used to determine a certain property of the material or structure which is affected by its crystalline quality (i.e. defect density). For example, using **HRXRD** one can measure the FWHM of an omega rocking curve which reflects the mosaicity of the material. Indeed, we will show that the peak width correlates with the layer quality (Figure 1a) and that it can be translated quantitatively into a threading dislocation density [2]. In principle, the technique can also be applied to nm scale fins by simply probing an ensemble of identical features simultaneously in order to achieve an acceptable SNR [3]. However, we will also show that this approach fails in case of strain-relaxed buffers and selectively grown fins where the dislocation density in the near-surface (i.e. active) region significantly varies from the bulk defect density.

Moreover, the sensitivity towards {111} defects such as stacking faults is limited. We will also discuss **RBS** channeling where the number of backscattered He^+ ions in a channeling configuration and the derived channeling yield reflect the number of displaced atoms and hence the dislocation density in the sample (Figure 1b). We will present results obtained on blanket films and fin structures and discuss the limitations of the technique when targeting low defect densities.

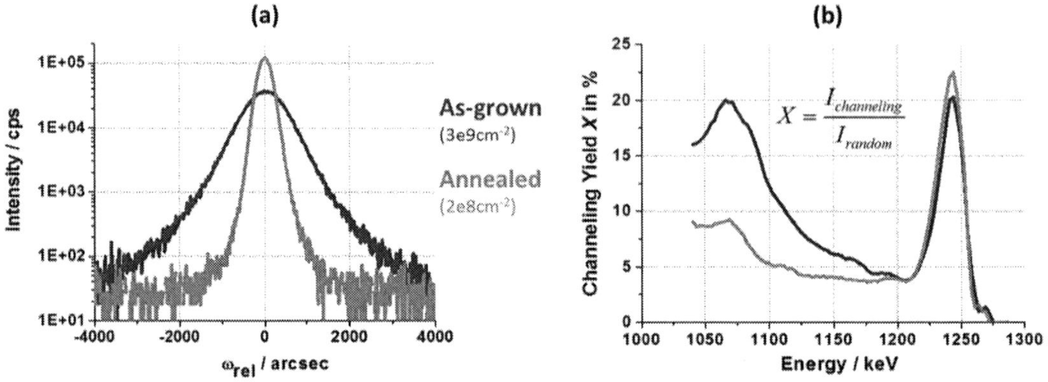

FIGURE 1. (a) Omega rocking curve obtained on blanket Ge layers of different threading dislocation densities (TDD). **(b)** RBS channeling yield measured on the same Ge layers.

Optical techniques appear very attractive for in-line applications as they are typically non-destructive in nature. Since extended defects can lead to states in the bandgap of a semiconductor, they modify the generation/recombination and transport properties of charge carriers. For this reason, photoluminescence (**PL**) decay can be used as a measure for the defect density in highly defective Ge layers (Figure 2a). However, it must be noted that sensitivity is ensured only as long as Shockley-Read-Hall (SRH) recombination via bulk defects dominates. This is typically given for highly defective material. For defect levels <1e7cm^{-2}, on the other hand, SRH recombination at surface/interface states or even radiative recombination (direct semiconductors) start to dominate and therefore limit the measured PL lifetime. As a consequence, the impact of bulk defects becomes negligible. The latter effect is particularly pronounced in narrow fins where surface/interface recombination is strongly increased due to the large surface-to-volume ratio. The latter in combination with an indirect semiconductor (e.g. SiGe fin) can even lead to a scenario where no PL at all is observed.

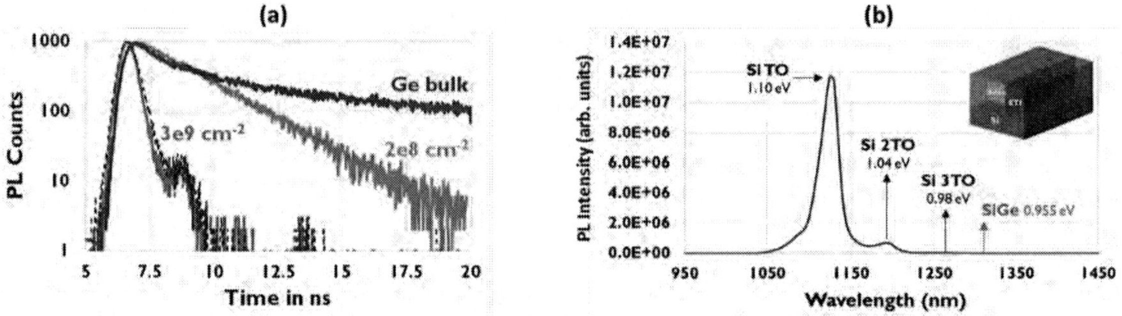

FIGURE 2. (a) Photoluminescence decay measured on thick relaxed Ge layers with different threading dislocation densities at 77K. **(b)** PL spectrum (77K) of an array of relaxed $Si_{0.3}Ge_{0.7}$ fins (225nm wide) selectively grown in a STI matrix.

SMALL PROBE TECHNIQUES

TEM is perhaps the most obvious small probe technique and can be used to study extended defects in great detail. However, TEM is destructive in nature and its reduced specimen size and limited field of view hamper the assessment of defect densities below ~1e8cm^{-2}. In contrast, **chemical etching** can be used to decorate extended defects on the layer surface. The obtained etch features can then be visualized using AFM (SEM, light scattering etc.) at lower magnification, thereby enabling the analysis of lowly defective materials with statistical relevance (Figure 3a). The destructiveness of the etching, however, prevents the technique from being used in high volume manufacturing.

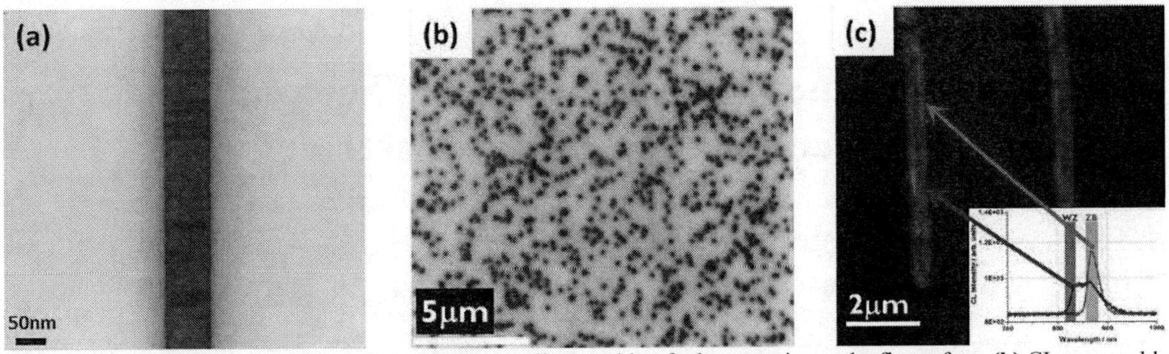

FIGURE 3. **(a)** AFM on InP fins after defect etching revealing stacking faults emerging at the fin surface. **(b)** CL map on blanket GaN layer. Threading dislocations act as non-radiative recombination centers and hence appear as dark spots. **(b)** CL map on InP fins. The relaxation of InP grown on Si is facilitated by the formation of a dense network of twinning planes leading to a Wurtzite crystal structure detectable by CL.

According to our studies, Cathodoluminescence (CL) and electron channeling contrast imaging (ECCI) can be considered the most promising defect metrology candidates. Both techniques are inherently non-destructive and can be performed using a (modified) SEM. In **CL,** the focused electron beam of the SEM is used to generate excess carriers. The light emitted after their recombination is recorded as a function of wavelength. In case of direct bandgap materials we observed strong luminescence for blanket layers but also fin structures. This enabled us to visualize threading dislocations as non-radiative recombination centers in GaN layers (Figure 3b). In case of InP fins we could identify the local presence of a Wurtzite crystal (Figure 3c) which we believe is linked to {111} defects present in the material. The application of CL towards aggressively scaled devices is challenging due to carrier diffusion and undesired interface recombination. In **ECCI**, backscattered electrons are used to derive information about the sample's defectivity. This is possible since the intensity of the backscatter signal depends on the alignment between primary beam and lattice planes of the sample. This alignment and hence the backscatter signal is modulated when the electron beam is scanned across an extended defect characterized by lattice distortions in its vicinity [4]. The induced channeling contrast allowed us to image linear as well as planar defects in blanket layers and fins (Figure 4). As ECCI is not based on the generation/recombination and transport of charge carriers, it is particularly interesting for confined structures made of direct as well as indirect semiconductors.

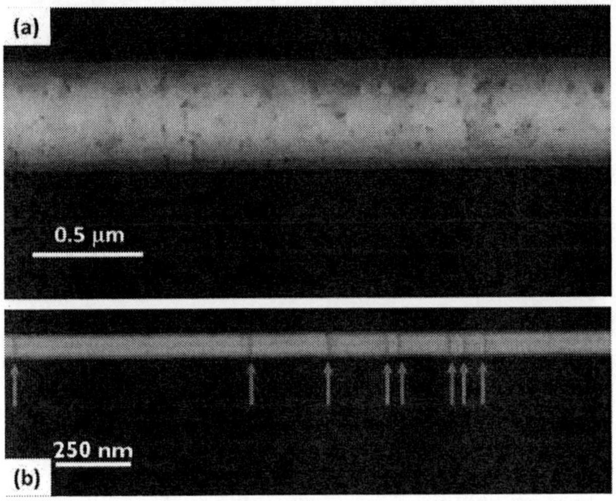

FIGURE 4. **(a)** Threading dislocations and **(b)** stacking faults in SiGe fins visualized by ECCI

REFERENCES

[1] M. M. Heyns et al., *ECS Trans.*, vol. 3, no. 7, pp. 511–518, 2006.
[2] J. E. Ayers, *J. Cryst. Growth*, vol. 135, no. 1–2, pp. 71–77, 1994.
[3] A. Schulze et al., FCMN, Dresden, 2015.
[4] S. Zaefferer et al., *Acta Mater.*, vol. 75, pp. 20–50, 2014.

KEYWORDS

Defect Metrology, Heteroepitaxy, RBS, Electron Channeling, Photoluminescence, Cathodoluminescence

High Brightness MetalJet X-Ray Technology for Semiconductor Process Metrology

J. Hållstedt[1], U. Lundström[1], E. Espes[1], B. A. M. Hansson[1], O. Hemberg[1], G. Johansson[1], M. Otendal[1], T. Tuohimaa[1], P. Takman[1]

[1] *Excillum AB, Torshamnsgatan 35, 164 40 Kista, Sweden*

INTRODUCTION

Recently the interest for x-ray source technology has been intensified due to the fact that several critical semiconductor metrology methods such as optical based scatterometry is starting to run out of steam [1]. Furthermore with introduction of advanced strain engineering and 3D structures in manufacturing new challenges for semiconductor metrology will emerge. Proposed solutions involve a variety of techniques including x-ray based methods, however a main problem with most of those is the very low throughput since conventional X-ray tubes are very limited in x-ray flux.

The limitation of a conventional x-ray tube is illustrated in Fig 1a, showing that X-rays are generated when highly energetic electrons are stopped in a solid metal anode. The fundamental limit for the X-ray power generated from a given spot size is when the electron beam power is so high that it locally melts the anode. A novel method of overcoming this problem is by using a liquid as metal anode [2]. The liquid-metal-jet anode (MetalJet) technology solves the thermal limit by replacing the traditional anode by a thin high-speed jet of liquid metal (see Figure 1b). Melting of the anode is thereby no longer a problem as it is already molten, and significantly (currently about 10x) higher e-beam power densities can therefore by used.

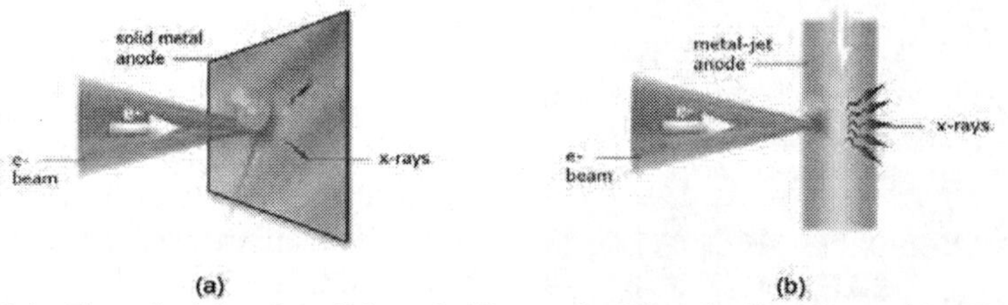

FIGURE 1. The principle of a solid anode X-ray tube (a) and a liquid-metal-jet X-ray tube (b).

SYSTEM DESCRIPTION

Figure 2 illustrates a complete MetalJet X-ray source. The upper part is the source head with the electron gun and the vacuum pump. This part is similar to any open-type X-ray tube apart from the metal-jet anode that is formed by ejecting a ~200 μm diameter metal jet through standard nozzle inspired by the nozzles used for water cutting. One key requirement to achieve 24/7 operation is that the liquid-metal-alloy can be 100% recirculated. The path of the circulating liquid alloy is illustrated by the arrows in Fig. 2.

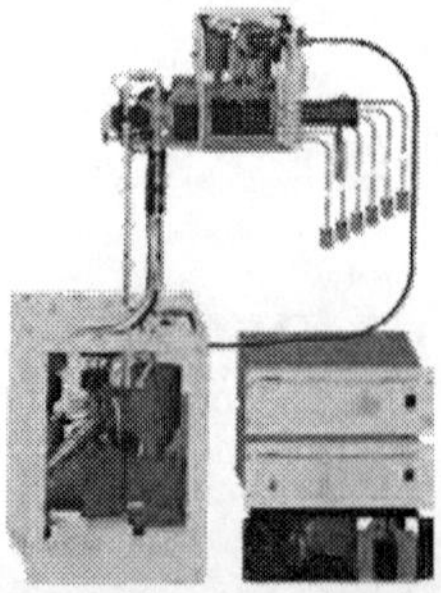

FIGURE 2. A complete MetalJet X-ray source with source head, pump box, support electronics and water-chiller.

The engineering aspects of closed-loop recirculation are significantly relaxed if anode alloys molten at room temperature or slightly above can be used. This limits the corrosiveness of the alloys, that tend to increase rapidly with temperature. The current generation of sources use either a mainly gallium based alloy or an alloy which also contain significant amounts of indium. The gallium Kα line is at 9.2 keV, which makes it an attractive high-brightness replacement of copper Kα at 8.0 keV. The indium Kα is at 24.2 keV making it an attractive high-brightness replacement of silver Kα at 22.2 keV.

TYPICAL SOURCE PERFORMANCE

The main specifications of the state of the art MetalJet sources are listed in Table 1. Table 2 summarizes typical source parameters for operation in four different configurations. Normally a 1:4 e-beam line focus is used which results in a round spot with angled viewing. For example, in X-ray diffraction the source is typically operated at 250 W with an approx. 20×80 μm^2 spot viewed at an angle to generate an effective ~20-μm-diameter spot corresponding to effective power densities of several megawatts per square mm. However, the spot shape may be freely tuned from larger sizes, with upper limit related to the size of the jet, down to 5-6 μm diameters, limited by the e-beam focus and electron scattering. The aspect ratio is also freely tunable. The brightness numbers in table 2 are given as the Kα (double) line brightness.

TABLE 1. Typical source specification.

Voltage	Power	Max. current	Min. focal spot size	Min. spot-object distance	Available beam angle
21-160 kV	0-300 W	4.3 mA	~5 μm	18 mm	18°

TABLE 2. Typical source performance.

Alloy	Acceleration voltage [kV]	Apparent Spot Size[a] [μm]	E-beam power [W]	Line	Kα peak brightness [photons/(s·mm2·mrad2·line)]
Gallium alloy	70	20	250	Ga K-alpha	$3x10^{10}$
Gallium alloy	70	10	125	Ga K-alpha	$5x10^{10}$
Indium alloy	70	20	250	In K-alpha	$3x10^{9}$
Indium alloy	70	10	125	In K-alpha	$4x10^{9}$

[a] *Actual e-beam spot has a 1:4 aspect-ratio line focus, but the projected diameter is essentially circular.*

APPLICATIONS

CD-SAXS

Research on critical dimension small angle X-ray scattering (CD-SAXS) show that this technology could potentially complement and replace optically based CD tools as dimensions become smaller and more complicated. CD-SAXS can be performed both in reflection and transmission geometry. For transmission geometry energies higher than 20 keV are needed to get enough photons through the wafer for non-synchrotron CD-SAXS and early results with the MetalJet source technology and indium K-α emission at 24 keV show great promise towards meeting the requirement needs of the semiconductor industry. [3,4]

XRD

Various X-ray Diffraction (XRD) [5] techniques can benefit significantly from the high brightness achievable with the MetalJet X-ray source technology. This is becoming even more important as structures are going from planar to 3D as well as when device design involves the introduction of a variety of carrier mobility enhancement engineering such as channel alloying e.g. SiGe and/or strain, one of the few ways to measure and enable control of this is via high resolution x-ray diffraction (HRXRD) and high resolution reciprocal lattice mapping (HRRLM). Current state of the art solid x-ray sources are far (several orders of magnitude) from achieving throughput even close to what the industry is requesting when metrology is needed on patterned wafers in production control [6]. This is truly highlighting the need of new types of high brightness sources such as the MetalJet X-ray source.

XRR

X-ray reflectivity (XRR) is a method used to determine thickness, density and roughness of thin films. The high brightness and small spot of the MetalJet enable faster throughput and/or smaller measurement area. The shorter wavelength of Gallium Kα compared to Copper Kα may also enable more accurate measurements of especially extremely thin layers for high-k gate stacks.

X-ray Microscopy

The highest available resolution in lab-based X-ray microscopes is achieved with zone-plate based projection microscopes [7] Such microscopes, however, typically use Cu Kα radiation which is not so well suited to see copper structures in silicon due to poor contrast between copper and silicon. As illustrated by Fig. 3, the Kα of gallium used in MetalJet sources, is just above the K-absorption edge of copper [8] and thus much better suited to create a sufficient contrast between copper and silicon. First x-ray microscopy instrument using the MetalJet for Cu interconnect inspection was recently presented [9]

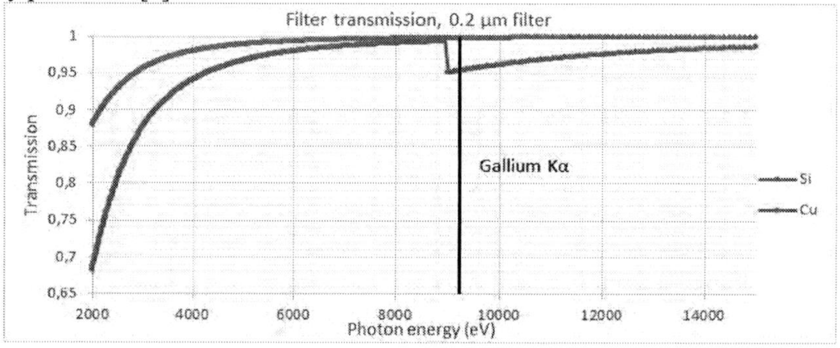

FIGURE 3. X-ray transmission trough a 0.2 µm filter of silicon and copper. The black line indicates the Kα of gallium [8].

XRF and TXRF

Recently a group at the Atominstitut, in Vienna, used the liquid-metal-jet technology to do Total reflection X-ray fluorescence (TXRF) investigations [10]. The high brightness and small spot-size, which results in a low divergence of the primary beam, results in a very low spectral background. In a non-optimal set-up, they still achieved detection limits in the high femtogram range. Another possibility that opens up with the use of a liquid anode is the unique characteristic lines coming from e.g. Ga, In, Sn which previously could not be used as x-ray source anode material which may improve sensitivities for e.g. 3d transition metals.

REFERENCES

1. M. Lapedus, 'Can We Measure Next-Gen FinFETs?', *Semiconductor engineering,* Nov 21 2016
2. H. M. Hertz et al, 'Method and apparatus for generating x-ray or euv radiation' Patent US20020015473 *
3. R. J. Kline, D. F. Sunday, D. Windover and W. Wu, 'Bringing CD-SAXS to the Fab', *SEMICON West* 2014, 2014.
4. M. Lapedus, 'Measuring Atoms And Beyond', *Semiconductor Engineering* Nov 21st
5. D. Bowen and B. Tanner, *X-ray metrology in semiconductor manufacturing*. Boca Raton: CRC/Taylor & Francis, 2006.
6. A. Schulze, 'X-Ray Metrology For The Semiconductor Industry', Int. Workshop on Compact EUV and X-ray Light Sources 2015.
7. T. Beetz, 'High-resolution X-ray Tomography Imaging Systems', CHESS Users' Meeting Ithaca, NY, 2008.
8. E. Gullikson, *Filter transmission*, http://henke.lbl.gov/optical_constants/filter2.html, visited on 2015-03-24.
9. Juana Rudati - Rapid integrated circuit inspection for reliability and security inspection, *International Conference X-ray Microscopy XRM,* Oxford, 2016.
10. A. Maderitsch, S. Smolek, P.Wobrauschek, C. Streli and P. Takman, *Spectrochimica Acta Part B* **99,** 67–69 (2014).

KEYWORDS

X-ray, tube, source, microfocus, metaljet, liquid-metal-jet, gallium, indium, TXRF, microscopy, imaging, CD-SAXS, metrology, XRR, XRD, HRXRD

Nanoelectronics Dimensional Metrology: Understanding the Differences between Secondary and Backscattered Electron Imaging [1]

Michael T. Postek, András E. Vladár, John S. Villarrubia

Engineering Physics Division
Physical Measurement Laboratory
National Institute of Standards and Technology
Gaithersburg, MD 20899

INTRODUCTION

In many fields of research and production, a great deal of dimensional metrology, characterization and process control is accomplished using scanning electron microscopes (SEM). The accuracy of these SEM measurements has always been important, but is often overshadowed by two other main measurement drivers: throughput and precision. It is slow and often tedious to achieve high degree of accuracy and, so it is often ignored, especially in production where measurements must be made very quickly. However, the accuracy of a measurement is becoming more important as the frontiers of nanoelectronics are being explored, and sub-10 nm semiconductor device structures are routinely produced. Hence, the metrology error budget has shrunk, and has become truly atomic scale. This presentation will discuss new measurement, signal collection and modeling methods applied to sub-10 nm metrology being pursued for all types of semiconductor nanostructures, nanomaterials and nano-enabled materials to ultimately achieve the needed accurate measurements.

SEM MEASUREMENTS

The quality of the SEM measurements depends on how the acquired image is influenced by vibration, drifts, sample contamination and charging. In addition, accounting for specimen-electron beam interactions and their contributions to the acquired image must also be considered. New acquisition methods and successful mitigation of the previously mentioned environmental and instrument-induced effects can alleviate some of the imaging uncertainties. However, another key element is the application of advanced electron beam-solid state interaction modeling such as the NIST JMONSEL model [2] to interpret and account for the physics of the signal generation and help to understand and minimize the various contributions to measurement inaccuracy.

EXPERIMENT

The first part of this work involved a fundamental comparison of secondary (SE), backscattered (BSE) and low-loss (LLE) electron signals acquired on a new instrument that was equipped with SE, high-angle BSE and energy-filtered LLE detectors. Early work indicated that the BSE and LLE signal could be advantageous to help to understand the metrology of semiconductor structures [3]. The LLE image is produced by high-pass energy filtered BSE. These have undergone only minimal inelastic interactions with a sample and therefore, carry high-resolution information, specific to sample geometry [4-7]. LLE imaging is difficult because the collected signal is produced not only by the most energetic, but also a small minority of backscattered electrons. Early work pointed to a potential measurement difference between the SE and the BSE signals [8], it was difficult to obtain the needed information because of the resulting poor signal-to-noise ratio and other instrument-specific geometric limitations. A measurement difference between the two signal modes was documented and was as large as 100 nm on the relatively large lines studied. Later, technology improvements facilitated the collection of conventional BSE and SE electron signal with a microchannel-plate electron detector. This approach was shown to be advantageous at low landing energies because of its improved geometry and signal collection capabilities [9]. In that early work, on different samples, collection and comparison of BSE and SE images of line structures again demonstrated that comparative threshold-based width measurements of the BSE images yielded values as much as 10 nm larger for the SE measurements than the BSE measurements on nominal 1.0 μm gold lines. Due to the enhanced emission of low-energy (typically less than 10 eV) electrons at the sides and corners, there are common circumstances in which the SE intensity increases more abruptly at an edge than the BSE intensity. If width assignments are based on an intensity threshold, as is often the case, SE images would then be interpreted as showing a wider, larger feature than the BSE image. It was anticipated that the LLE signal would provide results similar to the BSE results, but with higher edge fidelity. A Hitachi SU 8230 FESEM [10], equipped with an in-lens, high-angle energy-filtered BSE detector, was used to compare the SE, BSE and LLE signals for dimensional measurements of the NIST RM 8820 magnification calibration sample [11]. The design of the field emission SEM incorporating a new in-lens energy filtered detector improves the LLE and BSE signal-to-noise ratio and reduces the geometrical limitations of the early LLE detectors.

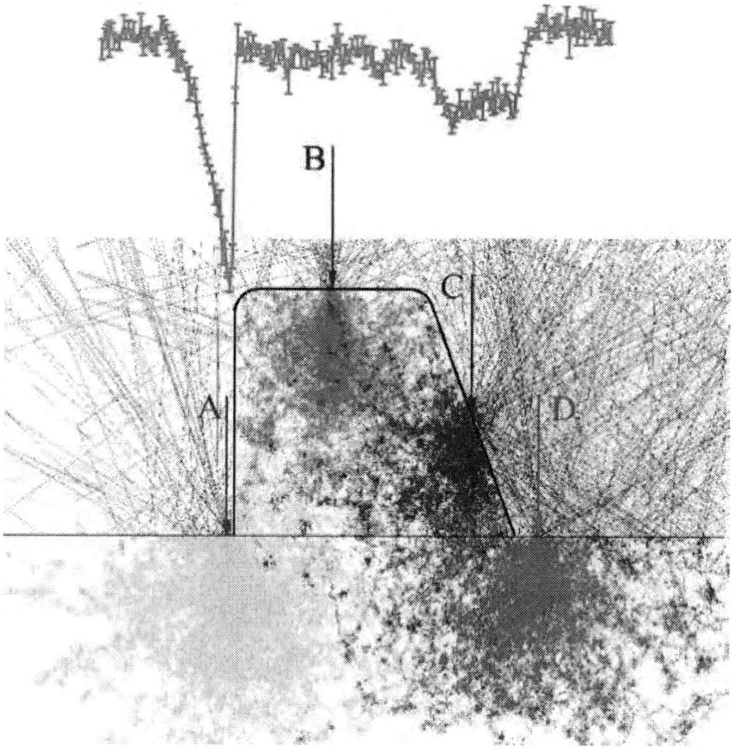

FIGURE 1. Model relationship of LLE signal to sample geometry. The upper (green) curve shows the modeled low-loss signal intensity from the line with near-vertical (left) and sloped wall (right) cross section shown in the lower portion. Electron trajectories at 4 landing positions, labeled A-D are superimposed on the sample geometry.

JMONSEL MODELING

In the second part of the work, we continued the analysis of the dimensional measurements using simulations with JMONSEL (Figure 1), an electron microscope simulator. JMONSEL indicated, as expected, that the nanometer-scale differences observed on this sample can be explained by the different convolution effects of a beam with finite size on signals with different symmetry (the SE signal's characteristic peak vs. the BSE or LLE signal's characteristic step). But, this effect is too small to explain the 10 - 100 nm discrepancies that had been observed in earlier work on the different samples [8, 9]. Additional modeling then indicated that those discrepancies could be explained by considering the much larger sidewall angles of the earlier samples, coupled with the different response of SE vs. BSE/LLE linescans and measurement algorithms to those wall angles. Clearly, serious measurement errors would be encountered in such cases if measurement algorithms were applied blindly without considering the underlying physics provided by applying model-based metrology.

CONCLUSION

In summary, this paper has: (1) demonstrated, for the first time, by simultaneous imaging that the previously observed bias between SE and LLE/BSE images is indeed real, not just an artifact of charging, drift, detector positioning, or some other instrument - or measurement-related error; (2) documented the measurement variation inherent in algorithm choice both on modeled and experimental data; (3) clearly pointed out that modeling of the image formation is necessary for highly accurate measurements, and (4) explained the previously observed mysterious size difference in the measurements with a simple phenomenological model and by a more complete Monte Carlo model.

REFERENCES

1. Contribution of the National Institute of Standards and Technology; not subject to copyright.
2. J. S. Villarrubia et al., Ultramicroscopy 154:15 (2015).
3. M. T. Postek, et al., SCANNING 23(5): 298 (2001).
4. O. C. Wells, Appl Phys Lett 16(4):151 (1970).
5. O. C. Wells, Appl Phys Lett 19(7): 232: (1971).
6. O. C. Wells, Scan Electron Microsc, 1, IITRI Chicago, 43, (1972).
7. O. C. Wells, Appl. Phys. Lett. 49(13): 764: (1986).
8. M. T. Postek, et al., SCANNING 10:10-18, (1988).
9. M. T. Postek, et al., Rev Sci. Instrum, 61(12):3750, (1990).
10. Certain commercial equipment is identified in this report to adequately describe the experimental procedure. Such identification does not imply recommendation or endorsement by the National Institute of Standards and Technology, nor does it imply that the equipment identified is necessarily the best available for the purpose.
11. *https://www-s.nist.gov/srmors/view_detail.cfm?srm=8820*

.

ACKNOWLEDGEMENT

The authors would like to thank Mr. Atsushi Muto and Hitachi High Technologies America for their assistance and use of the instrumentation used in this paper.

KEYWORDS

scanning electron microscope, SEM, JMONSEL, modeling, metrology, secondary electron, backscattered electron, low-loss electron

High-Throughput X-ray CD Metrology

R. Loewen, M. Feser, R. Ruth, M. Gifford

Lyncean Technologies, Inc.
44755 S. Grimmer Blvd.
Fremont, CA 94538
USA

ABSTRACT

Conventional techniques for Critical Dimension (CD) measurement face considerable challenges to keep pace with Roadmap requirements. New 3D device architectures, variety of materials, and instrument limitations drive the need for development of new tools that could offer a pathway to High Volume Manufacturing (HVM). Hard X-rays are a natural extension as an optical probe in which photons can penetrate materials to create scattering profiles that are highly sensitive to patterned structure dimensions. And unlike Optical CD metrology, X-ray scattering methods become easier as future node sizes shrink.

To date, mainly two X-ray scatterometry methods—transmission-based Critical Dimension-Small Angle X-ray Scattering (CD-SAXS) and reflection-based Grazing Incident-Small Angle X-ray Scattering (GI-SAXS)—have been evaluated for CD metrology. CD-SAXS has successfully demonstrated multi-parameter fitting of Fin structures but is currently too slow to implement for HVM. CD-SAXS requires the development of high-brightness X-ray sources at higher energies (15keV+) to make the method practical in the 10's of second timescales. Present laboratory sources are three to four orders-of-magnitude away from realizing this goal. GI-SAXS is traditionally done in the lab at lower energies (8keV) and produces more relative scattering signal than CD-SAXS but suffers from large spot sizes, is dominated by surface features, and requires more complicated analysis to extract structure parameters. Both methods are based on the diffraction of a more-or-less collimated beam incident on a grating-like array of scattering objects in which the scattering intensity depends on the phase shift of the incident beam through different paths in the material.

This paper describes a complementary, reflection-based X-ray method that employs a focused X-ray beam in which the sample itself produces interference effects that are highly dependent on the periodic structure of the patterned objects. Unlike diffraction methods, the sample itself modulates the intensity profile by relying on a superposition of optical path phase effects, which for a monochromatic beam, is entirely geometrical. By using higher energy X-rays (15keV+) in a specific reflection orientation, enhanced in-plane signals resulting from the interference of multiple interfaces can dramatically enhance the cross section of scattering. The in-plane scattered signal contains information about the absolute linewidth and spacing, and in multiple-pattern deposition applications, variations such as pitch walk and roughness.

Any path for high-throughput CD metrology also requires advances in X-ray source brightness, where brightness must quantify the amount of useful photons that can take part in the particular measurement. The combination of the proposed new X-ray CD method, which accepts a larger angular acceptance from the source than previous methods, together with a state-of-the-art X-ray generator such as the Lyncean Compact Light Source, offers a solution for X-ray CD metrology that is scalable today for HVM.

FIGURE 1. Photograph of the Lyncean Compact Light Source miniature synchrotron.

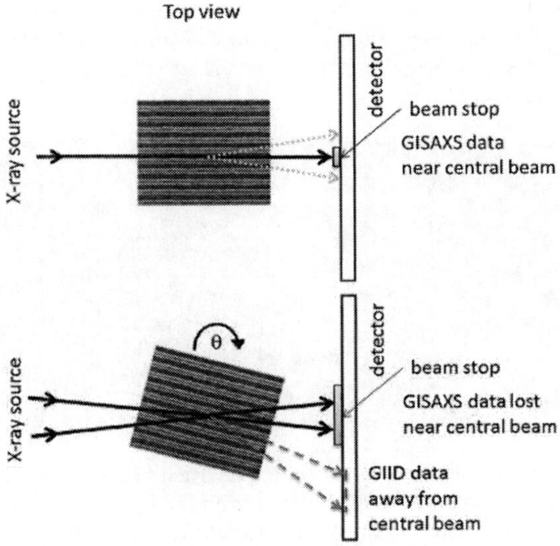

FIGURE 2. Top view of GISAXS vs. Grazing Incidence In-Plane Diffraction (GIID) configuration. On the left, GISAXS requires a collimated beam and fine alignment of features along the beam direction. A beam stop is used to prevent the detector from saturating in order to measure scattering. On right, rotating sample in-plane creates conditions for reflection from multiple GIID scattering sites in the sample. GIID can use a focused beam to significantly increase intensity on the sample.

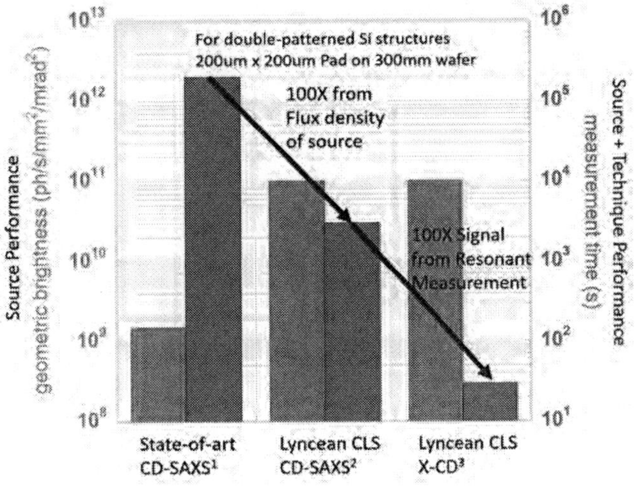

FIGURE 3. Comparison of raw source performance (left axis) and CD measurement time (right axis) for a 200um x 200um measurement pad on a 300mm wafer.

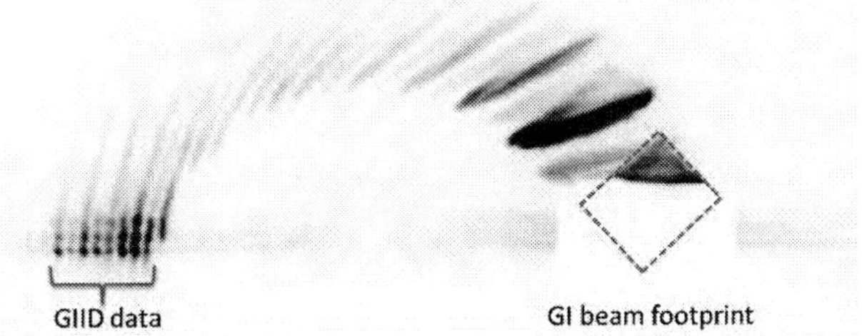

FIGURE 2. CCD detector image of data collected on a NIST Si Fin structure using the CLS to explore a new method for GI in-plane diffraction (GIID). The CLS beam was focused at the sample with a K-B multilayer (shown as a footprint of direct GI beam outlined in the box). The sample is rotated in-plane to reveal strong scattering resonances near grazing incidence. Note the sample itself is acting as an angular filter.

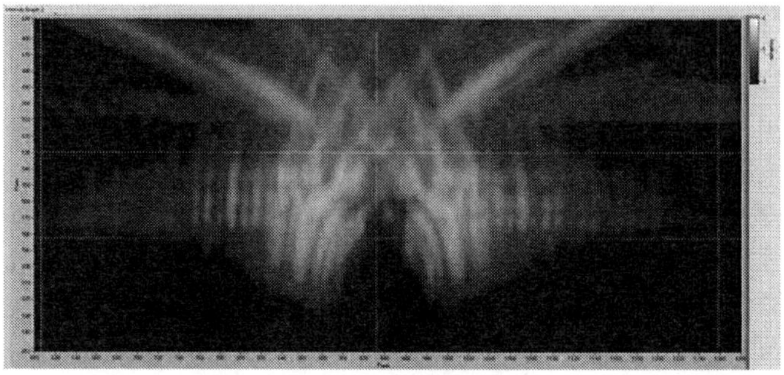

FIGURE 3. Example of a preliminary measurement on a double patterned Si Fin sample using the GIID method developed at Lyncean Technologies. The data set was collected in 60 seconds on the Lyncean CLS (2014) and shows many orders of resonance spaced at a particular pitch. Variations in intensity between resonances is sensitive to sub-nm variations of relative gap and line widths in the periodic structure. (Note: The in-plane signal is within the white rectangle, and the intensity scale is logarithmic-many orders of magnitude are displayed. More than 20 higher order resonances displayed.)

KEYWORDS

Keyword: CD Metrology, X-ray, SAXS, CD-SAXS, GI-SAXS, GIID

Electrical Property Characterization of Vacuum-Channel Nanoelectronics Via Scanning Capacitance Microscopy

Jin Woo Han

NASA Ames Research Center, Moffett Field, CA 94035, United States of America

Gerald Pascual, Byong Kim, and Keibock Lee

Park Systems Inc., 3040 Olcott Street, Santa Clara, CA 95054, United States of America

INTRODUCTION

From their invention at the beginning of the 20th century, vacuum tubes were used as basic electronics components that enabled the proliferation of devices which have had an incalculable impact on human society: television and radio, telephone networks, industrial process controls, and more. However, vacuum tube-based electronics were eventually rendered outmoded by the advent of solid-state devices powered by semiconductor devices. By mid-century, semiconductor devices, with their smaller footprint, greater efficiency, longer longevity, and lower cost, became the de facto wellspring for mass-produced electronics and would eventually grow into the multi-billion dollar industry it is today. However, with humanity now more than a decade into the 21st century, the challenge of expanding our space exploration efforts has revealed that semiconductor devices are vulnerable to the types of radiation that bombard spaceborne equipment. Efforts to harden components and systems have become a major engineering operation for space exploration agencies [1] and other organizations that seek to safeguard their electronic components and systems. Unfortunately, such processes are not only time-consuming but expensive; hardening electronics against radiation and heat lead to solutions that are both costlier and older than what is available to other types of consumers [2].

In order to address these drawbacks and produce a transistor technology that can achieve higher speeds and frequencies than any semiconductor device, research has begun focusing on reviving vacuum tube technology and applying its principles at nanoscale whilst still leveraging a silicon-based fabrication method. The result is a vacuum-channel transistor device which could potentially be manufactured at an industrial scale using already existing silicon fabs for solid-state transistor devices [2]. This investigation uses Scanning Capacitance Microscopy (SCM) to investigate the nanoscale electrical properties of a newly developed vacuum-channel device to both ascertain its viability as a transistor as well as to observe if the method used to fabricate its gate insulators can be controlled.

METHODOLOGY

To acquire data regarding the vacuum-channel device's nanoscale electrical properties, a Park NX20 Atomic Force Microscopy (AFM) system was used to run SCM scans of the device area containing the source-drain interface, vacuum channel, and insulated gates. Supplemental topographical data at the scan area was collected simultaneously through contact mode AFM using the same probe.

SCM with AFM

SCM with AFM is a powerful combination for investigating transistor devices—together, the two methods provide the user with a non-destructive process of characterizing both charge distribution and surface topography with high spatial resolution and sensitivity [3]. In SCM, a metal probe tip and a highly sensitive capacitance sensor augment standard AFM hardware. A voltage then is applied between the probe tip and the sample surface creating a pair of capacitors in series (in Metal-Oxide-Semiconductor devices) from (1) the insulating oxide layer on the device surface and (2) the active depletion layer at the interfacial area between the oxide layer and doped silicon. Total capacitance is then determined by the thicknesses of the oxide layer as well as the depletion layer which is influenced by the both how doped the silicon substrate is well as the amount of DC voltage being applied between the tip and device surface.

The principle of how capacitance is measured in vacuum-channel devices via SCM is similar. Again, a thin layer of oxide is used as insulation on the device [4], this time insulating the gate from the source-drain interface on the device surface. A DC voltage is applied between the probe tip and the sample surface as the tip scans across various device features. The data of the detected changes in capacitance are also supplemented by AFM data generated by recording the deflections of the probe's cantilever as the tip engages the device surface [5]. As the sample scan is being completed, a laser beam is reflected off the probe cantilever and onto a position-sensitive photodiode. The deviations of the laser's position are then processed with software to create a rendering of the device's surface topography.

RESULTS AND DISCUSSION

Functional Device Topography

The first and properly fabricated vacuum-channel device (Device 4) was scanned at an area measuring 500 x 1000 nm capturing the device's source-drain interface. Contact mode AFM reveals that the source and drain terminals are shaped into sharp tips. This design was implemented in order to intensify the electrical field generated at this location [4]. The topography data at this location also reveals the distance between the source and drain tips, a span which includes Device 4's insulated gates and vacuum-channel, is approximately 250 nm (Figure 1). For reference, the mean free path of electrons and gas molecules under normal atmospheric pressure is about 200 nm [2]. If the voltages running through the device were kept low enough, the electrons traversing from source to drain would not have energy to ionize any lingering gas molecules left in the channel [4]. Therefore, Device 4 could technically operate without incident under normal air pressure—the presence of a vacuum in the channel serves as an extra precaution to guard against ionized molecules from damaging the terminals of the source-drain interface. Inspection of the topography line profile (Figure 2) for this image reveals the source and drain terminals, at roughly 8-12 nm in height, represent an 5-7 nm elevation from the lowest portions of the vacuum-channel. Of further interest are two bumps, speculated to be the device's insulated gates, approximately 6-7.5 nm in height standing on opposing edges of the vacuum channel.

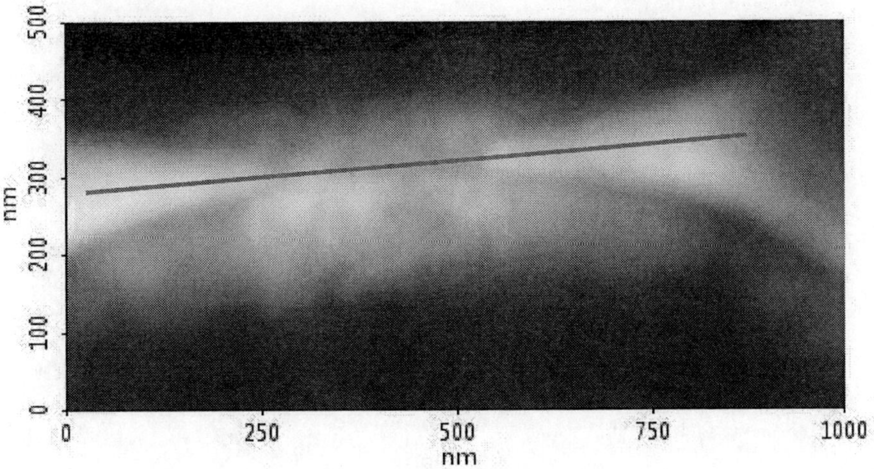

FIGURE 1. Contact mode AFM topography image of the first and properly fabricated vacuum-channel device's (Device 4) sensor-drain interface, vacuum-channel, and insulated gates region. The overlaid red line corresponds to the topography line profile displayed in Figure 2. Scan size: 500 x 1000 nm, image size: 380 x 190 px.

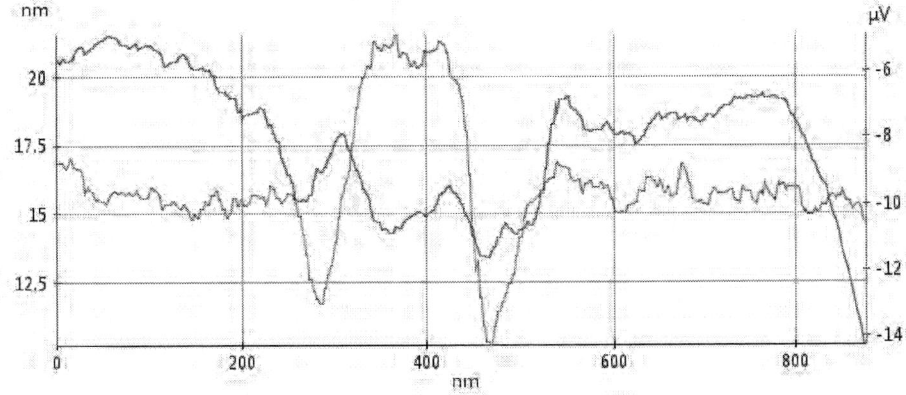

FIGURE 2. Line profiles of the AFM topography (red, left y-axis in nm) and the capacitance data (green, right y-axis in µV) of the Device 4 area scanned in Figures 1 and 3.

Functional Device Capacitance with Topography

The SCM data from was also gathered from the same 500 x 1000 nm scan. This image (Figure 3) shows each of Device 4's source-drain terminals as well as the insulated gates being darker in color or more negatively charged than surrounding portions of the device. Note the difference in shades between the terminals and gates—these features are both more negatively doped than the rest of the imaged area, however the gates are noticeably darker. This is corroborated by the capacitance line profile (Figure 2) which shows an average capacitance of about 1 µV in areas corresponding to the sections of the line covering the terminals, but that capacitance dips sharply to -4 µV in the 50 nm leading to the aforementioned 6-7.5 nm bumps revealed in the topography data.

Another important feature clearly seen in the capacitance image is Device 4's vacuum-channel which shows a brighter, more positively charged (a 7 µV) region. This particular region, approximately 175 nm in length, cannot be easily identified in the topography image but is easily spotted with the greater contrast afforded by using SCM. The capacitance change from the insulation gates to the middle of the vacuum-channel is observed to be 11 µV, the greatest change observed in this line profile.

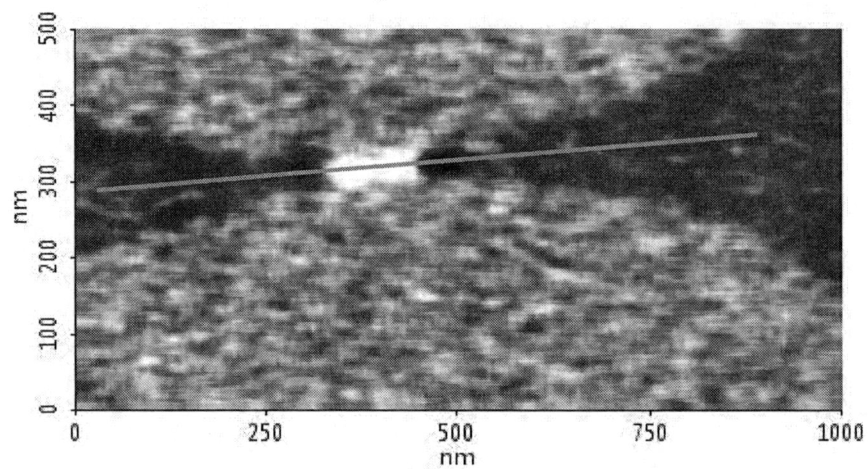

FIGURE 3. An SCM capacitance image of the region containing Device 4's sensor-drain interface, vacuum-channel, and insulated gates. Brighter colors correspond to relatively more positively charged areas on the device whereas darker colors correspond to relatively more negatively charged areas. The overlaid green line corresponds to the capacitance line profile displayed in Figure 2. Scan size: 500 x 1000 nm, image size: 380 x 190 px.

Malformed Device Topography

A second, purposefully malformed vacuum-channel device (Device 6) was also scanned at an area measuring 700 x 1000 nm centered on the device's source-drain interface. Devices 4 and 6 use an identical design, however the latter was purposefully malformed to render one of its two insulated gates nonfunctional. This investigation served as a check on the process used to fabricate both devices to observe whether this specific amount of control over fabrication was possible. As expected, the major observable features in the topography image (Figure 4) are the source-drain terminals which were again created with pointed tips. The span between the tips on Device 6 measure approximately 225 nm, comparable to the 250 nm observed on Device 4. A look at the line profile (Figure 5) reveals that Device 6 significantly departs from Device 4 in surface topography. The area between the source and drain terminals is now bout 2-2.5 nm taller than the terminals themselves (6-7.5 nm in height). This placement of material along this line profile in what was an empty space in the functional Device 4 suggests Device 6's vacuum-channel has disappeared.

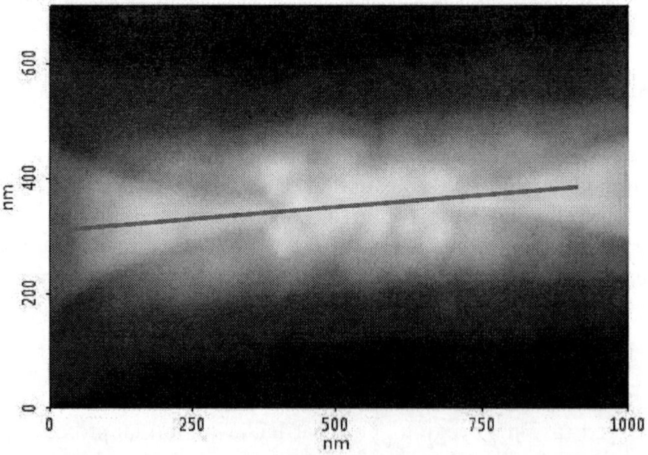

FIGURE 4. Contact mode AFM topography image of the second and malformed vacuum-channel device's (Device 6) sensor-drain interface, vacuum-channel, and insulated gates region. The overlaid red line corresponds to the topography line profile displayed in Figure 5. Scan size: 700 x 1000 nm, image size: 290 x 200 px.

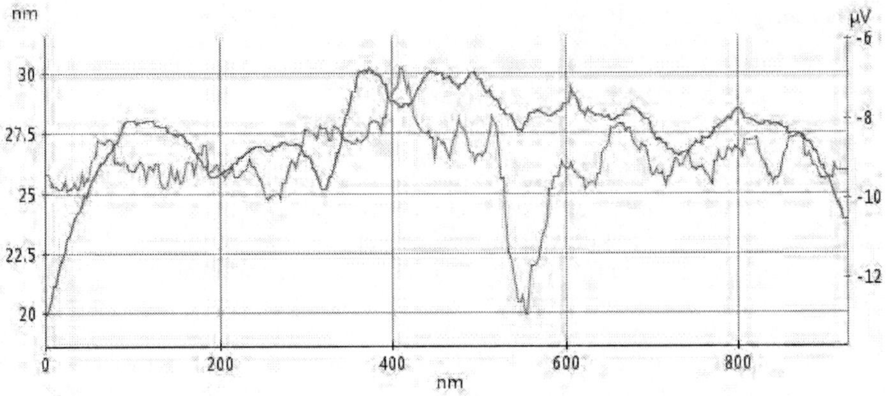

FIGURE 5. Line profiles of the AFM topography (red, left y-axis in nm) and the capacitance data (green, right y-axis in µV) of the Device 6 area scanned in Figures 4 and 6.

Malformed Device Capacitance with Topography

Device 6's SCM data from was also gathered from the same 700 x 1000 nm scan area. This image (Figure 6) shows each of Device 6's source-drain terminals, but only one functioning insulated gate on the right which carries a slightly more negative charge relative to the terminals and other surrounding areas of the device. Absent from the image are the second insulated gate which would have been on the left as well as the vacuum channel between the gates which would have carried the highest relative positive charge of any area within the scan. Looking at the line profile of the capacitance data (Figure 5), the right side of the device with the visible gate does exhibit a drop in relative charge from 1 µV to -3 µV across a distance of 50 nm that we observed on Device 4. Due to the deposition of excess material in what would have been the gap between the source and drain here on Device 6, this increase in negative charge cannot be correlated to the bump that would have been the insulated gate observed in Device 4.

The same excess material which obfuscates the functional insulated gate in the topography data also affects the device's electrical properties in the area which the vacuum channel would have normally been. There is nothing here on the malformed device's capacitance line profile comparable to the 175 nm span on seen on the functional device which carries a charge of 7 µV. The lack of a clearly defined length of relatively more positive charged vacuum-channel space and a second relatively more negatively charged insulated gate on the left have rendered this device inoperable as an effective transistor.

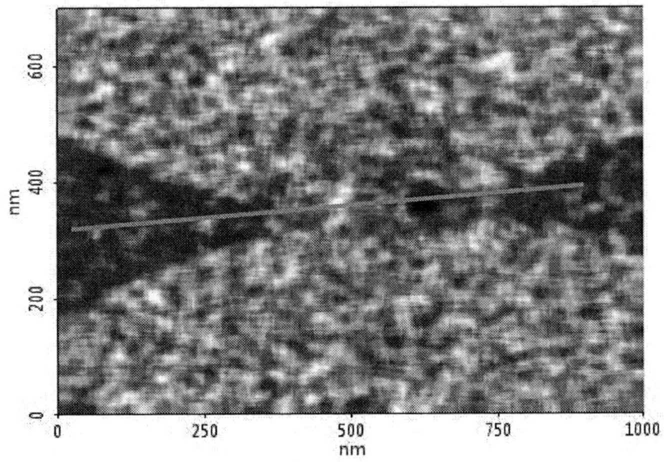

FIGURE 6. An SCM capacitance image of the region containing Device 6's sensor-drain interface, vacuum-channel, and insulated gates. Brighter colors correspond to relatively more positively charged areas on the device whereas darker colors correspond to relatively more negatively charged areas. Note the absence of electrical evidence for a functioning second

insulated gate and vacuum-channel. The overlaid green line corresponds to the capacitance line profile displayed in Figure 5. Scan size: 500 x 1000 nm, image size: 380 x 190 px.

CONCLUSION

SCM together with AFM successfully characterized both the spatial variations in capacitance as well as the topography of the functional and malformed vacuum-channel devices. By examining the line profiles of the topography and capacitance data acquired down an identical path on each of the device's source-drain interfaces, additional insight was gained by being pairing certain physical structures or locations with changes in capacitance. In the functional Device 4, the expected structures of a relatively negatively charged (-4 μV) insulated gate followed by a relatively positively charged vacuum-channel (7 μV) and another such gate (-4 μV) provided evidence that the sample could function as a transistor. However, in the malformed Device 6, the absence of one of the insulated gates and as well as the entire vacuum-channel gave evidence that the sample would not be able to function as intended. Lastly, while contact mode AFM confirmed that there was a discrepancy in the topographies between Devices 4 and 6, only SCM could provide direct empirical evidence of the latter's ineffective capacitive charge distribution.

REFERENCES

1. A.S. Keys and M.D. Watson, *Radiation Hardened Electronics for Extreme Environments*, Huntsville, AL: NASA Marshall Space Flight Center, 2007.
2. *Vacuum Nanoelectronics: Interview with M. Meyyappan and Jin Woo Han*, EEWeb, 2013. Retrieved from https://www.eeweb.com/blog/eeweb/interview-with-m.-meyyappan-and-jin-woo-han-on-vacuum-nanoelectronics
3. *Scanning Capacitance Microscopy (SCM)*, Park Systems, 2016. Retrieved from http://www.parkafm.com/index.php/park- spm-modes/94-electrical-properties/235-scanning-capacitance-microscopy-scm
4. J.W. Han and M. Meyyappan, *Introducing the Vacuum Transistor: A Device Made of Nothing*, IEEE Spectrum, 2014. Retrieved from http://spectrum.ieee.org/semiconductors/devices/introducing-the-vacuum-transistor-a-device-made-of-nothing
5. *How AFM Works*, Park Systems, 2016. Retrieved from http://www.parkafm.com/index.php/medias/nano-academy/how-afm-works

KEYWORDS

Vacuum, vacuum-channel, transistor semiconductor, silicon, atomic force microscopy, AFM, scanning capacitance microscopy, SCM, topography, capacitance, characterization

Assessing Scanning Electron Microscopy Stereophotogrammetry Algorithms with Virtual Test Samples

John S. Villarrubia, András E. Vladár, and Vipin N. Tondare
Engineering Physics Division,
Physical Measurement Laboratory,
National Institute of Standards and Technology,
Gaithersburg, MD 20899-8212, USA[1]

Contact:

John Villarrubia
Engineering Physics Division
National Institute of Standards and Technology
100 Bureau Drive, MS 8212
Gaithersburg, MD 20899
John.Villarrubia@nist.gov
301-975-3958

Key Words

Monte Carlo SEM modeling, nanometer-scale dimensional metrology, scanning electron microscopy, stereophotogrammetry

Assessing Scanning Electron Microscopy Stereophotogrammetry Algorithms with Virtual Test Samples

John S. Villarrubia, Vipin N. Tondare, and András E. Vladár

Engineering Physics Division, Physical Measurement Laboratory, NIST, Gaithersburg, MD 20899-8212, USA†

INTRODUCTION

Planar memory and logic devices have always had functional dependences on in-plane dimensions of their structures, but non-planar devices have added another dependence: on vertical dimensions. For example, the size of the conduction gate channel of a FinFET (a field effect transistor with raised channels, called fins, between source and drain) depends on the height of the fin. Optical and electron microscopy images have only two spatial dimensions—in the lateral plane. Their third dimension is an intensity. However, images from different viewing angles may be combined. Features extended along the insensitive vertical axis in one image will have a component in the sensitive lateral plane in one or more of the others. This permits in principle reconstruction of the 3D shape via stereophotogrammetry. Application of stereophotogrammetry to scanning electron microscopy (SEM) was described by Piazzesi in 1973.[1]

Since SEMs now have spatial resolution near 1 nm, the question naturally arises whether SEM-based stereo methods are sufficiently accurate for 3D nanometrology needs. Apart from the usual question of measurement errors that affect the inputs (e.g., the SEM images and the coordinates and angular viewpoints assigned to them) and how these errors then propagate to the result, there is the question to which we here address ourselves: whether important errors result from assumptions and approximations within the reconstruction software itself. For example, reconstruction of the position of a point A on the sample generally requires identifying its corresponding homologous image points A, A', etc., in two or more different views. Identification may be based on similarity of appearance as determined by correlation, but this is an approximation since appearance changes partly due to electron beam/sample interaction effects for which the reconstruction software does not account. Filtering may also be used to reduce errors caused by noise in the images. Filtering errors will also propagate to some extent into the reconstruction. Because of effects like these, even in the ideal case of error-free inputs, we might expect reconstruction errors. Also, of course, different strategies that might be selected by the algorithm developer to handle these or other issues may differ in their effectiveness.

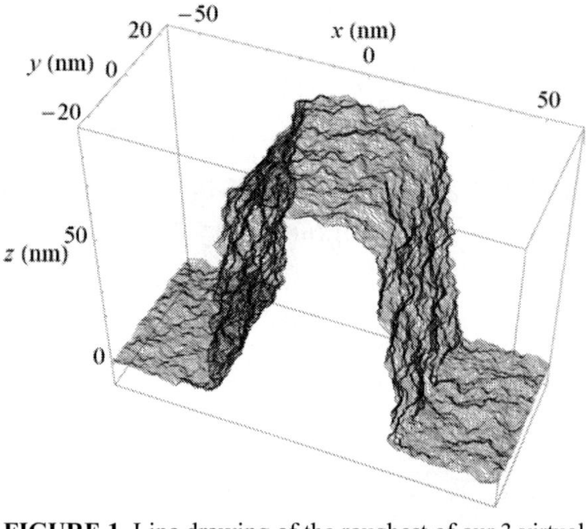

FIGURE 1. Line drawing of the roughest of our 3 virtual samples. The line is oriented to show the right edge. The left edge is obscured by absence of hidden line removal.

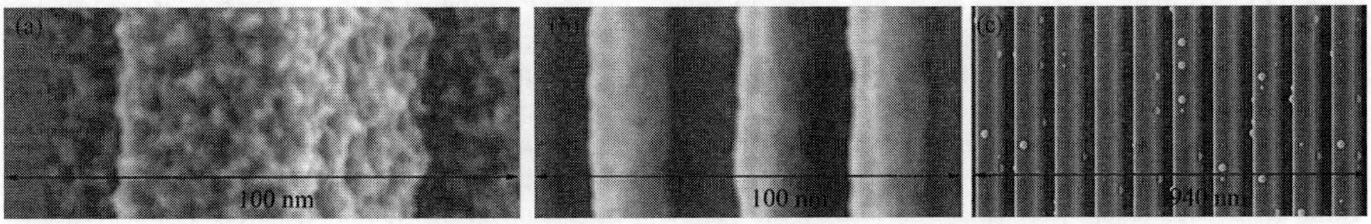

FIGURE 2. Simulated SEM images of 3 virtual samples. The samples were imaged with tilts at 5° increments from −85° to 85° from the normal (axis of rotation along the lines). The views shown here are at 20°.

Developers of most algorithms can test them by applying them to test problems with known answers. We construct such problems by starting with virtual samples. A virtual sample is a mathematical object, which is completely known. The virtual samples are "imaged" at varying tilt angles with an SEM simulator. These images are then input to commercially available stereo SEM reconstruction software, and the software's output is compared to the known virtual sample to assess errors.

PROCEDURES

We produce images of the virtual samples with the JMONSEL simulator.[2] The simulator employs models of electron elastic scattering, secondary electron generation, and scattering at boundaries to compute electron yield vs. position, capabilities that have been used for model-based metrology that agrees with transmission electron microscopy and critical dimensions small angle x-ray scattering measurements to better than 1 nm.[2]

We report results here for three virtual samples with height, *h*, width at mid-height, w_{mid}, sidewall angle, θ, and top corner radii, *r*. One was a near-trapezoidal line (h = 80 nm, w_{mid} = 50 nm, θ = 3°, r = 10 nm) around which a rough skin was wrapped. The sample is shown in Fig. 1 and its image in Fig. 2a. It has 1 nm root mean square (RMS) roughness and a 15 nm correlation length. Some results from this sample were given in an earlier report.[3] The second sample (image in Fig. 2b) had the same RMS roughness and similar design but with a longer roughness correlation length (30 nm) and multiple smaller lines (h = 30 nm, w_{mid} = 10 nm, θ = 2°, r = 2 nm) separated alternately by 18 nm- and 26 nm-wide trenches. The final sample (image in Fig. 2c) was the smoothest. It consisted of an array of 10 lines (h = 60 nm, w_{mid} = 60 nm, θ = 3°, r = 10 nm) on 100 nm pitch. The sample was decorated with hemispherical bumps on the trench floor, line tops, and line sides.

The images along with the tilt angles are input to the reconstruction software, which produces a number of different outputs, among them a profile across the sample.

RESULTS

Profiles across the middle of the Fig. 2a sample are shown in Fig. 3. The rough black line is the true profile. The smoother green and blue ones were outputs of two commercial software packages. For this very rough structure, the two packages made errors of 1.4 nm or less in the average height and less than 1 nm for the average width. These errors are small enough that we can not confidently attribute them to reconstruction error. They could be due to modeling or sampling errors (the latter because the software did not reconstruct the feature all the way to the edge of the image, so sampled less than the full profile). As we mentioned, the reconstructed profiles are obviously smoother than the true profile. This means the reconstruction is not useful for assessing surface roughness of this sample.

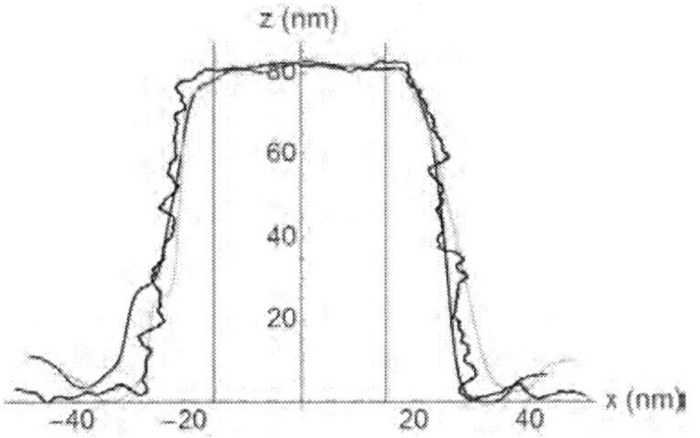

FIGURE 3. Profiles across the Fig. 1 sample from two different commercial software packages (the smoother blue and green lines) superimposed on the true profile (rougher black line).

The Fig. 2a sample was the roughest of the three. At the other extreme, the Fig. 2c sample has line tops and trench floors that are completely flat planes except for a sprinkling of hemispherical bumps. These proved difficult for all the tested algorithms. Even the best of the reconstructions had the average line almost 40 % shorter than its true 60 nm height, and although the lines are—apart from the hemispherical markers—all the same height, the reconstructed heights varied with standard deviation of more than 8 nm. Interesting in this context is that we could determine the height from the input images along a slice through one of the small hemispheres with much smaller error than this (~1 nm) by using some simple geometry and a calculator. The line shapes in this case were also strongly distorted and about 10 % too narrow. The sample with intermediate roughness (Fig. 2b) also suffered from significant errors on reconstruction, with heights smaller by 4 nm in the wider trench and 8.7 nm in the narrower trench than the true value of 30 nm. On this sample, six pairs of homologous points identified by eye on line tops and in the trenches produced reconstructed heights that averaged only 2.5 nm too low with standard deviation 0.4 nm.

DISCUSSION AND CONCLUSION

Roughness serves a useful function for stereo reconstruction algorithms. It provides a non-periodic surface texture that facilitates identification of homologous points in images from different viewing angles. Two of the samples we examined represent extremes relative to the amount of roughness one would ordinarily encounter in integrated circuit production. The sample of Fig. 1 and Fig. 2a is very rough whereas that of Fig. 2c is very smooth. The best of the tested commercial software algorithms reconstructed the very rough sample with height and width errors of about 1 nm. (This excludes errors due to vibration, noise, angular positioning, etc., for which the reconstruction algorithm bears no responsibility. In a real measurement, these other errors would of course be additional.) Errors on the very smooth sample, on the other hand, were a significant fraction of feature size. The disparate performance at these two extremes motivated a test with a sample (Fig. 2b) intermediate between the two, with roughness bearing a greater resemblance to that expected in practice. Errors in this case were again a significant fraction of feature size, e.g., height errors of 13 % to 29 % of the true height.

Manual line height determinations using simple geometry and homologous points identified by eye outperformed the automated algorithms on the smooth and moderately rough samples. This demonstrates that the larger errors are not *inherent* in these kinds of samples, but must rather be due to less than full use of available information by the tested algorithms. That is, a different algorithm design could perform better. Absence of test problems with known solutions has heretofore been an obstacle to algorithm development. Test problems based on simulated images of samples with known shape can be useful in this respect.

REFERENCES

1. G. Piazzesi, "Photogrammetry with the scanning electron microscope," *J. Phys. E: Sci. Instrum.* **6**, 392-397 (1973).
2. J. S. Villarrubia, A. E. Vladár, B. Ming, R. J. Kline, D. F. Sunday, J. S. Chawla, and S. List, "Scanning electron microscope measurement of width and shape of 10 nm patterned lines using a JMONSEL-modeled library," *Ultramicroscopy* **154**, 15-28 (2015), http://dx.doi.org/10.1016/j.ultramic.2015.01.004
3. J. S. Villarrubia, V. N. Tondare, and A. E. Vladár, "Virtual rough samples to test 3D nanometer-scale scanning electron microscopy stereo photogrammetry," *Proc. SPIE* **9778**, 977809-1 (2016).

KEYWORDS

Monte Carlo SEM modeling, nanometer-scale dimensional metrology, scanning electron microscopy, stereophotogrammetry

Development of a Nanometer Probe Helium Ion Microscope with Time of Flight Element Identification

Viacheslav Manichev [1, 2], Alexander Livernois[1, 3], L. C Feldman [1, 3] and T. Gustafsson [1, 3]

[1]*Institute for Advanced Materials Devices and Nanotechnology*
[2]*Department of Chemistry and Chemical Biology;*
[3]*Department of Physics and Astronomy, and Laboratory for Surface Modification, Rutgers University, Piscataway, NJ, USA*

and

D. Barbacci and A. Schultz

Ionwerks Inc., Houston TX

INTRODUCTION

Helium Ion Microscopy (HIM) [1] is one of the latest technologies in the imaging toolbox. The HIM utilizes ~ 35 KeV He ions focused on a small spot (a few ångströms in diameter). Interactions with a sample yield secondary electrons (SEs) which are straightforwardly detected; rastering across the sample provides a morphological image of the surface, similar to a Scanning Electron Microscope (SEM). Significantly smaller beam spot size, reduced multiple scattering and high SE yield gives important advantages over a SEM. The main advantage of HIM for imaging is the much deeper field of view and the unique capability to image non-conducting samples without a deposited metal overlayer.

The main uses of the apparatus are: 1) imaging via secondary electron emission, 2) ion beam lithography [2] and 3) ion beam machining on the nanoscale. In addition, various groups are developing element identification tools operable at the nanoscale [3, 4]. We describe recent progress in our laboratory in the development of a time of flight (ToF) energy analysis system which allows elemental identification on the nanometer scale in combination with the quantitative advantages of ion scattering technologies. The facility brings new capabilities for both characterization and metrology for a nano-electronics.

TIME OF FLIGHT DETECTORS AND ELECTRONICS

Our ToF system is based on a start signal associated with the emission of secondary electrons from an impinging He ion and a stop signal from the detection of the actual backscattered He ion. By measuring the ToF over a known path length we obtain the energy of the backscattered He ion, which, in turn, yields the mass of the sample atom associated with the scattering event. Hence elemental identification.

For the start signal, the system employs a customized set of electronic lenses, which first accelerate the secondary electrons to ~ 400 eV and then focuses the resulting beam onto a double stack channel plate. The electronics are specifically designed to handle the high incidence rates of order 10^5/s to 10^7/s. The lens system is also designed to minimize electron path length dispersion so as to retain high time resolution. A crucial advantage of using a He beam is that the secondary electron yield per incident He ion is larger than one, so that essentially every backscattered He ion can be detected.

The stop signal is generated by scattered He ions detected by a large (~4 cm diameter) two-channel plate array approximately 13 cm from the sample surface. This length corresponds to a flight time of ~110 ns for a 30 KeV He ion. The large area minimizes the total incident beam requirement for a statistically significant analysis. The channel plate detector is coupled to a ~15 ns X and Y meander array, which imposes a time delay dependent on the hit position of the ion. Knowledge of the position allows for corrections for slightly different flight paths associated with the large detector and for kinematic corrections inherent in ion backscattering.

The electron start signal and the four ion signals (two to define X and two to define Y) are coupled to an HPTDC8-PCI time to digital converter (TDC, 25 ps resolution) for energy/time analysis and position definition. A one dimensional schematic of the time sequence array is shown schematically below. Time delays with respect to the electron determine the ToF and time differences between the ion signals determine the position of the ion hit on the channel plate. The overall design time resolution is 0.4 ns. Early results in Figure 2, uncorrected for finite detector effects, show ~2 ns time resolution.

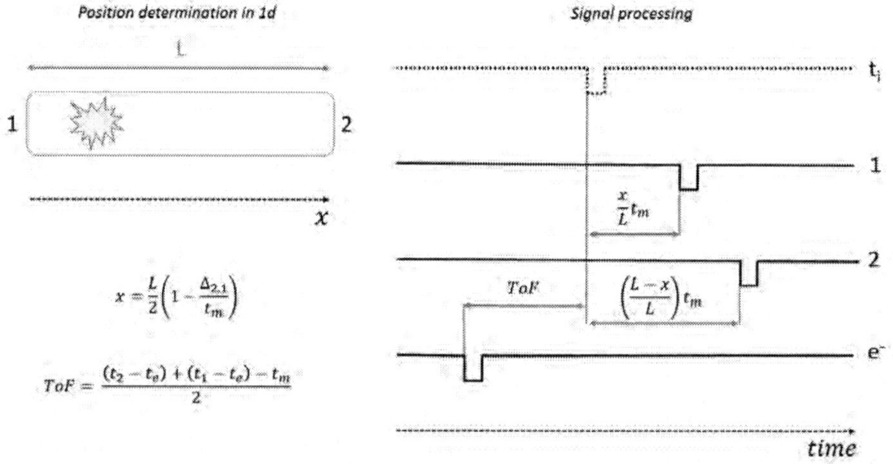

FIGURE 1. Position determination of ion hit in 1d case. t_i is the virtual signal denoting the real Time of Flight, channels 1 and 2 are two opposite outputs of meander wire. e^- is the electron start signal. t_m is the "meander time" associated with an imposed position dependent time delay on the actual ion arrival time. $\Delta_{2,1}$ is the difference in time between signals on channel 2 and 1.

ANTICIPATED CAPABILITES AND INITIAL RESULTS

Straightforward ion scattering kinematics for our ion detector size indicate that a time resolution of 400 ps with a 130mm flight path will yield an energy resolution of 0.22 KeV. The resolution can be improved further with a longer flight path. Current resolution would permit mass analyses that would separate a heavy metal from silicon, ^{69,71}Ga from ^{75}As, and ^{12}C from ^{13}C. It would correspond to ~ 1.0 nm depth resolution in silicon, better in higher Z materials. We note that the random count rate is not directly determined by the time resolution but rather the time limits of the TDC. As such, real to random rates have been demonstrated at better that 10:1 as shown below. Damage and sputtering effects of the sample provide a fundamental limit on sensitivity. Estimates suggest a single monolayer of heavy metal (Pt say) on Si can be detected using a 100 nm beam spot before sputter erosion of the overlayer. Equivalently this translates to a limit of a 30 nm beam spot, and hence a 30 nm spatial resolution, for scanning a ten monolayer film. These are conservative estimates based on bulk Pt sputtering coefficients and can be improved with a larger detector. Recent experimental studies also indicate a considerable greater tolerance to beam effects than stated above.

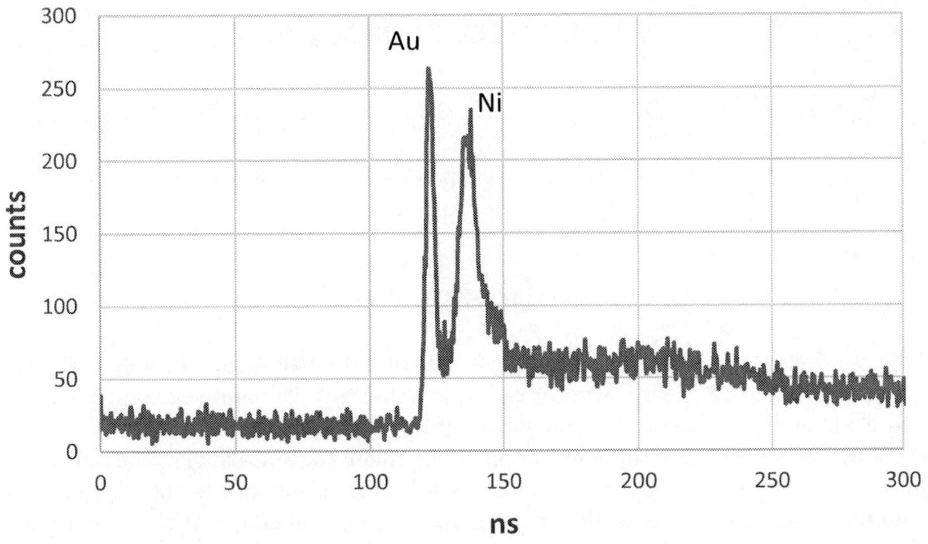

FIGURE 2. Time of flight spectrum of 0.7 nm Au on 2.7 nm Ni on Al_2O_3 taken with a 10 μm beam spot size and uncorrected for finite detector size and kinematic effects.

Recent applications of the HIM in our laboratory include imaging of the fine structure in "aged" (rat) kidney tissue and its reconstruction with new test drugs; inspection of calcification in ocean coral providing insights into ocean acidification mechanisms, investigation of the nano-scale pore structure in shale rock as an indication of mining productivity and nanoscale creation of pores and individual vacancies in 2D electronic materials such as MoS_2. All of these applications will benefit from the new elemental capability described above. In microelectronics we anticipate new investigations to examine nanoscale metal-semiconductor surface diffusion, trench structures and their metallization and semiconductor hetero-structures as to possible interdiffusion and interface control.

ACKNOWLEDGEMENT

This work was supported by NSF grant DMR-1126468 and by the Rutgers Institute for Advanced Materials, Devices and Nanotechnology.

REFERENCES

1. Scipioni, L. et al. *Advanced Materials and Processes* 166.6 (2008): 27.
2. Luo, F. et al., *Proc. SPIE 9779, Advances in Patterning Materials and Processes XXXIII*, 977928 (March 21, 2016)
3. Sijbrandij, S. et al., *J. Vacuum Science & Technology B*, 26, 2103-2106 (2008)
4. Klingner, N. et al., *Ultramicroscopy*, 162, 91-97 (2016)

KEYWORDS

Helium Ion Microscopy, Elemental Identification, Nanometer scale Materials analysis, Time of Flight.

Subsurface Nano-imaging Using Torsional Scanning Probe Microscopy

Violeta Navarro, Maarten H. van Es and Hamed Sadeghian

TNO, Stieltjesweg 1, 2628 CK Delft, The Netherlands

INTRODUCTION

The need for nondestructive imaging of buried structures with nanometer resolution is becoming more and more urgent in science and industry, for example in the field of nanoelectronics. Such imaging can be achieved by applying an ultrasonic excitation to a sample and analyzing the response of a micro fabricated cantilever with a sharp tip at its end in contact with the surface [1]. The mechanical response of the cantilever to the propagating wave contains information related to the viscoelastic subsurface properties of the material right below the area where the tip is placed [2]. If the excitation frequency is close to the contact resonant frequency of the cantilever (both for flexural and torsional modes), the sensitivity and signal-to-noise-ratio are enhanced.

TORSIONAL SUBSURFACE VS FLEXURAL SUBSURFACE

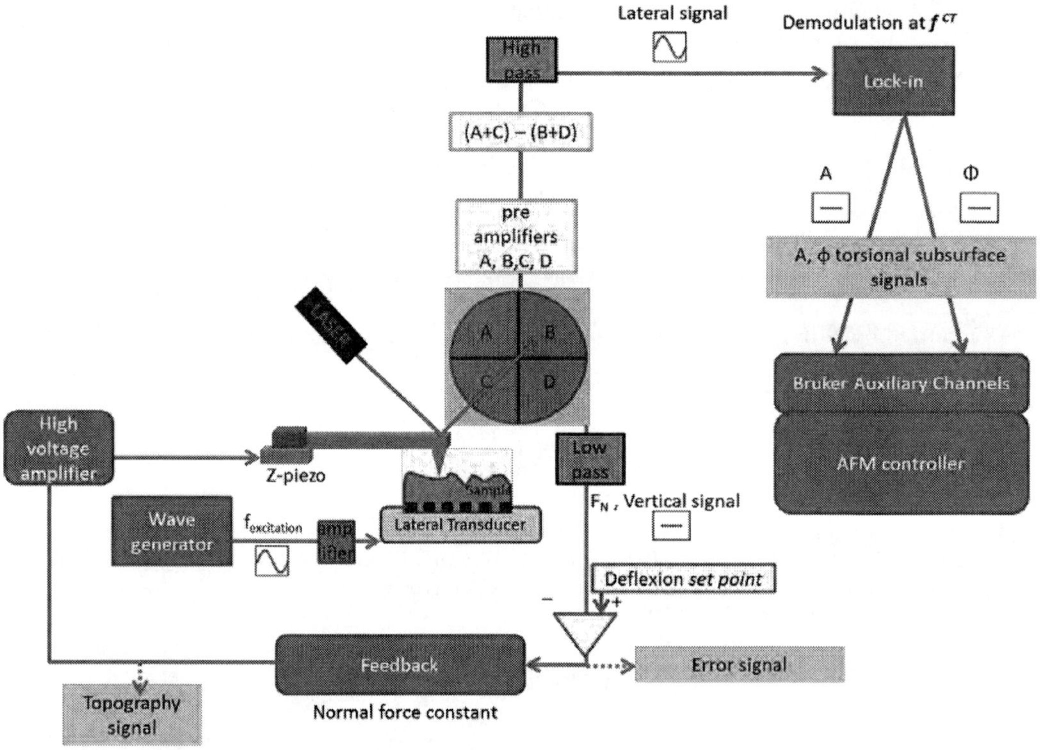

FIGURE 1. Schematic of the subsurface torsional imaging using scanning probe microscope.

When the excitation is perpendicular to the surface in the megahertz range (close to the contact flexural resonant frequency of the cantilever), the information obtained is the viscoelastic mechanical response of the sample of interest, below the surface and in the direction perpendicular to it. Complementary to the vertical excitation of the sample, a shear or lateral excitation of the sample parallel to the surface, induces the natural torsional oscillation of the cantilever when the tip is in contact. From the response the cantilever when the sample is excited close to its first contact resonant torsional excitation, we can obtain information about the shear mechanical response of the material at the surface, below the surface and in the direction parallel to it. Torsional subsurface is complementary to flexural subsurface since it may have enhanced contrast to structures and defects which are in a plane, parallel to the surface (as for example, stacking faults), which would otherwise be "invisible" for the flexural motion. Several modes can be excited simultaneously and demodulated at the different frequencies separately, such as the first resonant contact frequency for the torsion and for the flexural movement. The advances on the development of torsional subsurface technique will be presented.

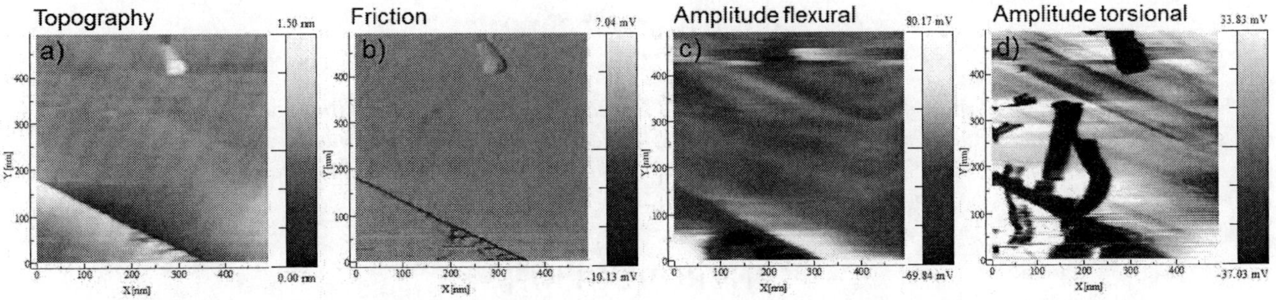

FIGURE 2. 492 × 492 nm^2 scanning probe images of the same area of a graphite sample displaying different signals. a) Topographic image taken at 280 nN applied normal (set-point) force; b) Friction (lateral force) signal image; c) Amplitude of the flexural response of the cantilever, demodulated at 1.65 MHz and d) amplitude of the torsional response of the cantilever, demodulated at 3.94 MHz. Dislocation system can only be distinguished at image d), by the dark lines.

REFERENCES

1. K. Yamanaka, H. Ogiso and O. Kolosov. *Appl. Phys. Letters* **64**, 178-180 (1994).
2. M. H. van Es *et al*, *Submitted* (2016).

KEYWORDS

Subsurface, atomic force microscopy, torsional mode, heterodyne, ultrasonic force microscopy, viscoelasticity, mechanical properties, ultrasound.

Engineering Ferroelectric Polymer Memories: Confounding Factors Which Obscure Polarization

Vasileia Georgiou[ab], Dmitry Veksler[a], Jason P. Campbell[a], Dimitris E. Ioannou[b] and Kin P. Cheung[a],

[a] *National Institute of Standards and Technology, 100 Bureau Drive, MS 8120, Gaithersburg, Maryland 20899, United States, kin.cheung@nist.gov,* [b] *Department of Electrical and Computer Engineering, George Mason University, 4400 University Drive, Fairfax, Virginia 22030, United States.*

INTRODUCTION

Ferroelectric memory based on polar polymers has attracted a lot of interest in the past few decades [1-4], due to the potentially easy low temperature processing as well as the application for flexible electronics. The most promising configuration of the memory cell is the FeFET (Ferroelectric field-effect-transistor) with the polar polymer incorporated in the gate dielectric stack [1-3,5]. This configuration allows a non-destructive read operation. The memory effect in these devices is originated from polarization of the ferroelectric polymer film and results in a hysteresis of the Id-Vg curve. However, many research groups have pointed out that the observed hysteresis may also be caused by charge-trapping and/or mobile ions. Therefore, development of the material processing and optimization of the device structure for maximizing the memory effect is not obvious. Decoupling of effects of different mechanisms was not systematically studied. One of the most common arguments [6], maintains that the steepness of the observed hysteretic Id-Vg characteristics unambiguously proves the occurrence of ferroelectric effect. However, this argument does not hold in general, because other mechanisms can also be responsible for the steep slope of Id-Vg curves. Another argument claiming that only the hysteresis caused by polarization effects saturates [7] is also incorrect. In reality, observations of ferroelectric switching could be unambiguously made from Id-Vg measurements only after careful consideration of all the possible mechanisms and their interactions.

DEVICE FABRICATION AND MEASURMENTS

We fabricated two types of devices: (1) metal-insulator-metal (MIM) capacitors with the CP1 polymer sandwiched between two metal layers; (2) FeFET devices in a form of depletion mode JFETs with 15 nm thick CP1 polar polymer film incorporated with 15 nm n-type poly-Si channel. The CP1 polymer has a glass transition temperature of around 260°C. Operating much below the glass transition temperature the ferroelectric effect should be significantly suppressed. To switch the memory state, the device needs to be brought above the glass temperature to allow polymer dipoles rotation, and then cooled down while maintaining an external electric field to "freeze" their position. The effects of charge trapping and mobile ions are parasitic effects, and will reduce the strength of polarization effect.

Dielectric Spectroscopy measurements were performed on the MIM capacitor devices at a wide range of temperatures. Measurements were executed using an Impedance Analyzer. At each temperature point, the imaginary part of the dielectric permittivity, ε'', vs. frequency, f, was extracted (Figure 1a). The peaks in the $\varepsilon''(f)$ spectra were associated with the specific corner frequencies of dipole rotations at given temperatures. At the frequencies above the corner, the dipoles are not able to follow the external electric field. No peaks were observed in the spectra at temperatures below 255°C, due to instrumental noise limitations. The obtained corner frequencies were then fitted into a model, $\log(f_c)=A\exp(-E_a/kT)$, to predict the value of the corner frequency at room temperature (star sign in Fig.1b). The projection to room temperature was obtained to be less than f_c = 100 μHz. Thus, at room temperature and for frequencies greater than 100 μHz, the dipoles in the polymer are "locked in place" and cannot follow the field, and no ferroelectric effect should be observed.

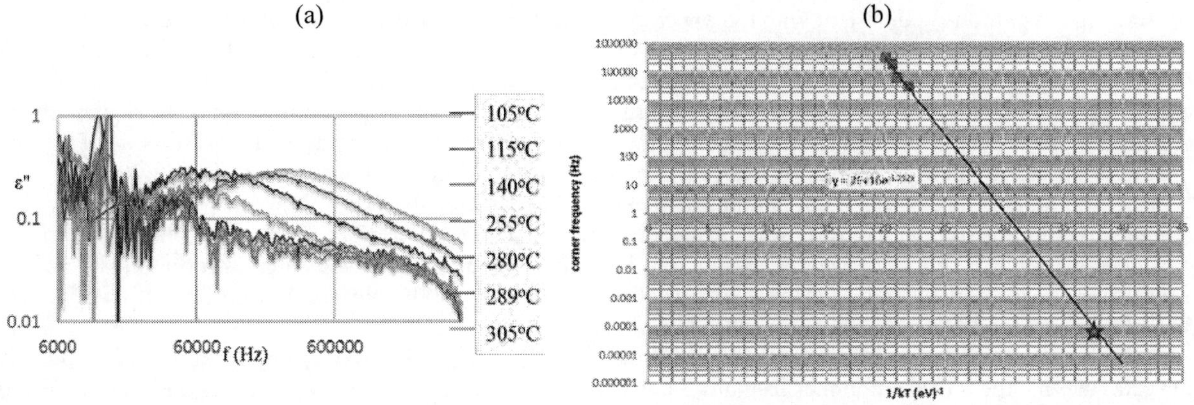

FIGURE 1. (a) Dependence of the measured imaginary part of the dielectric constant vs. frequency at different temperatures. MIM structure with 11nm thick CP1 was used. (b) Dependence of the peak frequency of the ε''(*f*) spectra vs. 1/kT and the theoretical fitting to the experimental data (solid line). Expected value of peak frequency at room temperature is shown by the star symbol.

Id-Vg characteristics of FeFETs (Fig. 2), measured at room temperature by applying sinewave voltage (V_{gs}) to the gate, still show significant hysteresis amplitude at frequencies $f \leq 10$ Hz. While this hysteresis cannot be caused by the polymer polarization effects (the corner frequency of this effect is supposed to much smaller), it is important to identify mechanisms responsible for this hysteretic behavior.

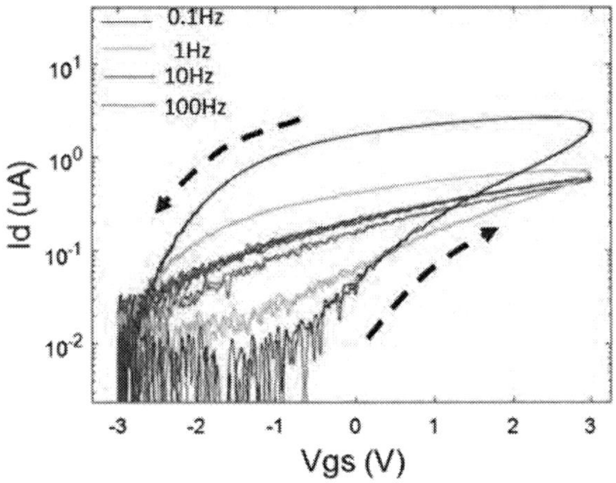

FIGURE 2. Experimental Id-Vg characteristics of polar polymer-based FeFET. 15 nm CP1 layer is incorporated as gate dielectric with 15 nm thick poly-Si layer used as the channel. The direction of observed hysteresis loop is schematically shown by arrows.

Fig. 2 shows that the Id-Vg loop measured at 100 Hz is practically hysteresis-free. One may conclude that 100 Hz is higher than characteristics frequencies of any possible involved mechanisms. The 10 Hz loop appears to be extended to the positive side of the 100 Hz loop, and to have a counterclockwise direction of the hysteresis. This observation alone allows to make preliminary conclusions about the mechanisms responsible for hysteresis in this case. Considering the thin polymer film used in our study, the gate dielectric thickness is comparable with the electron tunneling length at slow sweep rates. In this case, charge injection in the traps in polymer from both substrate and gate sides can contribute to the flatland voltage shift. The direction of the hysteresis caused by charge trapping can be either clockwise (electrons and holes injection from substrate) or counterclockwise (electron injection from gate). Thus the 10 Hz loop (counterclockwise) cannot be due to charge-trapping from the substrate side and leads to the conclusion that gate-side electron trapping and/or mobile ions in the polymer can be responsible for the observed hysteresis. Note that for the devices with much thicker polymer layers (> 100 nm) the charges injected from the gate have greatly diminished ability to affect V_{FB} and can in general be neglected. The frequency dependence of the hysteresis loop from 100 Hz to 10 Hz allows one to associate the hysteresis with injection of trapped charge from the gate, since the evolution is consistent with the expected shape of the hysteresis caused by this effect.

Further decreasing the voltage sweep frequency, hysteresis loops measured at 1 Hz and 0.1 Hz not only get wider, but also no longer stay on the positive side of the 100 Hz loop. This suggests that another factor should also be contributing to negative shift of hysteresis. Careful consideration of how mobile ions affect the electrical characteristic in the case of thick dielectric indicates that the entire hysteresis loop should be on one side of the mobile ion-free Id-Vg curve (there is almost always positive charge present: mobile charge is mainly due to Na^+, Li^+, K^+, H^+) and it should also be asymmetrical around the ideal Id-Vg characteristic. The 1 Hz sweep is sufficiently slow to allow positive mobile ions to move significant distances within a sweep period to contribute to the shape of the Id-Vg hysteresis loop. The effect of the positive mobile ions alone should cause the hysteresis loop to stay entirely on the negative side of the 100 Hz loop. However, our observations suggest that the combined effect of positive mobile ions and gate injection of trapped charges should be considered.

In conclusion, careful analysis of the direction of the hysteresis, dependence of the hysteresis on the gate voltage ramp rate, the thickness of the polymer insulating layer, the symmetry of the hysteresis curves and their dependence on the amplitude of the gate voltage can greatly aid in deconvolving complicated FeFET hysteresis. A large concentration of mobile ions is suspected to be present in the studied polymer, which suggests a path for process improvement, and may significantly help to minimize the parasitic effects which obscure observations of FeFET polarization effects.

REFERENCES

1. N. H. Van et al., "Ultralow-power non-volatile memory cells based on (PVDF-TrFE) ferroelectric-gate CMOS silicon nanowire channel field-effect transistor", Nanoscale, **7**, 11660-11666 (2015).
2. G. H. Gelinck et al., "Ferroelectric switching of poly(vinylidene difluoride-trifluoroethylene) in metal-ferroelectric-semiconductor non-volatile memories with an amorphous oxide semiconductor", Appl. Phys. Lett, **106**, 093503 (2015).
3. P. Heremans et al., "Polymer and Organic Nonvolatile Memory Devices", Chem. Mater., **23**, 341-358 (2010).
4. K. Asadi et al., "Organic nom-volatile memories from ferroelectric phase-separated blends", Nat. mat., **7**, 547 (2008).
5. Y. H. Chou et al., "Nonvolatile transistor memory devices using high dielectric constant polyimide electrets", J. Mater. Chem. C, **1**, 3235-3243 (2013).
6. R. C. G. Naber et al., "High-performance solution-processed polymer ferroelectric field-effect transistors", nature materials, **4**, 243 (2005).
7. C. A. Lee, et al., “Hysteresis mechanism and reduction method in the bottom-contact pentacene thin-film transistors with cross-linked poly(vinyl alcohol) gate insulator”, Appl. Phys. Lett., vol. **88**, no. 25, 252102 (2006).

KEYWORDS

Ferroelectric, FeFET, hysteresis, charge trapping, mobile ions

Nanoscale Chemical Imaging with Infrared Photo-induced Force Microscopy

Sung Park, Derek Nowak, and Thomas Albrecht

Molecular Vista, Inc., 6840 Via Del Oro, Suite 110, San Jose, CA 95119

INTRODUCTION

Real-space imaging and identification of chemical components in multi-component systems with nanometer scale spatial resolution remains a challenge for the nanoelectronics community.

PHOTO-INDUCED FORCE MICROSCOPY

In this talk, we introduce real-space and non-destructive chemical imaging capability of a new technique called infrared photo-induced force microscopy (IR PiFM) where the dipole-dipole force due to infrared absorption in nanoscale region of the sample is measured by an atomic force microscope (AFM). This dipole-dipole interaction is strongly affected by the optical absorption spectrum of the sample, thereby providing a significant spectral contrast mechanism which can be used to differentiate between chemical species. Due to its AFM heritage, PiFM acquires both the topography and spectral images concurrently and naturally provides information on the relationship between local chemistry and topology with sub 10 nm spatial resolution on a variety of samples. PiFM spectral images surpass spectral images that are generated via other techniques such as scanning transmission X-ray microscopy (based on synchrotron source), micro confocal Raman microscopy, and electron microscopes, both in spatial resolution and chemical specificity. The breadth of the capabilities of PiFM will be highlighted by presenting data on various material systems (organics, inorganics, 1D/2D, and nano-photonic materials). By enabling imaging at the nm-scale with chemical specificity, PiFM provides a powerful new analytical method for deepening our understanding of nanomaterials and facilitating technological applications of such materials.

ALPro System: An Electrical Profiling Tool for Ultra-Thin Film Characterization

Abhijeet Joshi and Bulent Basol

Active Layer Parametrics, Inc.
Blackstone Launchpad at UCLA, 200 De Neve Drive, Covel Commons, Room 203, Los Angeles, California 90024

INTRODUCTION

There is continuing need to improve contacts in devices employing ultra-shallow junctions. Various doping approaches are being developed to push the limits of dopant activation at the source and drain regions of MOSFET structures. Introduction of new materials to improve the channel mobilities is another area of high interest. In all of these applications, it is very important to characterize dopant activation and mobility fully through the ultra-thin layer under development.

The ALPro system provides direct measurements of resistivity and mobility depth profiles and the associated activated carrier depth profiles based on differential Hall effect-type approach. The method involves varying the effective thickness of a sample at a pre-assigned measurement region by oxidation or etching and carrying out Vander Pauw measurements to obtain the resistivity and mobility values as a function of material thickness removed from the circuit. Here, we discuss the status of our efforts towards developing a fully automated metrology tool for electrical depth profiling and present data on Silicon & Silicon-Germanium samples that detail the changes in carrier activation and mobility as a function of depth. Directly measured high-resolution mobility profiles from the ALPro system may also be compared with existing models (Thurber/ASTM Model [1] for Silicon and Cuttriss Model [2] for Ge) to provide additional information about defectivity [3,4]. It should also be noted that the primary data produced by this approach may be used to perform complete electrical studies to evaluate the impact of process variables on ultrathin layers and enable direct tuning of the activated region. Specifically dopant implant related damage that may persist after epitaxial growth/re-growth or annealing can have a large impact on mobility degradation, which may be eliminated by further tuning of the anneal or epitaxial re-growth process. Measuring the mobility distribution and comparing the measured mobility and carrier distribution with the carrier mobility curve for undamaged and unstrained materials can identify processes that result in the generation of significant concentrations of scatter defects [3,4].

ELECTRICAL PROFILE DATA FROM Si AND SiGe

There is a need for highly activated surface regions in current generation Si devices, and localized Ge and SiGe channels for advanced nodes. Advanced millisecond and laser annealing approaches are currently used to achieve *higher* than equilibrium dopant solubility activation, while simultaneously limiting the thermal budget to prevent excessive dopant diffusion [5]. In SiGe, direct selective epi-growth followed by oxide formation and 100% selective Ge-epi are some of the other methods used.

Achieving very high activation levels in current silicon device source and drain areas is essential to minimize the parasitic resistance in advanced CMOS devices [5]. High activation anneal is a topic of great interest and much work is being carried out to determine the precise recipe needed to achieve and retain activation near the surface areas of source and drain. Figure 1 shows data from Phos-doped silicon samples with doses of $1x10^{15}/cm^2$ (P1), and $3x10^{15}/cm^2$ (P3). The wafers were annealed in the Mattson Technology's Millios® MSA tool, which provides real-time measurement and control of the temperatures of both the front and the back surfaces of the wafer. As can be seen from this data ALPro technique was able to resolve the details of active carrier concentration within the top 8nm of the film. Agreement between the SIMS and ALPro profiles deeper into the surface is expected since the dopant concentration there is well within the solid solubility limit. It is clear that the process conditions of sample P(1) achieved higher dopant activation right at the surface.

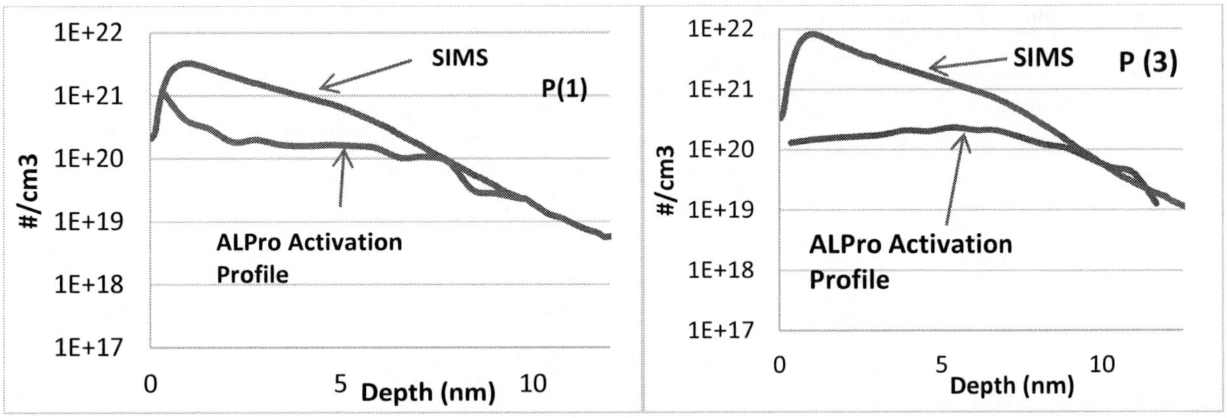

FIGURE 1. Samples: P1 = 1x10^{15}/cm^2 Phos-doped Silicon and P1 = 3x10^{15}/cm^2 Phos-doped Silicon. The post-anneal SIMS 'as-is' dopant profile, and ALPro Activation profile.

The ALPro System was also used to analyze SiGe layers that were deliberately strained by the addition of Ge into the silicon substrate. The study combined a very high dose Ge+B plasma implantation and selective surface laser melt annealing using a short wavelength laser. Structural information provided by XRD and X-TEM analysis provided information about the strain in the samples whose Ge content was 0%, 20% and 55%. ALPro measurements showed an increase in the mobility by 70% for the 20% doped sample (to 63 cm^2/V-s), and a sizable 4.3X increase in the 55% Ge sample. Note that 0% Ge has a mobility value of 38 cm^2/V-s.

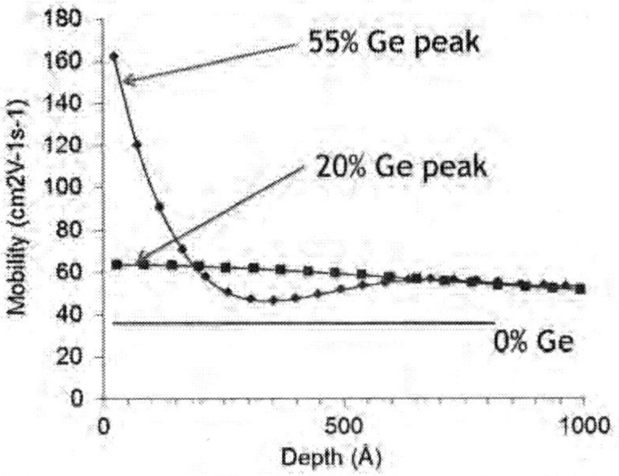

FIGURE 2. ALPro showed 70% increase in mobility for 20% Ge sample and 4.3x increase for 55% Ge sample 55% Ge sample concentration peak was at ~20 A (confirmed by AR-XPS)[6].

ACKNOWLEDGMENT

This material is based on partial support from the National Science Foundation under grant # 1632322.

REFERENCES

1. W.R. Thurber, R.L. Matthis, Y.M. Liu, and J.J. Filliben, NBS Special Publication 400-64, May 1981.
2. Cuttriss, D. B. "Relation between surface concentration and average conductivity in diffused layers in germanium." Bell System Technical Journal 40.2 (1961): 509-521.
3. A. Joshi, "Analysis of Defects in SiGe and Ge", 2015 FCMN, Dresden.
4. A. Joshi, S. Prussin, "Determining Defect & Strain Effects on Active Layer Mobility", IIT 2014, Portland, OR.
5. P. J. Timans, M. Hagedorn, C. Pfahler, A. Cosceev, M. Zwissler, L. Rubin and A. Joshi, "Dopant Activation and Deactivation Phenomena During Advanced Millisecond Anneal Cycles", Ion Implantation Technology 2016, Taiwan.
6. John Borland, Shu Qin, Peter Oesterlin, Karim Huet, Walt Johnson, Lauren Klein, Gary Goodman, Alan Wan, Steven Novak, Thomas Murray, Richard Matyi, Abhijeet Joshi, Si Prussin, "High mobility Ge-channel formation by

localized/selective liquid phase epitaxy (LPE) using Ge+B plasma ion implantation and laser melt annealing", Proc. 13th International Workshop on Junction Technology (IWJT), pp. 49 - 53. 6-7 June 2013.

KEYWORDS

Electrical profiling technique, complete characterization of Activated Layers, Strain-Effects and Defect Analysis, Defect Analysis of Germanium, Germanium epi-layers.

Precision of Micro Hall Effect Measurements in Scribe Line Test Pads

Maria-Louise Witthøft[a], Frederik W. Østerberg[a], Janusz Bogdanowicz[b], Andreas Schulze[b], Wilfried Vandervorst[b,c], Henrik H. Henrichsen[d], Peter F. Nielsen[d], Ole Hansen[a] and Dirch H. Petersen[a,1]

[a]Department of Micro- and Nanotechnology, Technical University of Denmark, DTU Nanotech Building 345 East, DK-2800 Kgs. Lyngby, Denmark
[b] IMEC, Kapeldreef 75, B-3000 Leuven, Belgium
[c] Instituut voor Kern- en Stralingsfysika, KU Leuven, Celestijnenlaan 200D, B-3001 Leuven, Belgium
[d]CAPRES A/S, Scion-DTU, Building 373, DK-2800 Kgs. Lyngby, Denmark
[1]email address: Dirch.Petersen@nanotech.dtu.dk

INTRODUCTION

Precise metrology for monitoring electrical properties in semiconductor manufacturing is of paramount importance [1]. However, obtaining high precision on measurements of electrical properties is difficult in the three dimensional device geometries of semiconductor devices being fabricated today. For nano-structured materials, electrical measurements become increasingly dependent on geometrical variability such as line width and roughness, thus decreasing the measurement precision on the electrical properties of the materials under test. In order to monitor electrical properties independent of geometrical effects, micro four-point probe sheet resistance measurements [2] or micro Hall effect measurements [3] in small scribe line test pads may still be the most practical solution, in combination with metrology, for precise measurement of nano-structure dimensions.

In micro Hall effect measurements, both sheet resistance, sheet carrier density and carrier mobility are measured simultaneously. The method relies on placing four electrodes on a thin conducting sheet in proximity of an insulating boundary with a magnetic field perpendicular to the sample surface. The measured four-point resistance includes both Hall effect and Corbino effect. Since the first description [3], several improvements have been made to reduce measurement time and increase measurement precision by application of geometrical error suppression methods, e.g., the far separation method [4], the short separation method [5], and most recently, the single-engage method using equidistant [6] or asymmetric electrodes [7], respectively.

In a previous numerical study, the single-engage method using equidistant electrodes was evaluated for semi-infinite sheets and square samples [8]. Here, we evaluate the precision of the single engage method on small rectangular pads using asymmetric electrodes [7]. Asymmetric electrodes can increase the signal to noise ratio by up to an order of magnitude. The assessment of measurement precision is performed numerically and will be supported experimentally by repeatability measurements on small test pads of B-doped SiGe. In this study, we examine how the size of a rectangular test pad affects the relative standard deviation of the extracted parameters. Furthermore, it is investigated whether the relative standard deviations can be improved by placing the probe closer to the edge of the pads.

MICRO HALL EFFECT MEASUREMENTS

Micro Hall effect measurements are performed by placing a micro 7-point probe (M7PP) parallel and close to an insulating boundary, cf. Fig. 1, and applying a magnetic field (B_z) perpendicular to the sample surface. A series of four-point resistance values are recorded using different combinations of probe pins for current and voltage electrodes. Utilizing that the resistance measured with different combinations of probes depend differently on the distance to the insulating boundary (y_0), it is possible to estimate this distance. Once the distance to the boundary is known it is possible to determine the sheet resistance (R_0) and the Hall resistance (R_H) of the sample. From these two parameters the Hall mobility (μ_H) can be calculated as, $\mu_H = R_H/(R_0ZB_z)$, where Z is the polarity of the charge carrier. The Hall sheet carrier density (N_{HS}) can be calculated as $N_{HS} = ZB_z/(qR_H)$, where q is the unit charge [3].

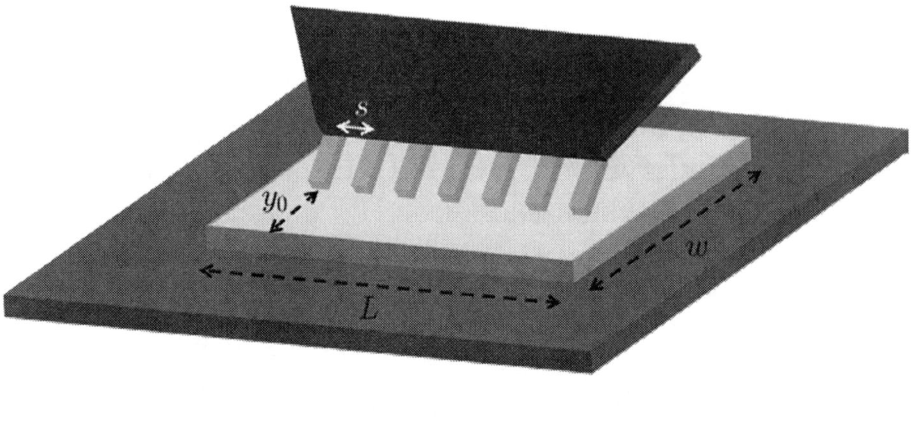

FIGURE 1. Illustration of an M7PP with 7 pins, separated by pitch s, placed on a rectangular test pad with dimensions $L\times w$, at a distance y_0 from the insulating boundary.

SIMULATION OF MEASUREMENT ERRORS

To assess the measurement precision on test pads of different sizes, Monte Carlo simulations including some of the most common sources of error, were performed. The main sources of error in micro Hall effect measurements are position errors and electrical noise. Static probe position errors are defined as displacement of contact points of the probe pins away from their ideal, equidistant positions. These position errors can be in-line and off-line. Static position errors are constant throughout a single engage, whilst using different configurations, but may change from engage to engage. Static in-line position errors can be suppressed using dual configuration measurements [4]. Further techniques for position error suppression are described in [5]. Electrical noise in the measurements will appear as random fluctuations and is described by the Signal-to-Noise-Ratio, SNR. The parameter most affected by the electrical SNR is the Hall signal, since this signal appears as a difference between two resistance measurements each with electrical noise proportional to the measured resistance.

In particular, in- and off-line errors, as well as electrical noise, were implemented in the measurement simulations. The in- and off-line errors were normally distributed with a standard deviation of $\sigma_{x,y} = 0.2s$, where s is the probe pitch, and applied to each pin in the probe. Furthermore, a SNR of 3000 was chosen, to account for electrical noise. The magnitude of these errors is based on experience, as well as previous simulations [8]. The Monte Carlo simulations were run 500 times for each parameter varied (w, y_0), using a 10 μm pitch M7PP. The results will be discussed in the following section.

RESULTS AND DISCUSSION

The results from the Monte Carlo simulations are displayed in Figs. 2 and 3. In Fig. 2, the relative standard deviation, σ_{rel}, for the sheet resistance, R_0, the Hall mobility, μ_H, and the Hall sheet carrier density, N_{HS} are plotted as a function of the width, w, of the test pad, for a pad length, $L=7s$ and a distance to the boundary, $y_0=0.4s$. As can be seen, the shape of the curves for the three different parameters is quite similar, although off-set from each other, with the sheet resistance assuming the lowest values and the Hall sheet carrier density the highest. The standard deviation remains low until the width of the test pad has been shrunk to around $w=3.5s$, at which point the geometry starts to affect the precision. In other words, it is possible, based on our simulations, to measure on test pads down to a size of $7s \times 3.5s$ without compromising the precision on the measurement significantly.

In Fig. 3, the same three parameters are plotted for the "limit" rectangular test pad size of $7s \times 3.5s$, as a function of probe-pin distance to the parallel boundary. Below y_0=0.4s, the relative standard deviation stays below 3 % for the worst case (N_{HS}), 2 % for μ_H and less than 1 % for R_0 which is acceptable for monitoring purposes. The precision can be improved on all parameters by moving the probe closer to the edge, e.g. y_0=0.2s. Moving the probe closer to the edge than 0.2s, may prove too difficult in practice and further away than y_0=0.4s will severely affect the precision and is therefore not recommended.

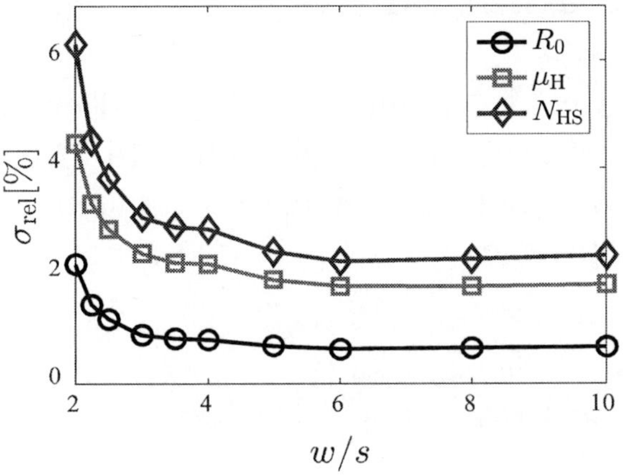

FIGURE 2. Relative standard deviation for R_0, μ_H and N_{HS} vs. width, w, of the test pad in question. The probe is placed at y_0 = 0.4s and the length, L, of the test pad is 7s. The relative standard deviation is seen to increase below w=3.5s.

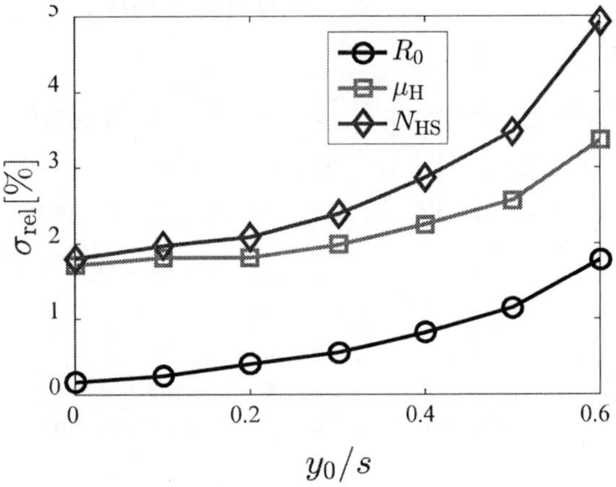

FIGURE 3. Relative standard deviation for R_0, μ_H and N_{HS} vs. distance to the edge, y_0, for w = 3.5s and L=7s. The probe should not be placed further from the edge than y_0 = 0.4s, which is the standard today, since a larger distance will lead to lower precision. Higher precision can be achieved by moving the probe pins closer to the boundary.

CONCLUSION

From this study it can be concluded that single engage micro Hall effect measurements can be performed with a micro 7 point-probe on rectangular pads down to a size of $7s \times 3.5s$ almost as accurately as on larger pads. The study also shows that if the sample parameters need to be determined more accurately, the relative standard deviation can be decreased by placing the probe close to the insulating boundary.

REFERENCES

[1] D. K. Schroder, Semiconductor material and device characterization, 3rd ed., Wiley-Interscience (2006).
[2]. S. Thorsteinsson, et al., Rev. Sci. Instrum. 80, 053902 (2009).
[3] D.H. Petersen, O. Hansen, R. Lin, and P.F. Nielsen, J. Appl. Phys. **104**, 013710 (2008).
[4] D.H. Petersen, et al., 16th IEEE International Conference on Advanced Thermal Processing of Semiconductors, RTP 2008 (2008) p. 251.
[5] F. Østerberg, et al., J. Appl. Phys. **110**, 033707 (2011).
[6] F. Wang, D.H. Petersen and O. Hansen, US Patent Appl. 20140015552 (2014).
[7] F. Østerberg, et al., *Manuscript in preparation.*
[8]. H.H. Henrichsen, et al., Precision of single-engage micro Hall effect measurements, 2014 International Workshop on Junction Technology (IWJT), pp. 1-4 (2014).

KEYWORDS

Four-point probe, Hall effect, sheet resistance, electrical characterization.

Coherent Extreme Ultraviolet Light Reveals The Elastic Properties Of Ultrathin Films

Travis D. Frazer[1]*, Jorge N. Hernandez-Charpak[1], Joshua L. Knobloch[1], Begoña Abad Mayor[1], Kathleen M. Hoogeboom-Pot[1,2], Damiano Nardi[1,2], Qing Li[1], Marie Tripp[2], Sean W. King[2], Erik H. Anderson[3], Weilun Chao[3], Henry C. Kapteyn[1] and Margaret M. Murnane[1]

[1]JILA and Department of Physics, University of Colorado, Boulder, CO 80309-0440
[2]Intel Corporation, 2501 NW 229th Ave, Hillsboro, OR 97124
[3]Center for X-Ray Optics, Lawrence Berkeley National Laboratory, Berkeley, CA 94720
**travis.frazer@colorado.edu*

INTRODUCTION

The continued scaling and improvement of nanotechnology has driven nanofabrication beyond most metrology techniques, with only more challenges to come as characteristic dimensions are reduced well below 10nm. Bulk material models' predictions no longer apply at these length scales, from thermal and elastic properties to electronic transport and beyond. This presents a unique opportunity for new metrology techniques to both inform new device design and advance the fundamental understanding of nanoscale materials. Ultrathin film systems are of particular interest, which can be fabricated with atomic layer resolution, but cannot be characterized with the same precision[1]. This impedes both technological advances, including the design, manufacture, and process control of nanoscale devices, and the fundamental understanding of how elastic properties change at the nanoscale as a function of thickness, fabrication technique, doping, or other structural changes. In this work, we present the further development of our novel nanometrology technique based on coherent extreme ultraviolet (EUV) light from table top sources that is both noncontact and nondestructive[2,3]. Using coherent EUV nanometrology, we characterize the full elastic tensor of isotropic ultrathin films down to 11nm in thickness. This technique represents a promising avenue to bridge the gap between fabrication and characterization of ultrathin films.

Most existing techniques for the measurement of elastic properties of nanoscale materials fall short when characteristic dimensions are below 100nm. Contact techniques, such as nano-indentation, can only access one component of the elastic tensor of thin film materials and struggle to decouple the effects of the substrate. Non-contact techniques based on visible light, such as picosecond ultrasonics, are diffraction-limited to hundreds of nanometers for direct measurements. In specific cases containing a high acoustic mismatch between the film and substrate material, Brillouin light scattering can characterize films down to ~20nm thick by detecting scattering between photons and acoustic phonons. However, this visible light technique is additionally limited by a challenging interpretation of the data and low intensity of the scattered light.

Coherent EUV nanometrology overcomes these limitations by using coherent EUV light from tabletop high harmonic generation sources. This light source is an ideal probe of nanoscale systems as its wavelengths and pulse durations match the inherent length- and time-scales of material dynamics at the nanoscale[3,4]. Coherent EUV light additionally offers high sensitivity to specific material properties, simplifying interpretation. Specifically, our coherent 30nm wavelength light is sensitive only to the surface profile of the sample, and is insensitive to contributions due to other transient effects such as hot electrons.

Using this technique, we characterize a series of sub-100nm a-SiC:H thin films with dielectric constants lowered through hydrogenation. The process of hydrogenation also decreases the average bond coordination in the film material. We are able to simultaneously extract the Young's modulus and Poisson's ratio of these films, and we observe, for the first time, a divergence in Poisson's ratio from the constant value assumption. For average bond coordination below a critical value of 2.5, the material transitions from brittle to ductile[5] as the Poisson's ratio increases toward incompressibility. See Figure 1. We additionally demonstrate the current capabilities of our technique by characterizing an 11nm thin film of SiOC:H, the thinnest film fully characterized to date[2].

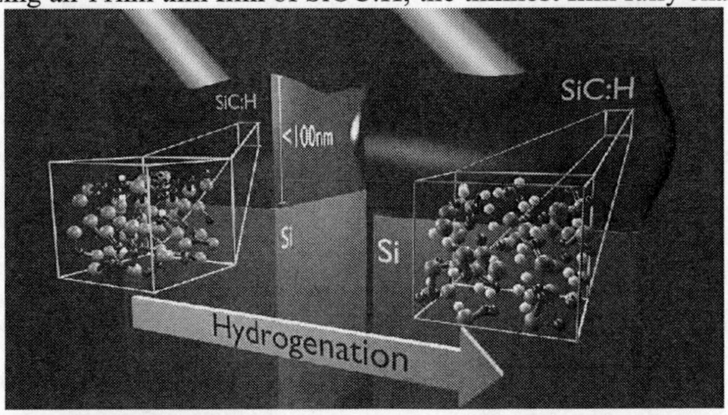

FIGURE 1. Artist's illustration of Coherent EUV Nanometrology. This study revealed increasing hydrogenation decreased average bond coordination and decreased compressibility in low-*k* SiC:H thin films[2].

COHERENT EXTREME ULTRAVIOLET NANOMETROLOGY

For our measurement, periodic gratings of nickel nanolines are deposited onto thin film samples, with periods varying from 45nm to 1500nm, and with a constant filling fraction of 1/3. These gratings act as transducers in a pump-probe experimental geometry. The 25fs pump pulse is predominantly absorbed by the metallic nanolines, causing them to impulsively expand. This excitation launches acoustic waves: longitudinal acoustic waves (LAWs) in the thin film and the nanolines, and surface acoustic waves (SAWs) in the substrate. The SAW wavelength and penetration depth are set by the period of the nanolines. With a set of different period gratings on each film, we are able to monitor both SAWs propagating mainly in the substrate and SAWs fully confined to sub-50nm thin films. This allows us to characterize thin film elastic properties independent of the substrate.

The coherent EUV probe pulses diffract from the dynamically changing sample surface onto an EUV sensitive CCD. The change in diffraction efficiency captured as a function of pump-probe delay time is sensitive to changes in the surface profile down to tens of picometers, and dynamics at sub-picosecond time scales. By monitoring the change in diffraction efficiency over time, we can extract the SAW and LAW velocities, which are related to the independent components of the elastic tensor for isotropic thin films. This method thus simultaneously extracts both Poisson's ratio and Young's modulus in a single measurement. We account for the mass loading by the nanostructures on the measured SAW frequencies using finite elements simulations. This further reduces our experimental uncertainty and produces a more general characterization of the system.

Characterization Of The Full Elastic Tensor Of Sub-100nm Thin Films

The results of our measurements on this thin film series is shown in Fig. 2. By extracting both the Poisson's ratio and Young's modulus simultaneously, we were able to observe a previously unknown trend in Poisson's ratio for these films. In previous studies, the Poisson's ratio was assumed to be constant. However, we found that this is not correct for films with Young's modulus lower than 30GPa, corresponding to films with bond coordination below the critical value of 2.5. This divergence in Poisson's ratio means the films go from more compressible and brittle, to incompressible and ductile.

In addition to this thin film series, we characterize a 10.9nm SiOC:H thin film to test the limits of EUV nanometrology. We launch SAWs in this film with periods as low as 60nm that have a penetration depth of approximately 19nm, giving a high, though incomplete, confinement of the surface wave to the thin film. Using our finite elements analysis procedure, and assuming an ideal interface between the film and substrate, we successfully extracted the Young's modulus (35GPa) and Poisson's ratio (0.39) of this film. The error in the Young's modulus (+19GPa, -9GPa) and the large allowed range of the Poisson's ratio (0-0.45) came from incomplete SAW confinement and experimental error. Future upgrades to our system will allow us to improve this measurement and determine if the departure from nominal values (20GPa Young's modulus) is caused by thickness-dependent material changes.

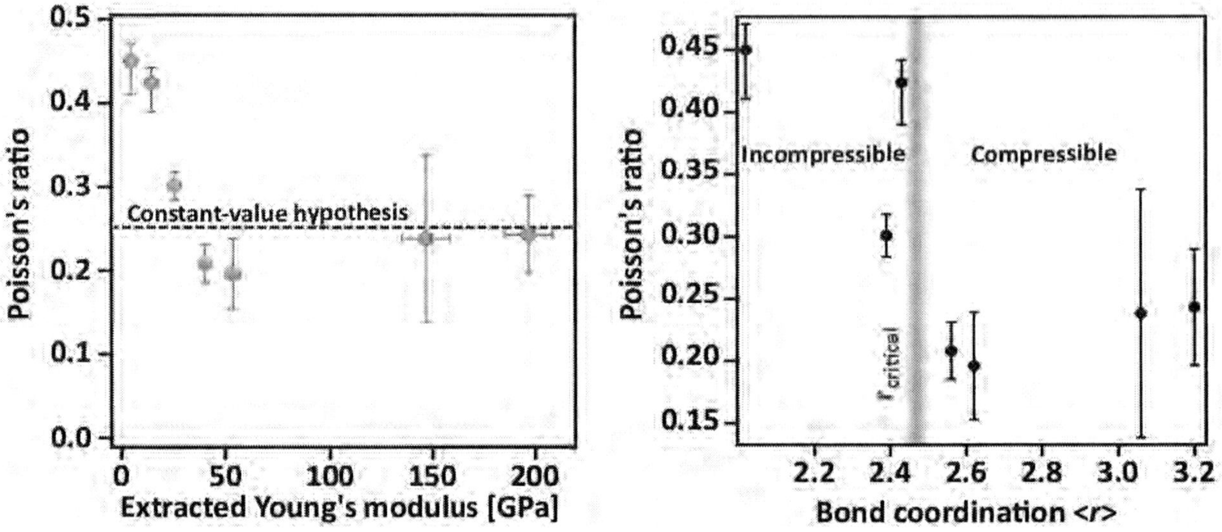

FIGURE 2. Extracted Poisson's Ratio of a Series of Sub-100nm a-SiC:H Thin Films. The Poisson's ratio versus the extracted Young's modulus (left) shows Poisson's ratio is consistent with the assumed constant nominal value for films with Young's modulus >30GPa, but for films with lower Young's modulus, the Poisson's ratio increases significantly. This corresponds to reducing the bond coordination below its critical value of 2.5 (right). A Poisson's ratio of 0 with a high Young's modulus corresponds to a compressible, brittle material, while a Poisson's ratio of 0.5 with a low Young's modulus corresponds to an incompressible, ductile material.

Beyond Elastic Properties

In addition to using this EUV nanometrology technique to extract elastic properties, we also use this same experimental design to measure the thermal transport from the nanolines to the substrate, allowing for deep nanoscale measurements of non-diffusive heat transfer. We have already discovered that the collective effect of closely spaced nanoscale heaters can return thermal transport towards the diffusive prediction[6], and we will in the future use this method to study the differences between 1D, 2D, and 3D confined heat sources, and the effect of different interfaces on thermal transport. Additional future work includes utilizing coherent diffractive imaging to reconstruct the entire dynamic 2D surface profile in phase and amplitude, characterizing anisotropic and engineered materials, and extending our technique to an EUV transient grating geometry, where we will eliminate the need for fabrication of nanoline transducers, allowing for generalization to other materials.

REFERENCES

1. S. W. King, H. Simka, D. Herr, H. Akinaga and M. Garner, *Appl. Phys. Lett. Mater.* **1**, 040701 (2013).
2. J. N. Hernandez-Charpak, K. M. Hoogeboom-Pot, Q. Li, T. D. Frazer, J. L. Knobloch, M. Tripp, S. W. King, E. H. Anderson, W. Chao, M. M. Murnane, H. C. Kapteyn and D. Nardi, *submitted* (2016).
3. J. N. Hernandez-Charpak, T. D. Frazer, J. L. Knobloch, K. M. Hoogeboom-Pot, D. Nardi, W. Chao, L. Jiang, M. Tripp, S. W. King, H. C. Kapteyn and M. M. Murnane, *Proc. SPIE* **9778**, 977801 (2016).
4. K. M. Hoogeboom-Pot, E. Turgut, J. H. Hernandez-Charpak, J. M. Shaw, H. C. Kapteyn, M. M. Murnane and D. Nardi, *Nano. Lett.* **16**, 4773-4778 (2016).
5. J. Cao and F. Li, *Philo. Mag. Lett.* **96**, 425-431 (2016).
6. K. M. Hoogeboom-Pot, J. N. Hernandez-Charpak, X. Gu, T. D. Frazer, E. H. Anderson, W. Chao and D. Nardi, *Proc. Natl. Acad. Sci.* **112**, 4846-4851 (2015).

KEYWORDS

Extreme Ultraviolet, High Harmonic Generation, Tabletop, Ultrafast, Elastic Properties, Ultrathin films

Thermoelectric Property of Silicon Nanowires Processed by Metal Assisted Directional Chemical Etching

David Rodriguez and Yong X. Gan

Department of Mechanical Engineering, California State Polytechnic University, Pomona, 3801 W Temple Avenue, Pomona, CA 91768, USA, Telephone: 909-869-2388, Fax: 909-869-4341, Email: yxgan@cpp.edu

INTRODUCTION

Thermoelectric (TE) materials offer the opportunity for direct conversion of heat into electrical power for any temperature differences. This phenomenon plays a crucial role in solving today's energy problems. Semiconducting materials are often used in such units because they have high power factors [1]. Among various semiconducting materials, silicon has been extensively studied because of the relatively low cost and high figure of merit [2]. In addition, silicon is compatible with oxide or metallic connecting electrode materials [3]. Nanostructured silicon in the form of thin film or fine pillar for thermoelectric energy conversion was investigated [4-6]. Such a nanostructured silicon is often processed by electron beam lithography [7, 8], and chemical vapor deposition [9]. Metal assisted directional chemical etching (MADCE) has recently been studied for making silicon nanowires [10]. In this work, we prepared silicon single crystal nanofibers using a silver induced self-catalysis chemical etching process. The thermoelectric energy conversion property of the silicon nanofiber has been characterized by measuring its Seebeck coefficient.

MATERIALS AND EXPERIMENTAL

The deposition of Ag nanoparticles (NPs) onto as-doped <100> n-type Si wafer pieces with the thickness of 500 micron and resistivity of 0.001~0.005 Ohm-cm was performed by chemical deposition. Before the deposition of Ag, Si pieces used were cleaned in acetone and ethanol, followed by a 1 min pre-soaking in a 4.8 M HF solution to remove any possible SiO_2 layers on the outside of the Si pieces. They were later rinsed with H_2O_2 followed by de-ionized water. Each piece was finally placed on a paper towel and left to air dry under room temperature. Chemical deposition of Ag onto the Si surface was performed on four of the five clean Si wafer pieces were chemically deposition with Ag by immersing the pieces in a solution of 0.01 M $AgNO_3$ and 4.8 M HF for 1 min. Afterward, three of the four Si pieces with Ag NPs were MADCE in a 0.2 M H_2O_2 and 4.8 M HF solution, fabricating the silicon nanowires (SiNWs). After the specimens were etched for 15, 30, and 60 mins, respectively, they were removed and post-treated with de-ionized water to wash away any remaining residue.

The Seebeck coefficient of the SiNWs/Si/SiNWs was measured at room temperature using an in-house built experimental apparatus consisting of a Talboys basic mini aluminum top hot plate, a UEi INF 165 digital infrared thermometer, and a CHI 400 electrochemical workstation. The CHI 400 electrochemical workstation was set to measure the open circuit voltages generated by the specimens for calculating the Seebeck coefficient of each sample. Figure 1 shows a schematic diagram of our experimental apparatus. One end of the sample was sandwiched in between two aluminum cylinders and the other end was free-standing, similar to a cantilever. The bottom aluminum cylinder was heated using a hot plate and an insulator block was added underneath the free-standing section of the sample to prevent any heat convection from the hot plate. Using a digital infrared thermometer, temperature measurements were taken at the start (T_H) and end (T_C) of the specimen, allowing to determine the temperature difference: $\Delta T = T_H - T_C$. This digital infrared thermometer measured the thermal radiation frequency distribution that each sample gave off. Using the measurement results, we can calculate the average temperature of the sample at an indicated point. It must be noted that the temperature measured is a combination of both the bulk Si and top SiNWs array. When T_H remained constant, an open circuit voltage (ΔV) was measured as a function of time for at least 90 seconds. The Al foil served purposely as an electrode for measuring voltage. The voltage measurement error is 0.1 mV. The temperature measurement error is around 0.1 degree in Celsius.

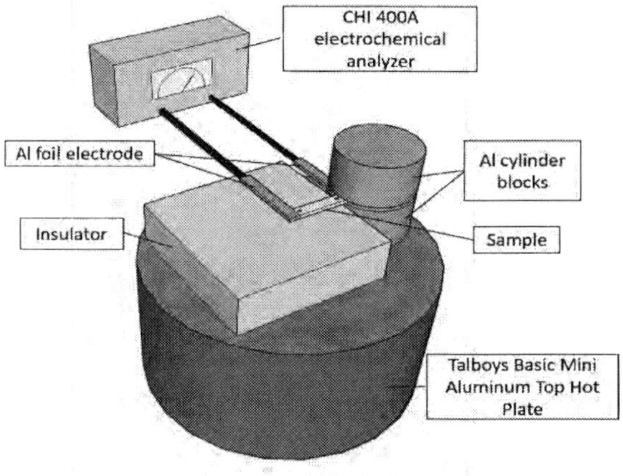

FIGURE 1. Schematic of the Self-built Seebeck Coefficient Measurement Apparatus.

RESULTS AND DISCUSSION

The Seebeck coefficient as a function of time is presented by plotting the trends of time-dependent Seebeck coefficient changes. Figure 2(a) shows the Seebeck coefficient of silver nanoparticles deposited on silicon, S_{SiAg}, with respect to time which has an average measurement of $0.09\frac{mV}{K}$ with a standard deviation of $0.04\frac{mV}{K}$. This result shows a 8-fold reduction in the S when a layer of Ag NPs, $S_{Ag} \approx 1.3\frac{\mu V}{K}$ [11], deposited onto the surface of bulk n-type silicon, $S_{bulk} \approx 0.74\frac{mV}{K}$ [12], $S_{bulk} \approx 8\ S_{SiAg}$. $S_{SiAg-15}$ and $S_{SiAg-30}$ have an average value of $0.98\frac{mV}{K}$ (standard deviation of $0.02\frac{mV}{K}$) and $0.91\frac{mV}{K}$ (standard deviation of $0.05\frac{mV}{K}$), respectively as shown in Figure 2(b) and 6(c). It is documented that increasing etching time improves the thermoelectric properties [13], contradicting our $S_{SiAg-30}$ result. The significantly lower S_{Ag} compared to S_{Si} and the Ag to Si counts on SiAg-30 sample as revealed by the EDS studies resulted in a 7% decrease from $S_{SiAg-15}$ to $S_{SiAg-30}$. Despite existing SiNWs, Ag dendrites and Ag^+ ions remained on the sidewalls and tips of the SiNWs causing the slight degeneration in *S*. Another possibility for the slight degradation has to deal with the Talboy's hot plate. The hot plate did not have the ability to keep its temperature stable, it functions by undergoing through heating and cooling cycles. Figure 2(c) shows an example of the hot plate entering a cooling cycle indicated by the downward trend of $S_{SiAg-30}$. In addition, the S_{SiAg^*} was measured (average of $0.71\frac{mV}{K}$ and standard deviation of $0.02\frac{mV}{K}$) and resulted $0.03\frac{mV}{K}$ below S_{bulk} which validates the non-existent NWs.

Figure 2(d) shows the measured $S_{SiAg-60}$ which resulted in a mean of $1.90\frac{mV}{K}$ and a standard deviation of $0.09\frac{mV}{K}$. There is a 3-fold improvement from S_{bulk}, $S_{SiAg-60} \approx 3\ S_{bulk}$ and a 2-fold improvement from SiAg-15 and SiAg-20, $S_{SiAg-60} \approx 2\ S_{SiAg-15} \approx 2\ Si_{SiAg-30}$. Table 1 shows the above S results. The improvement of electrodeposition from chemical deposition is associated with the adhesion of the nanoparticles (NPs) onto the Si substrate. In a previous study, it was found that Au, a noble metal like Ag, which was electrodeposited onto a Si substrate exhibits a much larger specific energy compare to a sputter deposition method [14]. Moreover, this same study found that the different behavior suggests that the mechanical interlocking between the Si and NPs contribute to a significant extent to the adhesion [14]. This same argument applies to this study. There existed a strong bond between the Ag NPs and the Si substrate which ensured that the proper etching process was performed. As a result, SiAg-60 produced an array of nanowires that was more uniformly dispersed as compared with SiAg-15 and SiAg-30, resulting in a higher Seebeck coefficient value, *S*.

Harvesting thermal energy more effectively needs to control the properties of thermoelectric nanomaterials properly. Energy-related nanoscience and nanotechnology research has caught significant attention [15]. Here we compare the Seebeck coefficient of Si nanowires fabricated with the assistance of Ag nanoparticles with different approaches. One approach is the self-catalyzed chemical deposition. The other is the electrodeposition method. It is observed that Si nanowires synthesized with the assistance of electrically deposited Ag nanoparticles show higher Seebeck coefficient as discussed based on the results in Figure 2. This point is meaningful for improving the thermal power of silicon nanomaterials.

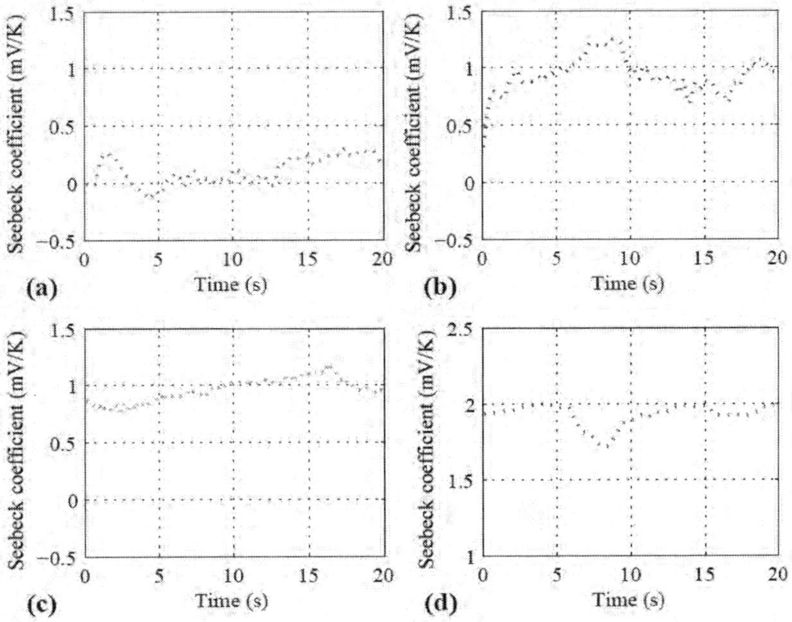

FIGURE 2. Seebeck Coefficient Measuring Results of (a) SiAg, (b) SiAg-15, (c) SiAg-30, and (d) SiAg-60.

In order to further validate this, more comparative experiments are performed. Specifically, we would like to address one of the important issues on how to decouple the effect of etching time on the Seebeck coefficient. This allows us to make sure that the improvement on Seebeck coefficient due to the use of Ag particles made by electrodeposition method is significant. The following comparative experiment results are presented in the last four rows of Table 1. The Seebeck coefficient value, S, of Si NWs fabricated by chemically deposited Ag NPs is compared with that of the sample using electrochemically deposited Ag NPs with the same etching time of 30 min. Si NWs fabricated by chemically deposited Ag NPs has the S value of 0.91 mV/K. While the Si NW sample using electrochemically deposited Ag NPs shows the S value of 1.33 mV/K. This is about 35% of increase. Similarly, Seebeck coefficient value of Si NWs fabricated by chemically deposited Ag NPs is compared with that of the sample using electrochemically deposited Ag NPs with the same etching time of 60 min. The Si NW sample using electrochemically deposited Ag NPs shows the S value as high as 1.90 mV/K. But the Si NWs fabricated by chemically deposited Ag NPs is much lower than this value. As shown in Table 1, the Seebeck coefficient of Si NW made using the chemically deposited Ag NPs is only equal to 1.07 mV/K. Again, this difference in the S values reveals the advantage of using the electrochemically deposited Ag NPs. Zhang et al. [12] found that the increasing in the etching time results in the slightly increasing in the S value. Our results as shown in both Figure 2 and Table 1 are in agreement with their claim. However, the etching time effect seems not as significant as the use of electrochemically deposited Ag NPs.

TABLE 1. Summary of Measured Thermoelectric Property Results

Sample	Ag Deposit Time	Etch. Time	ΔT (K)	ΔV (mV)	S (mV/K)
SiAg	1 min	0 min	2.0	0.18	0.09
SiAg-15	1 min	15 min	2.0	1.95	0.98
SiAg-30	1 min	30 min	1.2	1.09	0.91
SiAg-30	Electro. (2 cycles)	30 min	1.2	1.61	1.33
SiAg-60	1 min	60 min	1.8	1.93	1.07
SiAg-60	Electro. (2 cycles)	60 min	1.8	3.37	1.90

CONCLUSION

SiNWs were successfully synthesized by first depositing Ag NPs by two methods, via chemical deposition and cyclic voltammetry deposition, onto a piece of n-type Si wafer and secondly implementing the MaCE technique. The S of the bulk SiNWs/Si/SiNWs were investigated and our results show a significant increase in S when Ag NPs are electrodeposited due to the strong bond between the Ag NPs and the Si substrate, $S_{SiAg-60} \approx 3\, S_{bulk}$. Moreover, research needs to be done to determine the Seebeck coefficient, S_{SiNWs}, of samples similar to SiAg-60. Since $\Delta T = \Delta T_{SiNWs} + \Delta T_{Si}$, the values of ΔT_{SiNWs} and ΔT_{bulk} are determined only if the ratio of ΔT_{SiNWs}: ΔT_{bulk} was known. However, it is difficult to obtain these temperature differences experimentally or numerically because of the combination of heat conduction and convection, thus a finite element model will be ideal to approximate these values. Nevertheless, based on the results presented in this study, the thermoelectric performance improvement of the bulk SiNWs/Si/SiNWs, which are fabricated by using a combination of electrodeposition of Ag NPs and MADCE, is promising for the next generational thermoelectric devices.

ACKNOWLEDGEMENTS

This work is supported by the National Science Foundation (NSF) under Grant Number CMMI-1333044.

REFERENCES

1. F. J. DiSalvo, *Science*, **285**, 703–706 (1999).
2. P. Ci, J. Shi, F. Wang, S. Xu, Z. Yang, P. Yang, L.Wang and P.K. Chu, *Mater. Letters* **65**, 1618-1620 (2011).
3. Y.X. Gan and F.W. Dynys, *Mater. Chem. Phys.* **138**, 342-349 (2013).
4. D. Dávila, A. Tarancón, C. Calaza, M. Salleras, M. Fernández-Regúlez, A. San Paulo and L. Fonseca, *Nano Energy* **1**, 812-819 (2012).
5. Z.G. Chen, G. Han, L. Yang, L. Cheng and J. Zou, *Progress Natural Sci. Mater. Int.* **22**, 535-549 (2012).
6. A. Stranz, Ü. Sökmen, J. Kähler, A. Waag and E. Peiner, *Sensors Actuators A: Phys.* **171**, 48-53 (2011).
7. J.Y. Wu, C.L. Tseng, Y.K. Wang, Y. Yu, K.L. Ou and C.C. Wu, *J. Exper. Clinic. Medicine* **5** (2013) 12-16.
8. F. Wessely, T. Krauss and U. Schwalke, *Solid-State Electronics* **70**, 33-38 (2012).
9. P. Serre, C. Ternon, V. Stambouli, P. Periwal and T. Baron, *Sensors Actuators B: Chem.* **182**, 390-395 (2013).
10. X. Li, *Current Opinion Solid State Mater. Sci.* **16**, 71-81 (2012).
11. N. Cusack and P. Kendall, *Proc. Phys. Soc.* **72**, 898–901 (1958).
12. E. Krali and Z. A. K. Durrani, *Appl. Phys. Letters* **102**, 143102 (2013).
13. T. Zhang, S. Wu, J. Xu, R. Zheng and G. Cheng, *Nano Energy* **13**, 433–441 (2015).
14. J. Zhu, P. Henry, M. L. Weaver, M. L. Reed and G. Zangari, *J. Electrochem. Soc.* **160**, D507–D512 (2013).
15. X. Huang, G. Xing, Y. Li and E. Nannen, *J. Nanomater.* **2015**, 524095 (2015).

KEYWORDS

Silicon, nanostructure, metal assisted chemical etching, thermoelectric property, Seebeck coefficient.

Time of Flight Backscattering Spectrometry and Secondary Ion Mass Spectrometry in a Helium Ion Microscope - Chemical Analysis on the nm Scale

N. Klingner[1], R. Heller[1], G. Hlawacek[1], P. Gnauck[2], S. Facsko[1], and J. von Borany[1]

[1]Helmholtz-Zentrum Dresden-Rossendorf, Bautzner Landstr. 400, 01328 Dresden, Germany
[2]Carl-Zeiss-Microscopy GmbH, D-73447 Oberkochen, Germany

Helium ion microscopes (HIM) are commonly used high resolution imaging devices within several laboratories around the world. The latest generation of HIM devices (Zeiss Orion NanoFab) make use of both Helium and Neon ions, thus enabling additional possibilities for local surface modifications beside the sub nanometer resolution and an exceptional high depth of field in microcopy [1].

The image generation in a HIM is realized by evaluating the amount of secondary electrons (SE). The energy of the backscattered He or Ne projectiles was rarely taken into consideration so far. However, this energy contains information about the surface and sub-surface elemental composition and provides an additional contrast mechanism. Early attempts to measure BS energy spectra were carried out by Sijbrandij et al. [2] and gave evidence for the general feasibility but also revealed that a quantitative chemical analysis of thin layers would require the development of more sophisticated detection concepts.

In this contribution we present an experimental approach and the corresponding results of performing backscattering spectrometry (BS) in a HIM with a lateral resolution < 55 nm and an energy resolution < 2 keV [3] (see fig. 1). We show that pulsing the primary ion beam and measuring the Time of Flight (ToF) of the backscattered He/Ne enables BS in a HIM without disturbing its excellent imaging capabilities. Since our approach enables us also to perform Secondary Ion Mass Spectrometry (SIMS) by biasing the sample to a positive potential, elemental contrast in chemical analysis is further enhanced.

The combination of both techniques in conjunction with the excellent imaging capabilities of the HIM itself enables unique possibilities in the characterization of nano-electronic devices.

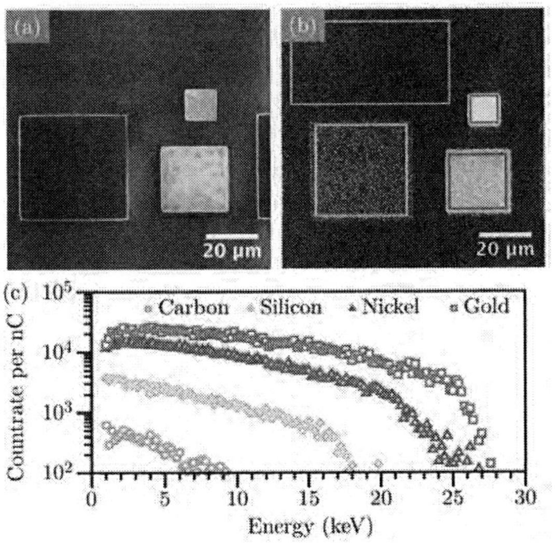

FIGURE 1: Images of a carbon sample covered with rectangular patches of Si, Ni and Au acquired in standard SE mode (a) and in ToF-BS mode (b). The color scale in (b) corresponds to the time of flight of the backscattered particles. In contrast to SE imaging this technique reveals well-defined elemental contrast. ToF-BS spectra from different regions in (b) - marked by rectangles - are plotted in (c). The color of the spectra equals the color of the rectangles in (b). The ToF-BS spectra allow a clear distinction between different elements. Taken from [3].

REFERENCES

1. G. Hlawacek, V. Veligura, R. van Gastel, and B. Poelsema, J. Vac. Sci. Technol. B **32**(2), 2014, 020801.
2. S. Sijbrandij, B. Thompson, J. Notte, B. W. Ward and N. P. Economou, J. Vac. Sci. Technol. B, **26**(6), 2008, 2103-2106
3. N. Klingner, R. Heller, G. Hlawacek, J. von Borany, J. Notte, J. Huangc and S. Facsko, Ultramicroscopy **162**, 2016, 91-97

KEYWORDS

Element mapping, backscattering spectroscopy, RBS, secondary ion mass spectrometry, SIMS, helium ion microscope, chemical analysis, laterally resolved element analysis, depth profiling, correlative spectroscopy

Atom Probe: Opportunities for CMOS Characterization

T.J. Prosa, I. Martin, Y. Chen, K.P. Rice, D.A. Reinhard & D.J. Larson

CAMECA Instruments, Inc. 5500 Nobel Drive, Madison, WI 53711 USA

INTRODUCTION

Integrated circuit and device technologies have evolved to the point of using highly advanced materials and complicated three-dimensional (3D) structures which continue to shrink as characteristic feature sizes in complementary metal oxide semiconductor field-effect transistors (CMOSFETs) shrink in accordance with Moore's Law. The functional properties depend on structure, elemental distribution, and interface roughness among other factors, at the nanometer or even the sub-nanometer scale. The distribution of just a few individual atoms, near the gate in a fin-shaped FET (finFET) or adjacent to a strained layer, often determines the performance of individual devices. As such, optimal metrology methods are needed to routinely measure these features in 3D near, or even at, the atomic scale [1].

Atom probe tomography (APT) is a time-of-flight mass spectrometry technique that also concurrently yields 3D spatial information [2]. Unique capabilities, such as detecting a low number of dopant atoms in nanoscale devices or measuring segregation at a buried nanoparticle interface, make this technique an important component of nanoscale metrology. In this presentation, we will discuss some aspects of APT relevant to CMOS characterization: FIB-based specimen preparation (controlling analysis direction), as well as some results from representative 14-nm finFET device structures and test structures.

APT CONSIDERATIONS

Modern APT analysis can require sophisticated approaches for sample preparation. Focused-ion-beam (FIB) methods are commonly used to manufacture APT specimens in the required nano-needle geometry with recipes that vary from simple to complex [2]. The standard lift-out process can be as simple as one, two, three: 1) extract wedge of material, 2) propagate material to multiple posts, and 3) sharpen. However, for the successful analysis of modern day CMOS devices, additional strategies to de-process, cap, isolate, encapsulate, fill, and/or reorient regions-of-interest (ROIs) may all become important to achieve successful analysis [3]. The overall goal is always the same: capture the ROI centered in the near-apex of a sharp tip, remove or avoid regions that are unimportant and might reduce analysis quality or yield, use capping or encapsulation to protect the ROI and control the field-of-view, and orient the ROI to maximize the opportunity to achieve desired analysis goals.

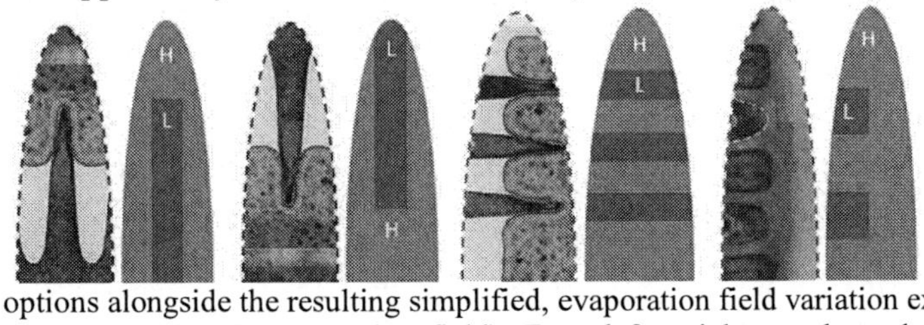

FIGURE 1. FinFET analysis options alongside the resulting simplified, evaporation field variation expected for four considered orientations with the fin as the ROI (blue=high field, gray=low field). From left to right: top-down, backside, parallel to the gate, and parallel to the fin.

The APT specimen preparation options for modern finFETs provide a good example of the various tradeoffs to be considered when utilizing this technique. The general finFET structure within an APT specimen is shown in Fig. 1. The finFET generally consists of multiple silicon fins along one direction and a series of high-k, multilayer metal gates making contact orthogonal to the fins. Each of these 3D structures is surrounded by electrical insulators that are periodically interspersed with electrical contacts that typically contain tungsten wrapped with multiple layers of other contact materials (not shown). For these small structures, it is certainly possible to include one complete repeating element within a specimen because they are so small; however, the presence of so many disparate materials and interfaces increases the chances for specimen failure and adds complexity to the reconstruction [2].

For successful analysis, one must decide in advance what aspect of the device is of interest and focus the preparation and analysis strategy around this goal. There are always tradeoffs as one considers the right strategy for a specific device. The approaches on the left in Fig. 1 may have higher survival rates (because most of the high-low-field interfaces are parallel to the analysis direction), but the reconstructions, especially near the interfacial regions, are likely to be highly aberrated for this same reason. So, choice of ROI orientation will depend on the analysis requirements.

Fig. 2 shows an example cross-section lift-out for a finFET device. Keeping track of the desired ROI during the various manipulations is essential for avoiding gallium damage and insuring proper placement of the ROI for analysis. This particular orientation might be useful for measuring composition within the inner part of the fin while also achieving higher analysis yield. The tradeoff is that the interfacial region around the fin is likely to have reduced spatial resolution because of the relative orientation of low-field fin with respect to the surrounding material.

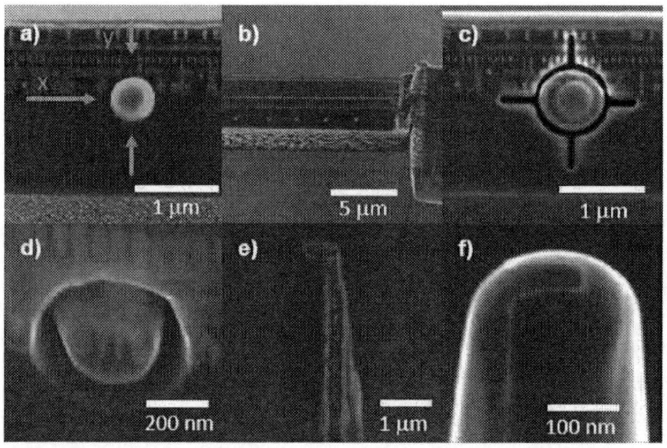

FIGURE 2. Illustration of the targeted cross-section lift-out method shown here after the first steps of the standard method have been completed (the wedge has been extracted, rotated, and the new top surface has been appropriately FIB-deprocessed). a) A fin located on a lift-out wedge is marked with Pt. b) Multiple fins are similarly marked and protected. c) Ion-beam milling is used to create additional markers. d) The wedge is aligned and the Pt is removed. The wedge is removed, capped, and then each ROI is transferred to a carrier post (not shown). e) Annular milling proceeds utilizing the ion-milled trench for proper centering. f) View of a final prepared specimen (after field evaporation showing the fin centered within the specimen tip.

CURRENT GENERATION CMOS TECHNOLOGY

One CMOS analysis example is shown in Fig. 3. This 14 nm finFET device was prepared by first taking the received device wafer and polishing down to metal-1 to reveal the various gate and tungsten contact locations. Then, cross-section lift-out was performed, analogous to that illustrated in Fig. 2 except that the orientation was parallel to the gate direction (also see Fig. 1).

The reconstructed data and 1D composition profiles through the various layers are shown in Fig. 3, highlighting the power of the technique to provide 3D compositional information. The silicon fin (gray center) is shown to be wrapped with layers of HfO, TiN, TiAlC, and TiN. As mentioned above, the spatial accuracy of APT can be limited by the evaporation field requirements of the various materials within a specimen. Here, analysis along the gate direction minimizes this effect by arranging layers of differing field orthogonal to the analysis direction. Because of this, the presence of carbon, for example, is shown to be limited to the TiAlC layer. The high-k metal stack is known to wrap around the fin and then follow along the bottom of the gate. Here, the bottom of the gate has been captured and is shown as the yellow HfO surface. As the layers wrap and become oriented parallel to the analysis direction, some blurring of the layer interfaces occurs because of layer field differences.

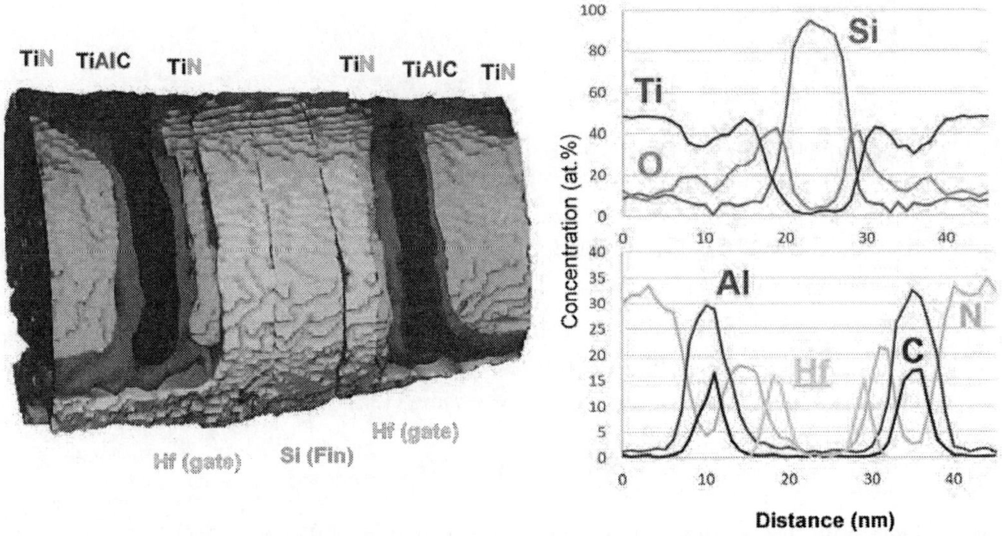

FIGURE 3. (left) Atom probe analysis of a 14-nm finFET device along the gate direction. Iso-concentration surfaces of arbitrary value were chosen, rather that the individual ion points, to clearly demark the various layers within this highly engineered structure. (right) Concentration profile of the central sub-volume of the device (gate-fin-gate). The accuracy of the measured compositions is affected by the relative field differences of neighboring layers.

Test structures are a required part of the device manufacturing process. They enable measurement of individual process steps and are part of the normal development of new technologies. For APT, simplified test structures (fewer material types with fewer variations in evaporation field) allow for faster time-to-knowledge by way of higher analysis success rates and more accurate characterization.

We have investigated two sets of test structures. The first consists of lines with patterned gate stacks that include source/drain regions. Both n-type and p-type structures were present on bulk silicon and silicon-on-oxide substrates, but lacking isolating oxides. The gates were arsenic-doped and the source/drain regions were phosphorous doped for n-type and boron doped for p-type.

The second test structure consists of an array of fins that were As-doped with plasma immersion ion implantation and processed with an oxide layer to allow a rapid thermal anneal without loss of the dopant. Measurements with multiple techniques were used to estimate the resulting dopant dose. APT estimates of dose were in good agreement with these complementary techniques.

For both examples, we utilized TEM to aid in selection of suitable APT reconstruction parameters. In this way, the localized information at the nanometer scale was accurately represented. The 3D information present in APT analysis can then be accurately extracted along any direction of interest to the analyst (e.g. parallel or perpendicular to the gate). In these structures, the dopant distributions and dose can be compared to other methods.

REFERENCES

1. M. Mayberry, *Pushing Beyond the Frontiers of Technology*, in Front. Charact. Metrol. Nanoelectron., E.M. Secula, D.G. Seiler, Eds., 2013, p. 21.
2. D.J. Larson *et al.*, *Local Electrode Atom Probe Tomography*, New York: Springer, 2013.
3. T.J. Prosa and D.J. Larson, *Microsc. Microanal.*, accepted (2017).

KEYWORDS

Atom Probe Tomography, FinFET, Microscopy

ICS X-ray Source for 3D Nano-Structured Metrology

Bradley J Nordell, Sudeep Banerjee, Grigory Golovin, and Donald Umstadter

Department of Physics and Astronomy, University of Nebraska-Lincoln, Lincoln, NE 68516
Corresponding Author: bnordell2@unl.edu

INTRODUCTION

As the semiconductor industry continues into the nanoelectronics era following Dennard scaling law, with half-pitch values decreasing below 10 nm (see Figure 1a), one of the significant challenges lies in the metrology of nanoelectronic devices and 3D interconnect architectures at these sub-nanometer domains [1]. In order to optimize 3-D in-line metrology of critical dimensions (CD) geometries for various nanoelectronic applications, industry requires a non-destructive rapid feedback monitoring system with angstrom resolution. Additionally, such monitoring systems will facilitate higher throughput and provide 3D spatial information for all aspects of an integrated circuit in order to image lithographic structures, impurities and defects—that in turn helps to diagnose and improve parasitic capacitance, mechanical failures, and signal speed for CMOS and MRAM devices. Therefore, due to high variability in critical dimensions and future pitch sizes in nanostructured devices, an all-purpose fab-ready metrology source remains a challenging problem. X-ray scattering approaches, specifically critical dimension small angle x-ray scattering (CD-SAXS), have shown great promise and success in meeting these challenges. One grand challenge for CD-SAXS is meeting both the x-ray energy and flux criteria (similar to a Synchrotron) while maintaining laboratory size. This paper will discuss the use of a laser-driven Inverse Compton Scattering (ICS) X-ray source that will not only meet the requirements for CD-SAXS but will allow for tunability necessary for meeting future semiconductor metrology requirements for a large range of applications. Lastly, this paper will emphasize various beamline designs for metrology applications (like Si FinFET 3D structures, tomography of Cu vias or buried interfaces, and pore metrology of low-*k* materials) and roadmap for an ICS X-ray source as the next-generation CD-SAXS industrial tool.

3D NANOSCALE METROLOGY NEEDS AND CD-SAXS REQUIREMENTS

The critical dimension is the smallest geometric feature for a specific technology node. These technology nodes are diverse and dependent on the scale of integrated circuit in question. According to the ITRS, such 3D architecture of nanostructured devices includes the ability to discern in-line topology of FinFET height and pitch, DRAM pitch, copper wire pitch, low-*k* dielectric thickness, and interfaces and nano-structures (like carbon nanotubes and graphene)[1,2]. Specifically, industry needs the ability to discern local features (like sidewall angle, top corner rounding, roughness, undercuts, and defects) with high spatial resolution. ITRS has also specified grand challenges in materials characterization metrology in imaging defect or fundamental pores (ranging from 1–5 nm) in low-*k* materials (see figure 1d), as well as 3D tomography of copper wire vias and defect monitoring in solder joints [1]. As seen in table 1, each of the technologies nodes or critical dimensions will range from 15–5 nm by 2023. The most problematic issue with metrology lies in the ability to have both high throughput while probing the local in-line properties with high spatial resolution at the angstrom level A clear solution is the use of x-rays through x-ray scattering and diffraction, which allows non-destructive probing at the atomic scale spacing for periodic lattices. At small diffraction angles (1-10°), the diffraction patterns allow for probing of periodic nanopattern instead of atoms in a crystal. Therefore, one optical approach that has shown great success is critical-dimension small-angle x-ray scattering (CD-SAXS). CD-SAXS uses a combination of scattering and diffraction of hard X-rays (>15 keV) to create 1D, 2D, and 3D images. The diffraction pattern from CD-SAXS is the k-space transform of the electron density distribution, which takes a 2D diffraction pattern (electron density from x-ray scattering off the three-dimensional space of nano-structure) on a screen and translates using Fourier analysis to render a complete three-dimensional image, which includes in-line periodic structures. One issue with CD-SAXS technology lies in the x-ray source. For CD-SAXS to be a usable metrology tool, it requires a source with x-ray energies greater than 20 keV, photon flux > 10^9-10^{10} ph/s, a small spot size < 100 μm, and most importantly, both fast scan time (< 1 min) and laboratory size are needed. Currently, synchrotron beamlines yield x-ray energies of 17–40 keV, with photon number of 8 x 10^{10} ph/s, spot size of 100 μm, and divergence of 60 μrad. The obvious problem with a synchrotron source lies in its size, yielding fab-ready tools improbable. Moreover, laboratory CD-SAXS sources like Mo Kα yields X-ray energy of 17.45 keV, photon number 10^6 ph/s, spot size of 300 μm, divergence of 300 μrad, but such sources suffer from lack of tunability, low brightness, and spot size [1,3]. The needs for improvements are resolution of lab source, energy greater than 20 keV for 450-mm wafers, reduce divergence, reduce spot size, increase brightness and scan time. As of yet, no CD-SAXS sources has been able to simultaneously meet all the stringent requirements.

APPLICATIONS OF X-RAY TUNABILITY OF ICS CD-SAXS SOURCE FOR 3D SPATIAL RESOLUTION OF FINFET AND LOW-K PORES

A viable solution to the CD-SAXS source problem is using all laser-based Inverse Compton scattering (ICS) x-ray sources [5, 6,]. Such sources are currently available but require greater research and development to reach optimal photon flux, lower energy resolution, angular spread, and emittance. Nonetheless, the ICS sources have the ability to create laboratory-scaled synchrotron radiation but with a much greater x-ray energy tunability at much smaller beam sizes [7]. One such source is currently available at the University of Nebraska-Lincoln, which houses the DIOCLES laser. The source consists of a femtosecond laser pulse (drive beam) focused on a plasma which accelerates electrons through wakefield acceleration mechanism; a scattering beam is then focused on the electron beam causing the electron beam to accelerate and leading to production of soft to hard x-rays. One advantage of ICS sources lies in their tunability, with the maximum photon energy $E_\gamma = 4\gamma^2\hbar\omega_L$, where γ is proportional to the kinetic energy of the electrons ($E_e = \gamma m_e c^2$) and ω_L is the frequency of the laser. Figure 1(b) represents the linear relationship between x-ray central energy and electron-beam central energy [8]. The high tunability results from the relationship between the electron-beam energy (gamma) and corresponding x-ray central energy. Figure 1(c) displays an ICS yielding 10^6 photons at 10 Hz operating single shot mode yields a photon flux of 10^7 ph/s pule per J of the laser beam.

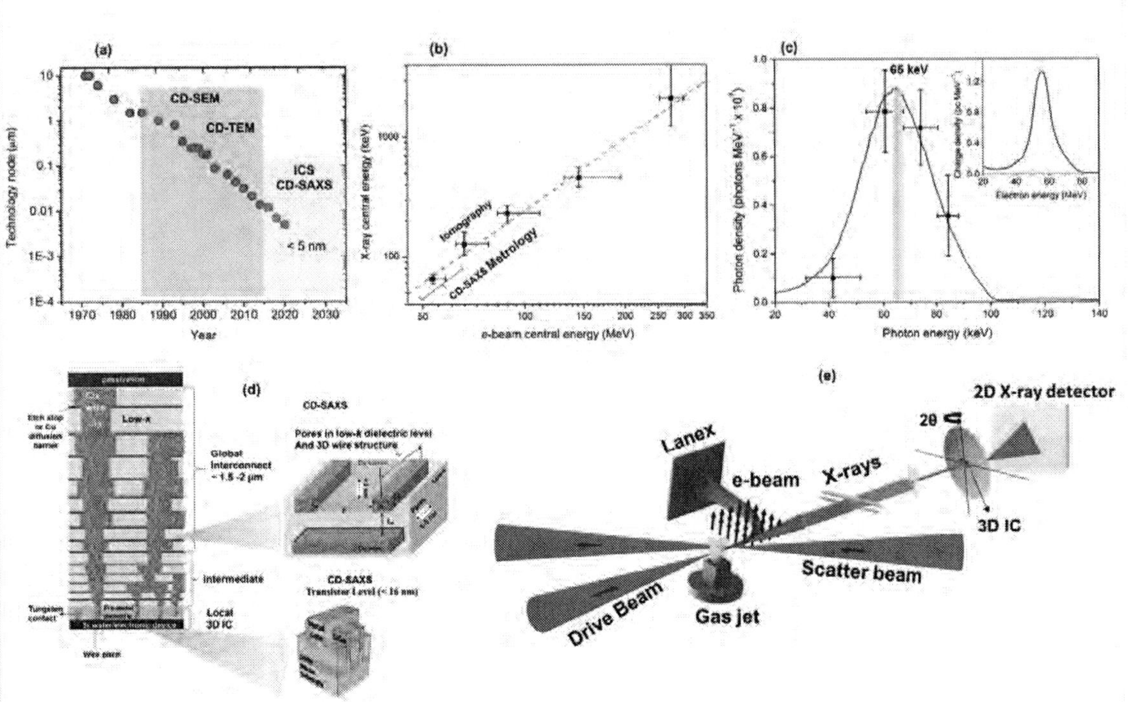

FIGURE 1. (a) Technology node versus year, indicating era of CD-SEM and future era of ICS CD-SAXS, (b) X-ray energy versus electron beam-energy [4, 6] (c) X-ray beam profile [4, 6] displaying the 65 keV band, (d) Various CD-SAXS and 3D tomography metrology applications in the integrated circuit, and (e) schematic layout of Inverse Compton scattering CD-SAXS beamline

We discuss the feasibility and design of multiple ICS beamlines with varying beam size, electron-beam energy, apertures, and laser energy to meet specific metrology requirements. For example, in table 1, one beamline consists of 20 keV with 10^{10} ph/s flux in order to image SOI FinFET, as seen in Figure (1e). To increase the photon flux for certain applications like low-*k* pore size, a second beamline may consist of a 100-mJ, 10-fs drive beam with a 1-J, 1-ps scattering beam with a Yb:YAG disk laser that can reach 1-kHz repetition rate [9]. We demonstrate that such a design can reach a line of around 10^{10}-10^{12} ph/s while simultaneously keeping the angular spread > 1 mrad and bandwidth around 2%. Another beamline will include x-ray energy >100 keV for 3D tomography and other applications to discern embedded interfaces and nanostructures [7]. Additionally, the size of the entire metrology system is 50ft^2, and this size will only decrease with future laser technology [4]. Further evolution and possible beamlines improvements of ICS sources will also be discussed, specifically in increasing photon flux and lowering bandwidth through applying new methods like coherence Thomson scattering and free electron lasers [4].

TABLE 1. ITRS 3D metrology critical dimensions as a function of time in conjunction with ICS X-ray source beamlines.

Beamline	**Nanostructure Type**	**CD [1-2] (nm)**				**X-ray Energy (keV)**	**Photon Flux (Ph/s)**	**Beam Size (um)**
		2017	2019	2021	2023			
	Si FinFET Half-pitch	19	15	12	9.5	20-70	10^{10}	< 100
1	High-k FinFET fin width	6.8	6.4	6.1	5.7	20-70	10^{8}	< 100
	Cu wire pitch	25	20	15.9	12.6			
2	Low-k pore size	5	3	1	<1	20	10^{12}	< 5
	Photoresist mask size	13	13	< 10	< 10	20	10^{10}	<10
3	3D tomography of Cu vias and sodder joints	<100	<50	<50	<25	>100	10^{12}	< 10

ACKNOWLEDGEMENTS

This material is based upon work supported by National Science Foundation under Grant No. PHY-153700 (ultra-low emittance electron beams); the US Department of Energy (DOE), Office of Science, Basic Energy Sciences (BES), under Award # DE-FG02-05ER15663 (ultrafast x-ray science); the Air Force Office for Scientific Research under award number FA9550-11-1-0157 (high-field laser-electron scattering); and the Department of Homeland Security Domestic Nuclear Detection Office, under competitively awarded contract HSHQDC-13-C-B0036 (low-dose x-ray radiography). This support does not constitute an express or implied endorsement on the part of the Government.

REFERENCES

[1] ITRS, Metrology, 2015.
[2] ITRS, Interconnect, 2013
[3] A. Vigliante and W.J. Lin, ECS Transactions, **52** (1) 865-871 (2013).
[4] D.P. Umstadter, Contemporary Physics, 1 (2015).
[5] S. Chen, N. Powers, I. Ghebregziabher, C. Maharjan, C. Liu, G. Golovin, S. Banerjee, J. Zhang, N. Cunningham, and A. Moorti, *Phys. Rev. Lett.* **110**, 155003 (2013).
[6] N. D. Powers, I. Ghebregziabher, G. Golovin, C. Liu, S. Chen, S. Banerjee, J. Zhang, and D. P. Umstadter, *Nature Photonics* **8**, 28 (2014).
[7] S. Banerjee, S. Chen, N. Powers, D. Haden, C. Liu, G. Golovin, J. Zhang, B. Zhao, S. Clarke, S. Pozzi, *et al*, *Nuclear Instruments and Methods in Physics Research Section B: Beam Interactions with Materials and Atoms* **350**, 106 (2015).
[8] W.S. Graves et al., *Physical Review of Special Topics – Accelerators and Beams* **17**, 120701 (2014).

KEYWORDS

Critical-dimension small-angle x-ray scattering, Inverse Compton Scattering, 3D nanometrology

Monitoring Accuracy And Robustness In On-Product Diffraction-Based Overlay Metrology

Arie den Boef[1], Kaustuve Bhattacharyya[1], Marc Noot[1], Aiqin Jiang[2], Martin Jak[1]

[1]*ASML Netherlands B.V., P.O. Box 324, 5500 AH Veldhoven, The Netherlands*
[2]*ASML, 320 Ushers Road, Ballston Lake, NY 12019, USA*

SUMMARY

We present progress that has been made on recipe creation in Diffraction-Based Overlay metrology (µDBO) and methods that are used to monitor the measurement quality during High Volume Manufacturing (HVM). This progress directly addresses the process robustness needs in µDBO for overlay critical layers in sub-10 nm nodes in HVM.

INTRODUCTION

Over the last couple of years µDBO has become a Tool-of-Record on many critical layers in logic and memory devices [1],[2]. In µDBO overlay is determined from an intensity difference ΔI between the +1st and -1st diffraction order of light that is diffracted by 2 overlapping gratings [3]. For small overlay numbers ΔI is linearly proportional to the overlay:

$$\Delta I = K \times OV$$

K is the overlay sensitivity and depends on the stack and measurement settings ("recipe") like wavelengths and polarization. In practice overlay sensitivity is unknown and in order to determine OV 2 biased pairs of overlapping gratings are used yielding 2 measured intensity differences $\Delta I_{\pm}=K\times(OV\pm d)$. From these 2 measured intensity differences we can determine overlay:

$$OV = d\frac{\Delta I_{+} + \Delta I_{-}}{\Delta I_{+} - \Delta I_{-}}$$

Various processing steps like etch, CMP and spin coat can introduce Bottom Grating Asymmetry (BGA) or an unequal processing of the 2 grating pairs called Grating Imbalance (GI) as schematically shown in figure 1.

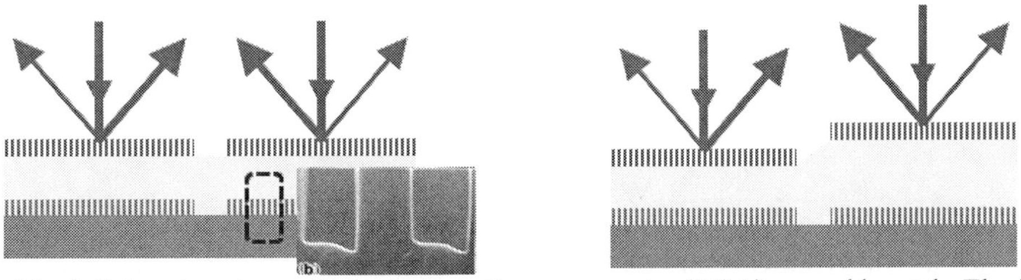

FIGURE 1. The left drawing shows the bottom grating asymmetry (BGA) caused by etch. The inset is only intended as an example to demonstrate an asymmetric grating and is taken from ref [4]. The right drawing schematically shows grating imbalance (GI)

BGA results in a small offset ΔI_{BG} in the measured intensity difference and yields an overlay error $\Delta OV=\Delta I_{BG}/K$. Grating imbalance results in a difference ΔK of the overlay sensitivity between the 2 grating pairs and the resulting overlay error is given by $\Delta OV=d(\Delta K/K)$. In practice these processing effects tend to dominate the measurement accuracy and it is therefore imperative that impact of these effects is minimized to the sub-nm level during process development but also during high-volume manufacturing (HVM). In this abstract we summarize methods that ASML has developed to deal with these sources of error:

1. selection of the best possible wavelength at setup phase
2. a method to monitor the impact of these processing effects run-time during HVM.

We will first present a simple model of the μDBO signal formation that shows the importance of wavelength in the signal quality. We will then show how this is applied to the selection of a recipe that is robust and immune against BGA and GI. Finally, we will describe methods that we have developed to monitor the quality of the measurements in a HVM environment.

SIGNAL FORMATION AND WAVELENGTH SELECTION IN μDBO

signal formation

The overlapping gratings in μDBO have a pitch P and form a pair of optically coupled gratings. The $\pm1^{st}$ diffraction orders of the top (resist) grating have an amplitude $A\exp(\pm j\alpha)$ where $\alpha=2\pi OV/P$. The $\pm1^{st}$ diffraction orders of the bottom grating have an amplitude $B\exp(j\beta)$. The phase β depends on the film stack between the gratings and the grating design. In a simplified approximation one could argue that β accounts for the time that the light from the bottom grating needs to reach the top of the stack. In such a simple model β is given by $4\pi T/\lambda$ where T is the optical stack thickness and λ the wavelength. This model is far from rigorous but it predicts the observed signals in μDBO to a level that is sufficient for practical applications. Adding the amplitudes of the top and bottom grating orders yields for the total detected intensities of the $\pm1^{st}$ diffraction orders:

$$I_1 = \left|A\mathrm{e}^{\mathrm{j}\alpha} + B\mathrm{e}^{\mathrm{j}\beta}\right|^2 = A^2 + B^2 + 2AB\cos[\beta+\alpha]$$

$$I_{-1} = \left|A\mathrm{e}^{\mathrm{-j}\alpha} + B\mathrm{e}^{\mathrm{j}\beta}\right|^2 = A^2 + B^2 + 2AB\cos[\beta-\alpha]$$

The intensity difference ΔI between the $+1^{st}$ and -1st order is $4AB\sin(\beta)\sin(\alpha)$. In practice, the overlay is much smaller than the pitch so $\sin(\alpha)\approx\alpha$ and we can write for the relation between ΔI and OV:

$$\Delta I = \frac{8\pi AB}{P}\sin\left(4\pi\frac{T}{\lambda}\right)OV$$

This expression shows a clear swing curve variation of the overlay sensitivity K with wavelength. An example of a calculated K as a function of wavelength on a multi-layer film stack is shown in figure 2.

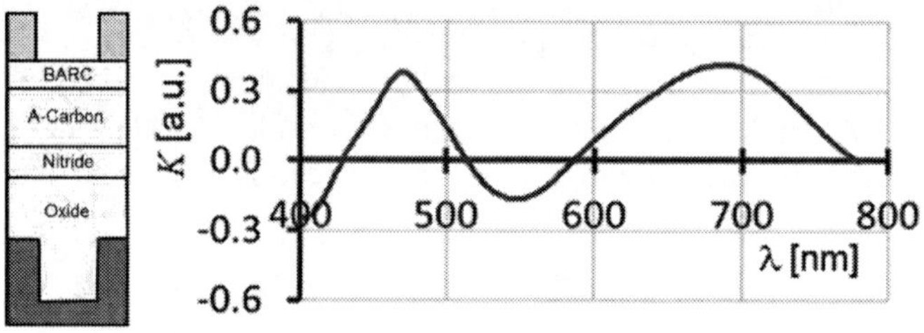

FIGURE 2. Simulated example of a swing curve

Wavelength selection

In order to minimize the BGA and GI induced error we should maximize K while also minimizing the sensitivity of K against thickness variations ΔT. For a simple model as presented in the previous section these conditions are met for a wavelength at a local extremum of the swing curve. This is one of the criteria that is currently used in the μDBO recipe selection. For recipe set-up many targets on one (or optionally more) wafer is measured for all available wavelengths. The resulting variation of K versus wavelength (the swing curve) and the variation of K across the wafer is analyzed and a wavelength where K is large and the across-wafer variation of K is small is selected as a potentially robust and accurate recipe. An example of such a measurement result is shown in figure 3.

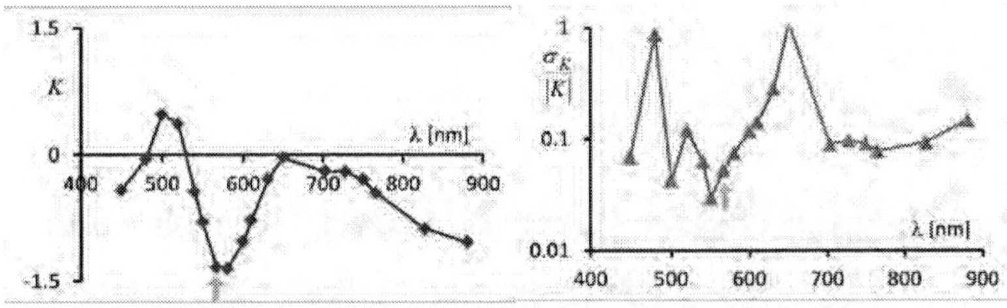

FIGURE 3. The left graph shows a measured K as function of λ on a product wafer and the right graph shows the normalized 1-σ variation of K across the wafer. The arrow indicates a wavelength with a high K and low σ_k/K

MONITORING IMPACT OF BGA AND GI

During HVM we need to be able to monitor changes in BGA and GI. BGA can be monitored by measuring the intensity unbalance ΔI_{BG} on a separate "BGA target" that only contains the bottom grating [5]. By applying the signal formation theory as outlined in the previous section we can convert this into an approximate nm-number that is of direct interest in HVM monitoring applications. Figure 4 shows some first results of this new approach where we measured the actual overlay and the predicted overlay error on a BGA target.

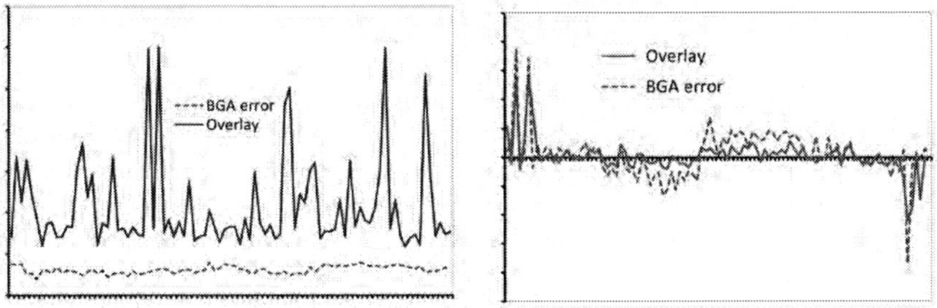

FIGURE 4. measured results of BGA-induced error (green curve) and measured overlay (blue curve) on product wafers. The left graph shows a case where grating asymmetry is quite stable (jumps in overlay are not related to BGA) and the right graph shows how a varying bottom grating asymmetry impacts the measured overlay.

For monitoring GI we use the average intensity I_m of the $+1^{st}$ and -1st diffraction order as a key performance indicator (KPI). At the optimum measurement setting (so with a large K and small σ_k/K) this KPI is independent of the actual overlay and only depends on the amount of grating imbalance. Moreover, at this optimum setting the impact of grating imbalance on the actual overlay is negligibly small. Monitoring this KPI on product wafers is currently in progress and first results will be reported in [5].

REFERENCES

1. K. Bhattacharyya et al., A study of swing-curve physics in diffraction-based overlay, Proc. SPIE 9778, Metrology, Inspection, and Process Control for Microlithography XXX, 97781I (March 24, 2016); doi:10.1117/12.2222040
2. K. Bhattacharyya et al., Holistic approach using accuracy of diffraction-based integrated metrology to improve on-product performance, reduce cycle time and cost at litho, Proc. SPIE 9424, Metrology, Inspection, and Process Control for Microlithography XXIX, 94241E (March 19, 2015); doi:10.1117/12.2085678
3. Arie den Boef, Optical wafer metrology sensors for process-robust CD and Overlay control in semiconductor device manufacturing, IoP publishing, Surf. Topogr.: Metrol. Prop. 4 (2016), doi:10.1088/2051-672X/4/2/023001
4. Marc Schaepkens and Gottlieb Oehrlein; "Asymmetric microtrenching during inductively coupled plasma oxide etching in the presence of a weak magnetic field" ; Appl. Phys. Lett. (72) 11 1293-1296
5. K. Bhattacharyya et al. SPIE 2017 to be published

KEYWORDS

overlay, μDBO, process robustness

Assessing Quantitative Optical Imaging for Realizing In-die Critical Dimension Metrology

Bryan M. Barnes, Mark-Alexander Henn, Hui Zhou, Martin Y. Sohn, and Richard M. Silver

Engineering Physics Division, National Institute of Standards and Technology, Gaithersburg, MD 20899-8212 USA

INTRODUCTION

The need for accurate critical dimension (CD) measurements as dimensions decrease for semiconductor manufacturing is now more acute with the ubiquitous challenge of reconstructing 3-D nanostructures. While optical methods are uniquely positioned as the high-throughput solution for process control due to their relative high speed and low cost, methods for sidestepping optical diffraction limits must be further advanced to meet industry needs. Our group at the National Institute of Standards and Technology (NIST) has recently demonstrated the capability of measuring finite sets of features of 15 nm-scale CD within a high-magnification imaging platform using $\lambda = 450$ nm light. This measurement technique requires a thorough understanding of the tool, rigorous electromagnetic modeling, as well as tailoring of the metrology target. This paper briefly reviews these recent accomplishments, target optimizations, and ongoing research into the challenges and extensibility of this scatterfield microscopy methodology.

QUANTITATIVE PARAMETRIC FITTING VIA IMAGING

Comparisons between experiment and simulation are made successfully in scatterometry for arrays of lines that are often larger than the incident spot size with pitches smaller than the wavelength. Much smaller line arrays (Fig. 1a) that would fit within the field-of-view of a microscope are desirable such targets will yield multiple scattering frequencies. These frequencies are inseparable within the real-space image, greatly complicating experiment-to-simulation comparisons. A new measurement approach has recently been published[1] that describes the fitting of simulation to experiment for intensity profiles obtained through-focus; two key components are highlighted here. First, the comparison is enhanced by correcting known errors from the instrument through Fourier Domain Normalization. When simulating the scattered field, the individual scattering frequencies are separable for each incident plane wave. By mapping the empirical imperfections of the tool through tool functions[2] (Fig. 1b), these known imperfections can be mapped upon each of the simulated, discretized Fourier components. By combining modified plane wave simulations to represent the finite aperture of the experiment, the resultant simulated image more closely takes on the imperfections of the experimental data.

Second, after investigating the possible Type "B" error components for the experimental data, potential error sources were determined to have correlated effects among the nominally independent measurements (the intensities at each pixel). These relationships were included within the covariance matrix (Fig. 1c) before performing the linear regression between simulated and experimental data to determine not only the best model parameter values but also their parametric uncertainties. Selected comparisons between Fourier Domain Normalized simulations and experimental profiles at several focus positions are shown in Fig. 1d.

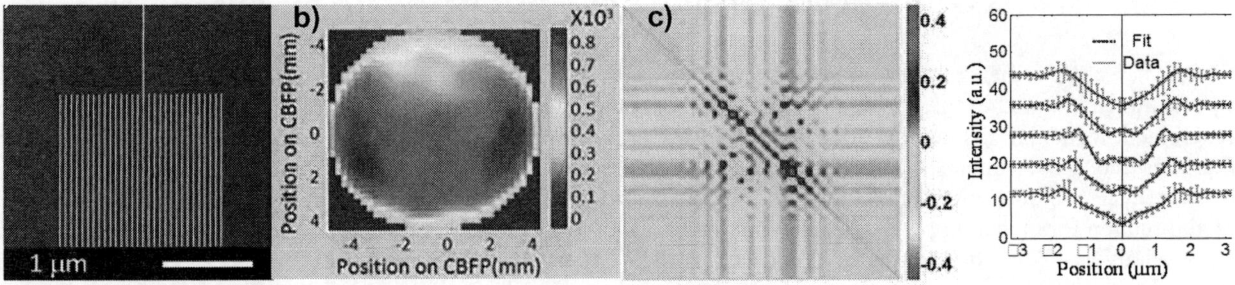

FIGURE 1. a) Scanning electron micrograph of a 30-line metrology target b) Example of an illumination path tool function (in arbitrary units) for one linear polarization. An aperture was rastered in the conjugate to the back focal plane (CBFP) of the objective lens and the light measured near the sample plane. Tool functions for the collection path were also required. c) Illustration of one part of an off-diagonal covariance matrix. d) Simplified example of theory to experiment fitting at five focus heights with multiple error bars removed for visual clarity. Actual fits were performed at 21 focus positions and two polarizations. Panels a), c) from Ref. 1, b) from Ref. 2.

OPTIMIZATION VIA SIMULATIONS

The metrology target in Fig. 1a has a patterned area of about 2 µm width and 6 µm length, excluding the protrusion for atomic force microscopy measurements. The total required area is also dependent upon the proximity of nearby objects, especially to the left and right of the target; objects should be at least 10 λ away from the sides of the target as shown in Fig. 2a. This observation is informed by simulations at $\lambda = 450$ nm for Ref. 1 of the target using rigorous coupled-wave analysis (RCWA), which implies a periodic structure. These 30-line targets were modeled as a repeating pattern of a finite set of lines surrounded by relatively vast unpatterned regions. The domain size was increased until the scattering from the target converged in amplitude, which occurred when the target was optically isolated from its periodic copies for a 10 µm domain. Simulations at 193 nm reinforce this 10 λ observation.

The specific target shown in Fig. 2a is impractically sized then for insertion into the active region of a device. Through simulation, however, optimization of the size and design of the target has been performed in addition analysis of the number of arrayed lines and number of focus measurements required to yield acceptable parametric uncertainties[3, 4]. The simulation study, unlike the fitting above, assumed a two-parameter model (height and mid-width) for simplicity and also application of a finite-element (FEM) Maxwell's equations solver to facilitate both two- and three-dimensional targets. It has been shown that using combinations of reduced line lengths and numbers of lines can potentially lead to a four-fold decrease in target area as compared to cutting-edge scatterometry targets[4]. In fact, as little as 10 lines can lead to uncertainties comparable to the ones obtained from measuring the larger 30-line target, see Fig. 2b. The restriction to less focus positions in the measurement can furthermore reduce the total time needed to acquire the necessary image data. Figure 2c shows little difference between taking images at four different focus positions and eleven focus positions. For favorable quantification of even smaller deep-subwavelength features, these optimization methods should be repeated to enable more thoughtful choices in geometrical layout and experimental design.

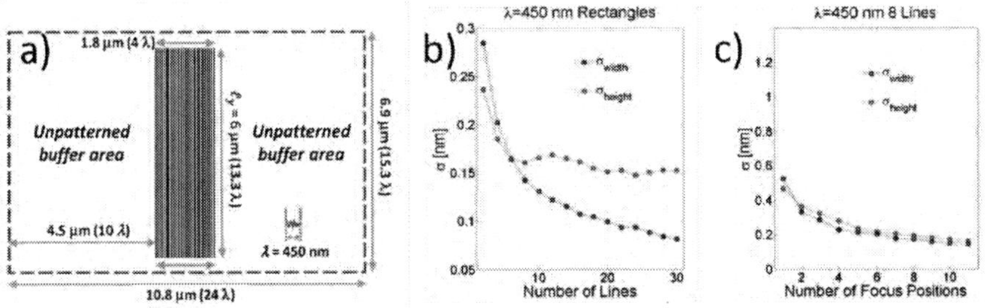

FIGURE 2. a) Initial proposed in-die target design (to scale) based upon the measured target from Ref. 1. The target has 30 lines with 60 nm pitch and line length (l_y) of 6 µm. A three-dimensional electromagnetic scattering model was used to determine the minimum line length for maintaining accuracy. b) Dependence of the estimated uncertainties on the number of lines. c) Dependence of the estimated uncertainties on the number of focus positions. Panel a) reprinted from Ref. 3.

CONSIDERATION OF POSSIBLE SYSTEMATIC BIASES

With increasing dependence upon accurate electromagnetic modeling for determining CDs, the limitations and simplifications assumed in these modeling methods become more relevant and may yield systematic biases in such measurement-to-model fits and yield erroneous parametric values. Two of the most prominent issues are the limitations due to the finiteness of the line arrays and the presence of roughness (Fig. 3a) in the measured targets. In order to reduce computation time and memory resources, arrays of lines are often assumed to be "infinite" along one spatial direction for modeling purposes, allowing the use of two-dimensional modeling codes. To test this assumption for the relatively short line lengths of proposed "in-die" targets, three-dimensional FEM-based simulations were compared against FEM-based two-dimensional solutions, shown in Fig. 3b. As shown, the simulated images diverge from the infinite model with decreasing line length, which if unchecked would lead to a systematic bias. The second source of systematic errors is the over-simplification of assuming perfectly smooth lines, whereas all actual arrays have some degree of line-edge width variation and roughness (LEWR). Simulation of such rough edges increases the computational challenges significantly, Fig. 3a shows an example of a rough grid used in simulations. The effect of the rough lines on the imaging can be observed in Figs 3c-d. A systematic bias that exceeds the expected random noise is introduced.

For both the finite vs. infinite comparison and the LEWR evaluation, two different ways to confront these problems are presented. The first one is to simply adapt the measurement set-up, i.e. choose measurement (including target) configurations for which the effects are minimal, while the second is to attempt to include resultant systematic biases into the actual modeling process.

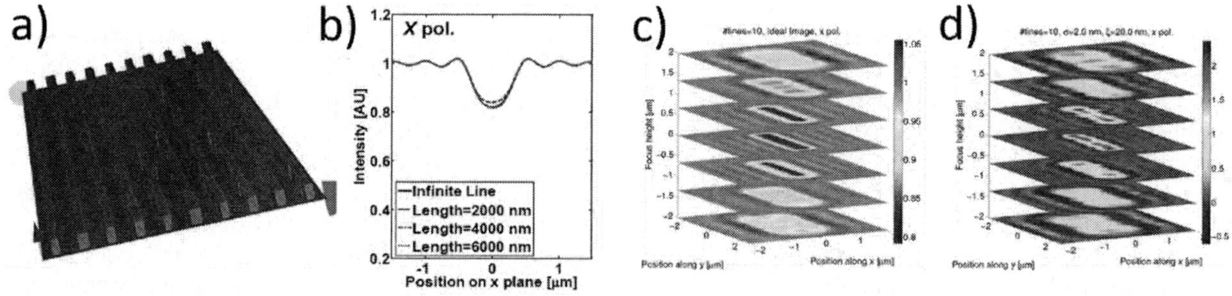

FIGURE 3. a) Single realization of a rough 10-line structure. b) Comparison between the profiles of the infinite and finite lines for three different line lengths l_y for a 10-line target at $\lambda = 450$ nm. c) Images of the ideal, i.e. non-perturbed 10-line structure at several focus positions. d) Relative difference between the image of ideal and rough 10 line structures at several focus positions.

REFERENCES

1. J. Qin, R. M. Silver, B. M. Barnes, H. Zhou, R. G. Dixson and M. A. Henn, *Light-Sci. Appl.* **5**, e16038 (2016).
2. J. Qin, R. M. Silver, B. M. Barnes, H. Zhou and F. Goasmat, *Appl. Opt.* **52** (26), 6512-6522 (2013).
3. B. M. Barnes, M. A. Henn, M. Y. Sohn, H. Zhou and R. M. Silver, " *Proc SPIE* **9778**, 97780Y (2016).
4. M.-A. Henn, B. M. Barnes, H. Zhou, M. Sohn and R. M. Silver, *Opt. Lett.* **41** (21), 4959-4962 (2016).

KEYWORDS

computational microscopy; light scattering; metrology; quantitative nanoscale microscopy; sub-nanometer uncertainties; metrology; scattering measurements; three-dimensional microscopy; linewidth; spatial frequency

Transmission and Reflection-Mode Imaging of Nanostructures with 13nm Illumination via Tabletop Ptychography CDI

Christina L. Porter[1], Michael Tanksalvala[1], Dennis F. Gardner[1], Michael Gerrity[1], Giulia F. Mancini[1], Xiaoshi Zhang[3], Galen P. Miley[2], Elisabeth R. Shanblatt[1], Benjamin R. Galloway, Charles S. Bevis[1], Nara Noeur[1], Robert Karl Jr., Daniel A. Adams[1], Henry C. Kapteyn[1], Margaret M. Murnane[1]

1. JILA, University of Colorado, Boulder, Colorado 80309-0440, USA
2. Department of Chemistry, Northwestern University, 2145 Sheridan Road, Evanston, Illinois 60208, USA
3. KMLabs Inc., 1855 S 57th Ct., Boulder, CO 80301-2811, USA

INTRODUCTION

EUV lithography is a promising candidate for reaching upcoming technology nodes in the semiconductor industry, but with this promise comes the need for reliable metrology techniques that allow mask defects to be found and repaired. In particular, there is a need for actinic mask inspection in which the imaging wavelength matches that of the intended lithography process, so that the most relevant defects are detected.

Here, we demonstrate reflection and transmission-mode ptychographic imaging of extended samples using 13.5nm illumination produced via tabletop high harmonic generation (HHG). We achieve the first sub-wavelength EUV imaging (0.9λ) in transmission mode, with the highest spatial resolution using any 13.5nm illumination source. We also present the first reflection-mode image obtained on a tabletop using 12.7nm light, as well as the first 12.7nm reflection-mode image using any source of a general, extended sample (that is, one not fabricated on a multilayer mirror).

We perform ptychographic coherent diffraction imaging (CDI)[1], a lensless imaging technique in which a beam is area-by-area scanned over an extended sample with overlap between scan positions. The overlap provides redundancy in the data that allows for robust reconstruction of the amplitude and phase of the sample and the illumination. The amplitude image provides quantitative information about the reflectivity or transmissivity—and thus the material composition—of the sample, while the phase image provides both material and height (or thickness) information.

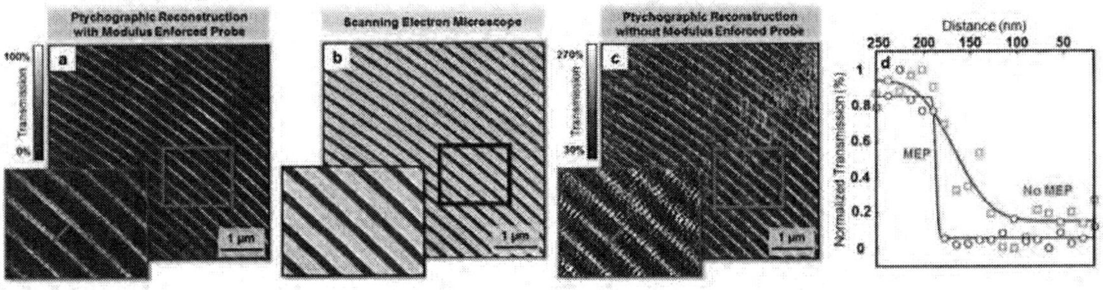

FIGURE 1. Ptychography CDI transmission image (a) reconstructed using the MEP technique, compared with an SEM image (b) and ptychography reconstruction obtained without the MEP technique (c). Lineouts (d) show 12.6nm resolution for the MEP reconstruction.

TRANSMISSION-MODE IMAGING

We generate high harmonics by focusing an 800nm Ti:Sapphire KM Laboratories Dragon laser with 2mJ pulse energy into a hollow-core glass waveguide filled with helium gas. The EUV beam is then passed through an in-vacuum iris and reflected from two ZrO_2 coated Si mirrors at high incidence angle as well as a ZrO_2 filter to reject the remaining 800nm light. The EUV beam was then reflected from two multilayer mirrors to select the 13.5nm harmonic. The second multilayer mirror was curved and focused the beam to a 2μm spot on a Fresnel zone plate sample placed normal to the beam, consisting of nearly-periodic PMMA rings deposited on silicon nitride. The sample was scanned with 121 positions collected in a rectilinear pattern, with a step size of 0.8μm. Diffraction patterns were collected during 4.25s exposures using an Andor Ikon-L, 16 bit, 2048 x 2048 pixel CCD with 13.5μm pixels, placed 22.6mm from the sample.

For this work, we used a novel new technique that we call Modulus Enforced Probe (MEP) ptychography CDI. In this method, we use an image of the undiffracted beam at the detector plane to constrain the probe guess in the sample plane. At the probe-update step in the extended ptychographic iterative engine (ePIE) algorithm[2], the probe guess is propagated to the detector plane and replaced with the measured amplitude of the undiffracted beam. This allows for faster convergence, improved robustness, and the retrieval of the quantitative transmissivity of the sample. Using this technique, we achieved 12.6nm resolution (see Fig. 1), the highest resolution image to date using any 13nm light source.

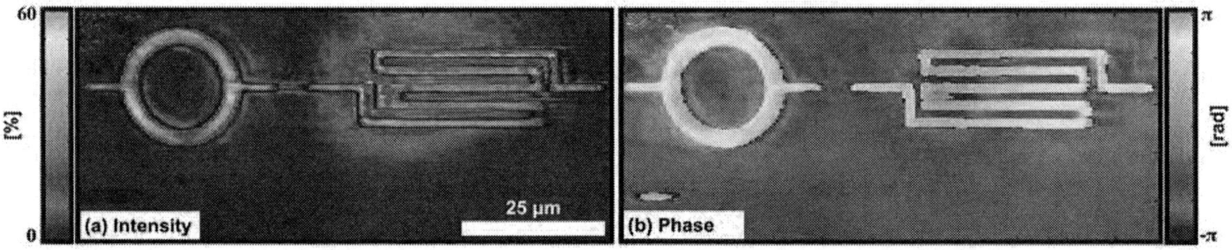

FIGURE 2. Reflection-mode ptychography CDI intensity (a) and phase (b) images of 29nm tall nickel structures on a silicon substrate.

REFLECTION-MODE IMAGING

For the reflection-mode experiment, our sample consisted of 30nm tall nickel structures deposited on silicon. To image this sample, we used a skimming incidence angle of 9.5° from the sample surface. Nearly all materials are significantly reflective to 13nm and shorter wavelengths between skimming incidence and 20°, making it important to develop high-resolution imaging techniques that work in these geometries.

The reflection-mode experiment was similar to the transmission experiment, but used a KM Laboratories Wyvern with 2.66mJ pulse energy. Also, in this case both multilayer mirrors were flat and selected the 12.7nm harmonic. The EUV beam was focused with an off-axis ellipsoidal focusing optic to a spot size of 4 x 11μm. Due to the skimming incidence angle, however, in the plane of the sample the projection of the beam was 4 x 65μm. The sample was scanned with 2μm steps using a Fermat spiral pattern consisting of 121 scan positions. Diffraction patterns were collected during 40s exposures and a sample-to-camera distance of 84.4mm.

We obtain a high resolution reconstruction by employing several novel algorithms. Because of the large size of the projected beam in the horizontal direction (65μm) compared to our sample to camera distance, our data does not fulfill the traditional oversampling criteria of ptychography (which requires that the diffracted intensity is Nyquist sampled) in all regions on the detector. In fact, because we are at high numerical aperture, we see spatially-dependent oversampling on the detector due to conical diffraction. That is, close to the DC peak the data is undersampled, but near the edges of the detector conical diffraction spreads out the diffraction peaks so that they are Nyquist sampled.

To process this data, we first upsample it before running tilted plane correction (TPC)[3], which re-maps the data onto a linear spatial frequency grid such that the fast Fourier transform may be used to propagate between the detector and sample planes. This upsampling ensures that after TPC, the Nyquist sampled data that starts out near the edges of the detector does not become undersampled when it is resampled into a narrower region of frequency space. This properly sampled data may then be reconstructed using ePIE. However, the data close to the DC peak that started out undersampled does not benefit from upsampling. Instead, we run sPIE[4] *only* on this central data, in which each detected pixel is split into several smaller ones and the algorithm solves for their values, ensuring that their sum agrees with the collected pixel value. Using a combination of ePIE on the properly sampled data and sPIE on the undersampled data near the DC, we obtain higher fidelity images (see Fig. 2) of our sample than using ePIE or sPIE alone.

We also use a technique we term "bit depth relaxation" to improve our reconstructions. Instead of imposing the ptychographic amplitude constraint at all pixels, we relax the constraint to apply only if a pixel's guessed value falls outside of some tolerance away from the measured value. This allows the algorithm to solve for a diffraction pattern less noisy than the one collected on the detector. Finally, because TPC asymmetrically condenses the diffraction pattern in the horizontal direction, we use super resolution ptychography[5] to solve for missing data on one side of our diffraction pattern.

CONCLUSION

Using advanced ptychographic algorithms, we obtain high fidelity images of general, extended samples at high NA in both transmission and reflection geometries. In transmission, we demonstrate 12.5nm resolution using a 13.5nm beam, achieving the first subwavelength EUV image of a nearly periodic sample. In reflection, we use a skimming incidence angle and obtain 77.5nm resolution in the vertical direction and 265nm resolution in the horizontal direction. Because the projection of our beam in the sample plane is very wide (65μm), we obtain a large field of view image (25 x 55μm = 1375μm^2) having scanned only a 95μm^2 area. This is the first tabletop 13nm reflection-mode image demonstrated to date. We believe that high incidence angle imaging will be beneficial for inspection of a variety of semiconductor industry samples due to its broad applicability for general, extended samples. Indeed, we believe ptychography CDI in reflection and transmission at 13.5nm is a promising method for sample metrology and actinic mask inspection that will compliment traditional metrology techniques as EUV lithography becomes the industry standard.

REFERENCES

1. J. Miao, T. Ishikawa, I. K. Robinson and M. M. Murnane, *Science* **348**, 530-535 (2015).
2. A. M. Maiden and J. M. Rodenburg, *Ultramicroscopy* **109**, 1256-1262 (2009).
3. D. F. Gardner *et al.*, *Optics Express* **20**, 19050-19059 (2012).
4. D. J. Batey *et al.*, *Phys. Rev. A* **89**, 043812 (2014).
5. A. M. Maiden, M. J. Humphrey, F. Zhang and J. M. Rodenburg, *J. Opt. Soc. Am. A* **28**, No. 4, 604-612 (2011).

KEYWORDS

Actinic Inspection, Coherent Diffraction Imaging, Ptychography, Reflection-Mode Imaging, Transmission-Mode Imaging, 13nm Imaging, High Harmonic Generation, Mask Inspection, Defect Inspection, Sub-Wavlength Imaging

An FEL source for CMOS lithography and characterization

Paul van der Heide*, Obert Wood, Erik Hosler

GLOBALFOUNDRIES, 400 Stone Break Road, Malta, NY, 12020, U.S.A
Paul.vanderheide@globalfoundries.com Ph: 518-305-6922

INTRODUCTION

With the advent of new technologies (examples include the GLOBALFOUNDRIES dual 2018 roadmap of 14 nm → 7 nm FinFET and 22FDX → 12FDX based devices) comes the requirement for enhanced characterization capabilities (both lab and fab based). This primarily concerns improved repeatability and faster turnaround times, all with the expectation of near 100% uptime and 24/7 utilization. These enhanced capabilities are generally satisfied through development of existing technologies (examples can include: further implementation of automated Transmission Electron Microscopy (TEM)/Scanning TEM (STEM) flows, faster STEM Z-contrast and Energy Dispersive X-ray (EDX) tomography, STEM based Precession Electron Diffraction (PED), X-Ray Diffraction (XRD), X-ray Photoelectron Spectroscopy (XPS), etc.) to the introduction of new capabilities (examples can include: Atom Probe Tomography (APT), Atomic Force Microscopy (AFM) based Infra-Red (IR) spectroscopy, etc.). The existing driving force behind these characterization needs comes from:

a) The reduced dimensions encountered on moving to subsequent device nodes
b) The transition from planer (2D) to 3D structures (FinFETs, 3D NAND, 3D ICs, etc.)
c) The introduction of new materials (SiGe, HKMG, along with various new barrier layers, etc.)

Future (2020 and beyond) device performance enhancements will require:

a) Examining structures supporting less than 3 degrees of freedom, i.e. films, wires, and dots, as well as structures supporting alternative information transport mechanisms such as photonics, spintronics, etc. These will require new materials which may include: organics, chalcogenides, ferromagnetics, III-Vs, etc.
b) Further developments of existing characterization capabilities (one example may include: Critical Dimension Small Angle X-ray Spectroscopy (CD-SAXS)), further developments in hybrid metrology, etc.
c) Introduction of additional characterization capabilities to those that already exist on site that allow for, among other things, a more detailed understanding of the electronic band structure around the Fermi edge (examples could include: x-ray scattering and/or adsorption, spin polarized photoemission, etc. with the valence band structure made more visible through the use of soft x-ray → UV sources)

Coincidently, the feasibility of an onsite Free Electron Laser (FEL) source to support the next generation device Extreme-UV (EUV) lithography needs has been investigated with the realization that a 13.5 nm wavelength FEL source could theoretically provide sufficient brightness to power 10 EUV lithography stations with up to 10 kW each, and with close to 100% uptime [1]. Although both linear accelerators (linacs) and Energy Recovery Linacs (ERLs) can provide the needed electron beams for injection into the undulator where FEL photons are generated, the latter provides two advantages, albeit at the cost of complexity, these being that 1) synchrotron radiation is produced from the induced deflection of electrons exiting the undulator, and 2) energy savings realized on re-injection of deflected electron beam (circular path) back into the undulator. Synchrotron radiation can then be used to support the soft x-ray source requirements of the additional characterization capabilities.

FEL-EUV SOURCE

EUV is currently expected to be inserted into High Volume Manufacturing (HVM) at the 7 nm or 5 nm FinFET nodes. This insertion is needed to contend with the increased complexity, and hence cost, realized on extending 193 lithography to ever smaller nodes, i.e. well beyond that implied by the diffraction limit (~50 nm half pitch). Surpassing this limit has required the introduction of new approaches such as: immersion lithography (193*i*), Phase Shift Masks (PSM), Optical Proximity Correction (OPC), Litho-Etch-Litho-Etch (LELE) techniques, Self-Aligned Double Patterning (SADP), etc. To move to the 7 nm FinFET node without the use of EUV lithography will likely require the extension of LELE, implementation of Self Aligned Quadruple Patterning (SAQP), etc.

The generation of effective Laser Produced Plasma (LPP) EUV sources have, however, been plagued by low power and uptime. Although improving, an alternative that involves the use of FELs for generating a coherent photon beam has been examined that can theoretically provide a sufficient brightness to power up to 10 EUV lithography stations with up to 10 kW each, and with close to 100% uptime [1]. The electron beam energy required for such an FEL would be of the order of 800 MeV. If contained within an ERL, this would provide what would be considered as a low energy synchrotron source (high energy sources are well into the GeV range).

The brightness realized from an FEL source is made possible through the fact that on passing a bunched electron beam through an undulator (a region in which the magnetic field alternates along the electron beam path), photons are produced that then seed the production of additional photons that are not only of the same energy, but are in phase with the seeding photons. This seeding process, otherwise referred to as Self Amplified Spontaneous Emission (SASE), then results in a significant enhancement in brightness relative to non-coherent sources. A conceptual design is illustrated in figure 1. Combining this with an ERL then provides additional capabilities not presently realized at an HVM site. Although this will require significant upfront costs, the longer term costs benefits along with the added capabilities (see next section) present this as a proposition of interest.

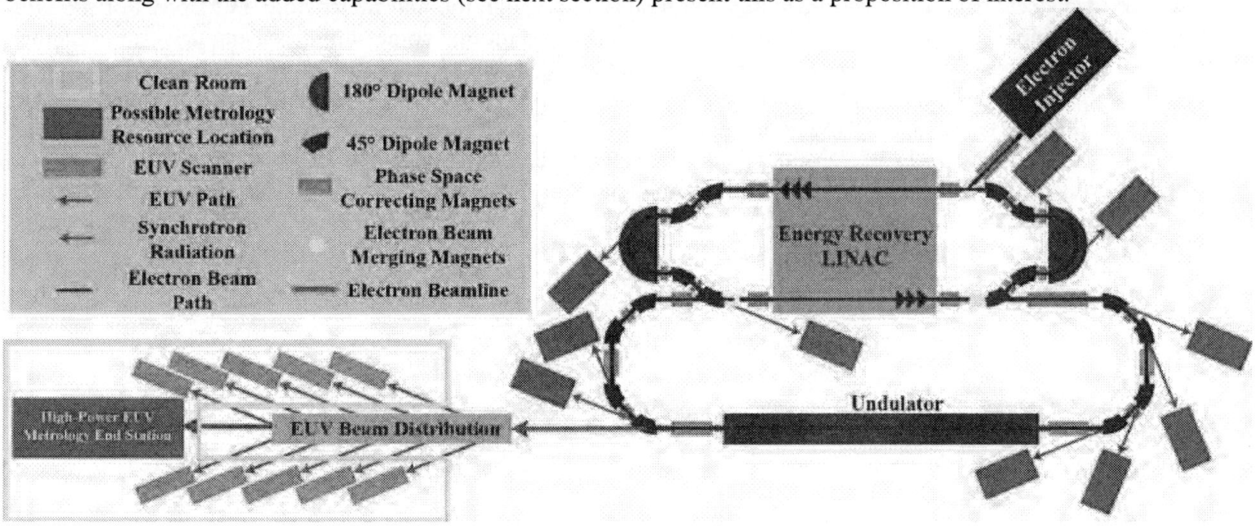

Figure 1. A conceptual schematic of an onsite ERL FEL design at GLOBALFOUNDRIES.

CHARACTERIZATION POSSIBILITIES

Although numerous characterization possibilities can be envisaged with the conceptual onsite FEL supported by an ERL at GLOBALFOUNDRIES (see figure 1)[1], two specific examples of the many possibilities are discussed. One would utilize the "High Power EUV Metrology End Station", while the other would use of the synchrotron radiation produced within the ERL section of which numerous stations referred to as "Possible Metrology Resource Locations" can exist (an arbitrary number of 10 are shown in figure 1).

For the first: A high brilliance coherent photon source of 13.5 nm (91.8 eV) presents unique characterization possibilities such as coherent EUV imaging. This could be carried out via a technique analogous to x-ray diffraction imaging, i.e. lenseless imaging using Coherent Diffractive Imaging (CDI) otherwise referred to as ptychography[2]. To avoid the introduction of additional patterning within the die area, such imaging would be carried out within a pre-specified test structure located in the scribe line. The fact that this should in principle be sensitive to pm depth scale displacements (this has been demonstrated using a High Harmonic Generation (HHG) source[3]) opens up the possibility of recording thermally induced variations induced during the EUV lithography process itself (spatial resolution would be wavelength limited). An EUV reflection based microscope[4] with no apertures/lenses situated between the sample and the detector would also ensure the fastest possible data acquisition times. The feasibility and value add of such opportunities are presently under investigation.

For the second: An ERL supporting 800 eV electrons will produce photons of energies that extend well into the VUV (10-100 eV) region. Photons of such energies are well suited to the elucidation of bonding in solids and surfaces; to the investigation of electron–electron correlations in solids. Indeed, XPS inclusive of Ultra-Violet Photoelectron spectroscopy (UPS) are techniques that utilize photons for producing photoelectrons from the surface of the aforementioned solid. Although this is more commonly referred to as Photo Electron Spectroscopy (PES) in synchrotron circles, this technique is able to provide detailed information of the bonding and the valence structure, inclusive of the band structure. Angle-Resolved PES (AR-PES) takes this to the next level as this allows for observations of the Fermi surface, i.e. provides the distribution of electrons, or more precisely, the density of single particle electronic excitations in reciprocal space from a solid's surface. AR-PES is realized by rotating the sample stage such the angle of incidence of the incoming photon (relative to the surface normal) can be adjusted. Likewise, Spin Resolved PES (SR-PES) provides for the capability of understanding the magnetization in ferromagnetic thin films. SR-PES is realized by passing the emitted photo-electrons through a Mott spin detector following energy filtering. Note: Synchrotron sources provide the added benefit that the signal of interest can be enhanced by appropriately tuning the energy of the incoming photons.

CONCLUSIONS

With the ever decreasing dimensions, increasing materials complexity, and increasing HVM costs of the latest generation CMOS devices comes greater characterization needs. With this comes an increasing drive to implement EUV lithography (expected to be inserted into High Volume Manufacturing (HVM) at the 7 nm or 5 nm FinFET nodes). EUV lithography is, however, presently plagued by multiple problems, primarily centered on uptime and brightness. To this end GLOBALFOUNDRIES has initiated a feasibility study of an onsite FEL based EUV source. Although the initial costs would be substantial, the long term benefits of multiple EUV lithography stations with near 100% uptime would serve to quickly close the financial gap. Such a source would also open up the possibility of additional characterization capabilities not typically accessible to a CMOS HVM site. Two specific examples are discussed, and there are many more possibilities.

REFERENCES

1. E.R. Hosler, O.R. Wood II, M.E. Preil. SPIE newsroom 10.1117/2.1201601.006323, (2016)
2. J. Miao, et al., Science, **348**, 6234, (2015)
3. K. Hoogeboom-Pot, et al., FCMN conference proc, 170, (2015)
4. M. Seaberg, et al., Optica, **1**, 39 (2014)
5. S. Hufner, Photoelectron spectroscopy: Principles and applications, 3rd edition, Springer, Berlin, (2003)

KEYWORDS

Keywords: FEL-EUV source, synchrotron source, AR-PES, CDI

Automated AFM for Small-Scale and Large-Scale Surface Profiling in CMP Applications

Ardavan Zandiatashbar, Byong Kim, Young-kook Yoo, Keibock Lee
Park Systems Inc., 3040 Olcott St., Santa Clara, CA, USA 95054

Ahjin Jo, Ju Suk Lee, Sang-Joon Cho, and Sang-il Park
Park Systems Corp., KANC 4F, Iui-Dong 906-10, Suwon, Korea 443-270

INTRODUCTION

With the low scaling of devices and scale up in production, metrology attracted more attention than before. The need for high functionality and value of metrology tools has been increasing. To keep up with the demand, the metrology tools are expected to improve four major specifications: precision, repeatability, throughput, and cost of ownership. In terms of precision, the tools should be compatible with the fabrication process for smaller nodes sizes and larger wafers. Therefore higher resolution has been required more significantly than before. However, higher resolution comes at the cost of throughput which includes the time required for sample preparation. As the result, lower cost of ownership while maintaining measurement repeatability has been becoming the challenges for the latest generations of high resolution metrology tools. This is true for surface characterization as well.

The chemical mechanical planarization (CMP) process is used to polish the surface of a wafer by using a physical pad and chemical active slurry, to remove topography and control the surface flatness to sub-nanometer. The CMP process has a significant role in process of shallow trench isolation (STI) and trenched metal interconnection (Damascene). However, CMP is a blind technique which makes it harder to know if desired amount of material is removed. Therefore, metrology tools such as surface profilers and AFM are required for monitoring the CMP process.

Surface characterization has been performed using two major tools: surface profilers, and atomic force microscopes (AFM). Surface profilers have been in use for a longer period of time and the tools has been used for both surface profiling and wafer stress measurement. The typical in-plane range of measurement is of few hundred mm scale with and out-of-plane range of up to 1 mm. The in-plane resolution of tens of nm and out-of-plane resolution of angstrom level are typical specifications of surface profilers. Therefore, it would be challenging for surface profilers to provide sub-angstrom surface roughness measurement over dielectric or poly-silicon, perform deep trench measurements, or detect defects with few nm in lateral dimension.

AFM systems has been introduced to and used in fabs for over a decade. Their major application has been surface roughness, step height, and critical angle measurements for monitoring etch, deposition, and CMP processes. The typical lateral dimension of below 70 um with out-of-plane dimensions of around 10 um has been measured by AFM. The noise level in the fab environment has been one of the major challenges of automated AFM systems. The systems were traditionally using piezo tube tip scanning systems. The piezo tube scanners were associated with a background out-of-plane motion that had to be compensated or filtered from the images. Tapping mode has been used as the standard imaging mode. As the result of using tapping mode, the tip life has been short and it lead to challenges in repeatability of the system.

In order to achieve the higher resolution of AFM and the larger measurement range of surface profilers, a hybrid tool, atomic force profiler (AFP), has been introduced. The conventional AFP has the associated limitations of conventional tube based AFM systems. In this work we introduce the latest in-line AFP solution that we have developed to address the aforementioned limitations associated with traditional AFMs. This system is based on decoupled XY and Z scanner design to eliminate the cross-talk between the scanners and provide better positioning capabilities. As a result of the improved positioning capability, the system is also used for automatic defect review (ADR). The decoupled XY scanner allows additional capabilities such as improved optical vision especially for wafers after CMP process and unpatterned wafers with small defects. This product has a commercial name of Park NX-WAFER and can perform measurements for 300 mm wafers and EUV reticles.

DECOUPLED SCANNERS DESIGN

To address the limitations in lateral scan range of conventional AFM and the significant out-of-plane background, the decoupled XY scanner is used in the new design. The background out-of-plane motion for the XY scanner is below 2 nm for 100 um scan range. The reduced background out-of-plane motion of XY scanner allows for surface profiling of up to 100 um ranges using only the XY scanner. Figure 1 shows an example of small-scale surface profiling using only the AFM XY scanner.

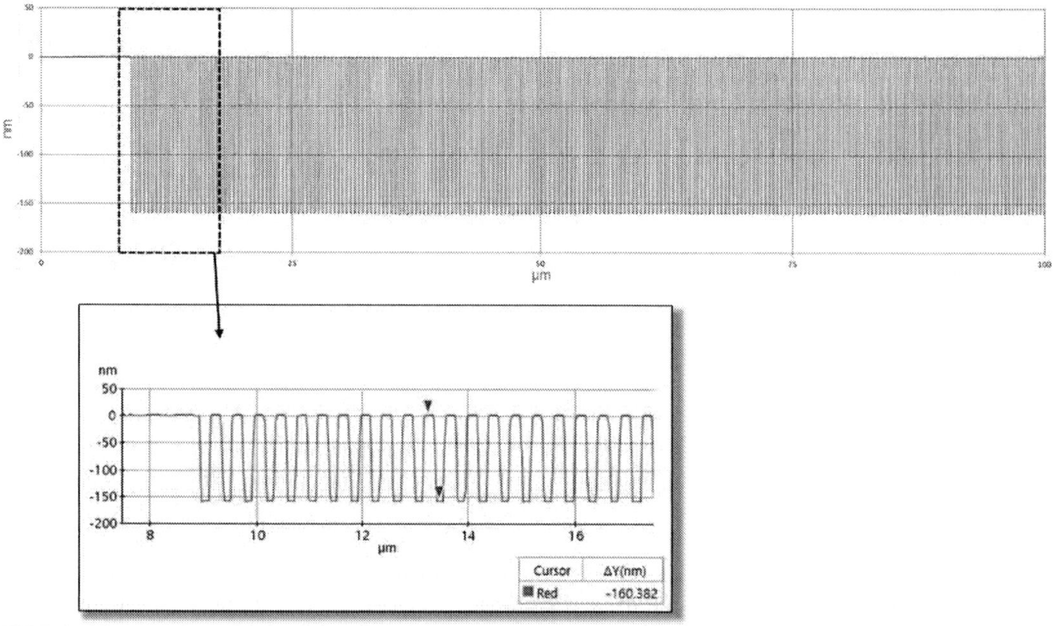

FIGURE 1. The small-scale profiling using XY scanner. The range is 100 um and the step height is 160 nm. The inset shows a magnified section of the same profile.

ENHANCED VISION

By using decoupled XY scanner, the sample can be moved accurately at nm-scale. On the other hand, the reduced volume of Z scanner (comparing to the larger volume of tube scanners) allows having an on-axis optical microscope with direct view of the sample. Hence, the frames of the sample before and after the accurate movement of its displacement by XY scanner can be recorded. Using differential frame averaging, it enables producing higher contrast comparing to standard optical images of the sample. This enhanced vision capability is highly useful when features are highly polished and are almost invisible in the optical camera. Another important application is for defect search on a blank unpatterned wafer [1].

NON-CONTACT MODE IMAGING

The advanced in-line AFP system utilizes non-contact mode for surface imaging to improve data consistency and preserve tip life. This imaging mode has been developed for inline measurement and is also a key factor in automatic defect review and surface roughness measurement capabilities of the system. An example of surface roughness measurement on a hard drive media with more than 500 repeats is shown in Figure 2. It is shown that for roughness values of 0.15 nm, the sigma of the measurement is below 2%. This is an essential requirement for tool matching and fleet management.

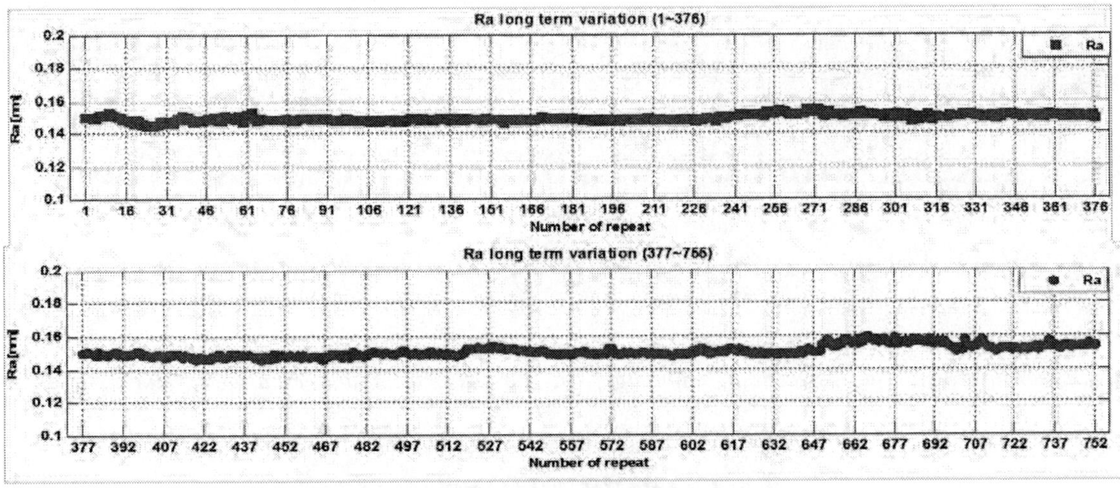

Average Roughness (Ra): 1.498 A
Long Term Variation (1σ): 0.024 A (< 2% of Ra)

FIGURE 2. The plot indicates surface roughness measurement using non-contact mode imaging on a hard drive media for 755 repeated measurements. The repeatability of below 2% is achieved..

LONG RANGE PROFILER

Long range profiling capability has become possible by addition of a slider underneath the XY scanner in the system. The slide provides surface profiling using the decoupled Z scanner over mm scale ranges. The decoupled Z scanner has a guaranteed engaging orthogonal angle with less than 0.015% deviation. Unlike tube scanners, no special algorithm is required to maintain the orthogonality of the Z scanner. Few examples of long range profiling including erosion and dishing are going to be demonstrated.

CONCLUSION

We demonstrate a new generation of atomic force profiler (AFP) using decoupled scanners. The system is capable of performing non-contact mode imaging for AFM applications such as surface roughness measurement. Enhanced vision enables the system to perform better positioning on highly polished patterned wafers and also is used in automatic defect review. The decouple scanner has minimized out-of-plane background motion and can be used for small-scale surface profiling up to 100 um. The long range profiler is used for profiling larger ranges than AFM and similar to surface profiler. AFP has been demonstrated as the valuable metrology tool in the fab.

REFERENCES

1. A. Zandiatashbar, P. A. Taylor, B. Kim, Y. –k. Yoo, K. Lee, A. Jo, J. S. Lee, S.-J. Cho, S. –i. Park, "Studying Post-etching Silicon Crystal Defects on 300 mm Wafer by Automatic Defect Review AFM", Proc. SPIE 9778, Metrology, Inspection, and Process Control for Microlithography XXX, 97782P (March 8, 2016); doi:10.1117/12.2220369

KEYWORDS

AFM, AFP, Non-contact, Profiler

Characterization Of A Lab Based CD-SAXS Tool

W. D. Thompson[1], R. Joseph Kline[2], and Osman Sorkhabi[1]

1. Lam Research Corporation,4650 Cushing Parkway, Fremont, CA 94538
2. National Institute of Standards and Technology, 100 Bureau Drive, Gaithersburg, Maryland 20899

INTRODUCTION

In accordance with Moore's Law, the reduction of critical dimensions(CDs) and increased densities of semiconductor devices continue to push the limits of current nanofabrication methods. Novel techniques for overcoming the limitations of standard 193 nm immersion lithography include multiple patterning techniques such as self-aligned double patterning (SADP) and self-aligned quadruple patterning (SAQP). These multiple patterning techniques can achieve CDs and pitches far below that of standard optical lithography, but are subject to compounding detrimental effects, such as pitch walk. Metrology techniques for monitoring the CDs and pitches of future technology nodes would be limited by current techniques such as optical scatterometry (OCD), which is reaching its fundamental limits, and CD scanning electron microscopy(CD-SEM), which only samples small areas and has difficulty producing detailed 3D information. Three dimensional structural information attained through destructive cross sectioning transmission electron microscopy(TEM) and SEM methods, are not practical for in-line metrology. A non-destructive, relatively large area, 3D metrology technique that shows promise for meeting future metrology needs is the CD small angle x-ray scattering (CD-SAXS) technique.[1] The technique has previously been shown to be capable of resolving CD, pitch, and pitch walk with sub-nanometer precision.[2] Previous reports on the results of the CD-SAXS technique have been generated from measurements performed at large synchrotron x-ray sources, but newly developed compact high brightness x-ray sources have made the development of viable lab based tools possible. A lab based CD-SAXS tool capable of providing the quality of information previously demonstrated by the CD-SAXS technique would be invaluable for satisfying the future metrology needs of the semiconductor industry. In this work we report on the results of continued efforts to calibrate and characterize a lab based CD-SAXS tool.

CHARACTERIZING THE LAB BASED CD-SAXS TOOL

For the CD-SAXS technique it is desirable to maximize the x-ray flux through the sample area, while minimizing the x-ray beam spot size at the sample and the divergence of the x-ray beam. Optimizing each of these properties is nontrivial and it is necessary to understand how each might be varied, as well any interdependencies that may exist between them. A liquid metal jet x-ray source[3] was used in this work, operated using indium Kα x-rays at 24.2 keV. X-rays were generated using a variety of electron beam sizes, accelerating voltages, and powers. The CD-SAXS system is also equipped with two sets of horizontal and vertical variable aperture GaAs scatterless slits. The resulting x-ray beam divergences, intensities, and spot sizes at the sample were examined for a variety of electron beam settings and aperture configurations.

X-Ray Beam Characterization Results

In order to determine the x-ray beam divergence, the x-ray beam spot on the detector was imaged at several x-ray source to detector distances. At each distance, the sets of horizontal and vertical slits were varied to better understand the dependence of the divergence on aperture size. An image of the x-ray beam on the detector at a source to detector distance of 418 cm is shown in Figure 1. Images such as this were analyzed to determine the full width at half max(FWHM) and to gather intensity information as a function of electron beam and aperture configuration. The resulting data were compiled, as in Figure 2., and the beam divergence was determined for each set of conditions.

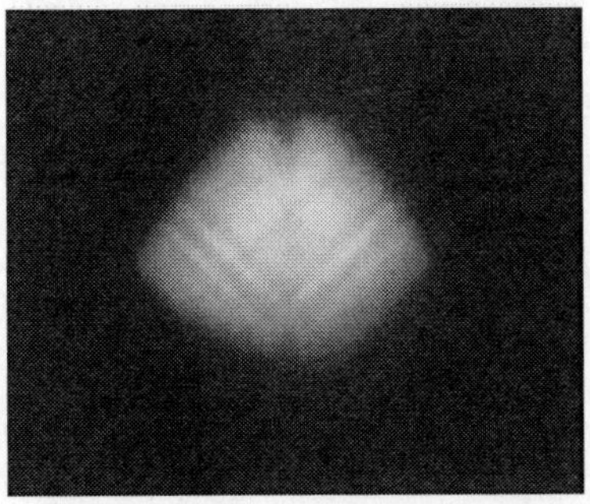

FIGURE 1. Image of the x-ray beam at a source to detector distance of 418 cm.

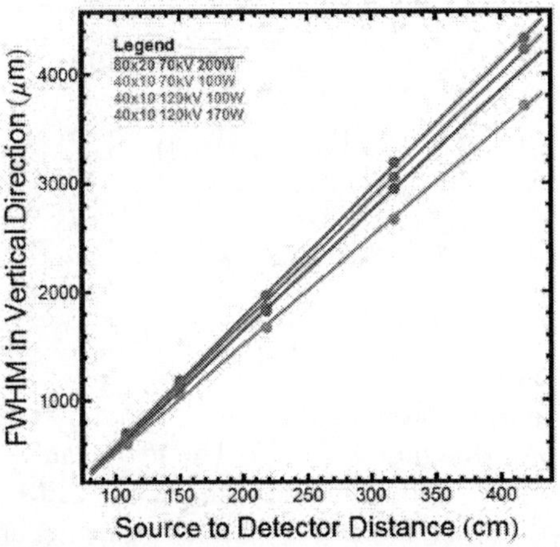

FIGURE 2. The x-ray beam size vs. the source to detector distance for a variety of electron beam configurations, from which the beam divergence was determined.

The x-ray spot size (FWHM) at the sample position was measured for a variety of electron beam and slit configurations. The x-ray beam intensity in the region of the FWHM was also determined. These data allow for maximizing the x-ray flux through the sample during the CD-SAXS measurement, when there are constraints presented by a limited sample target area.

In conclusion, significant progress is being made in characterizing a lab based CD-SAXS tool and is providing information that will aid in optimizing future CD-SAXS measurements on semiconductor samples. Examples of how these results aided in performing CD-SAXS measurements on silicon gratings will be presented.

REFERENCES

1. Jones, R. L.; Hu, T.; Lin, E. K.; Wu, W.-L.; Kolb, R.; Casa, D. M.; Bolton, P. J.; Barclay, G. G. Small Angle X-Ray Scattering for Sub-100 Nm Pattern Characterization. *Appl. Phys. Lett.* 2003, 83, 4059.
2. Sunday, D. F.; List, S.; Chawla, J. S.; Kline, R. J. Determining the shape and periodicity of nanostructures using small-angle X-ray scattering. *J. Appl. Cryst.* 2015, 48, 1
3. Hemberg, O.; Otendal, M.; Hertz, H. M. Liquid-Metal-Jet Anode Electron-Impact X-Ray Source. *Appl. Phys. Lett.* 2003, 83, 1483.

KEYWORDS

CD-SAXS, 3D Metrology

Quantum Noise Effects in e-Beam Lithography and Metrology

T. Verduin, S.R. Lokhorst, C.W. Hagen and P. Kruit

Delft University of Technology, Dept. Imaging Physics, Lorentzweg 1, 2628CJ Delft, The Netherlands
e-mail: c.w.hagen@tudelft.nl

Increasing the throughput in lithography, by using more sensitive resists and low exposure doses, also increases the Line Edge Roughness (LER). The main cause for this is quantum noise (or shot noise). LER is usually measured using top-down SEM images using a CD-SEM (Figure 1). To theoretically study the effect of shot noise on LER we developed a fast simulator of electron-matter interaction. A novel power spectral density based method is used to accurately determine the line edge positions and LER from very noisy top-down SEM images without the need for filtering. This approach allows very low-dose SEM imaging of resist patterns, thereby reducing the risk of resist shrinkage during imaging [1].
Top-down SEM images were simulated of 3D-resist patterns on silicon with a predefined side-wall roughness (SWR) (Figure 2). We then compare the LER as obtained from the top-down SEM images to the standard deviation of the 3D sidewall roughness σ-3D, and conclude that LER is smaller than σ-3D, i.e. LER underestimates the 3D-SWR [2].
The same simulator is also used to release acids from photo acid generators (PAG's) in a chemically amplified resist (CAR) [3]. The acids are allowed to diffuse during a post-exposure bake, and a threshold is set to determine the boundary between exposed and unexposed resist. In reality, there is also a development phase, which we so far have ignored in this study. We acknowledge that this is a simplified view of post lithographic processing. The exposed resist gives rise to a three dimensional feature (Figure 3), the LER of which is then analysed using the SEM-image simulation as described above.
We will discuss the electron-matter simulator and how it is used to obtain the results mentioned above.

REFERENCES

1. T. Verduin, P. Kruit, C.W. Hagen, *J. Micro/Nanolith. MEMS MOEMS* **13** (2014) 033009
2. T. Verduin, S. R. Lokhorst, P. Kruit, C.W. Hagen, *Proc. of SPIE* **9424** (2015) 942405-1
3. T. Verduin, S. R. Lokhorst, C.W. Hagen, P. Kruit, *Proc. of SPIE* **9778** (2016) 97781Z-1

KEYWORDS

Lithography, metrology, Monte Carlo simulations, Line edge roughness

FIGURE 1. A typical top-down SEM image of line edges, recorded by J. Jussot from CNRS-LTM/CEA-LETI using a Hitachi CD-SEM. This image is recorded with a probe current of 6.0 pA at an acceleration voltage of 500 V. The number of integration frames is 16. The magnification is such that the field size is 2.76 μm in length (1024 px) and 450-nm wide (1024 px). The resulting pixel size is approximately 2.7 nm × 0.44 nm.

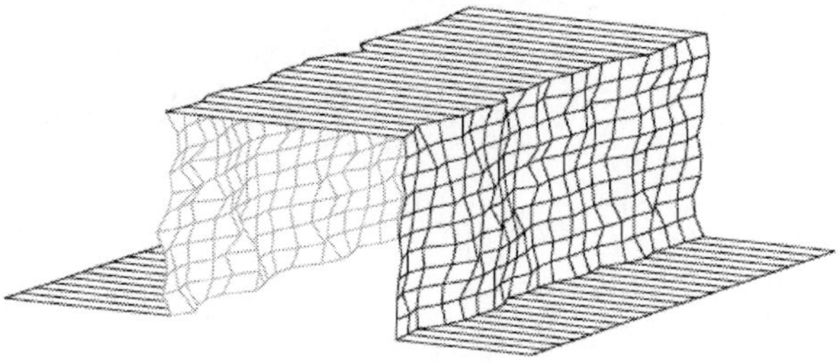

FIGURE 2. A randomly generated rough line of which a small section is shown here. At first, a flat line with a length of 2 μm, width of 32 nm and a height of 32nm is constructed. The sidewalls of the flat line are replaced with randomly generated rough surfaces. Note that the top of the line and the substrate, on which the line is located, remains flat.

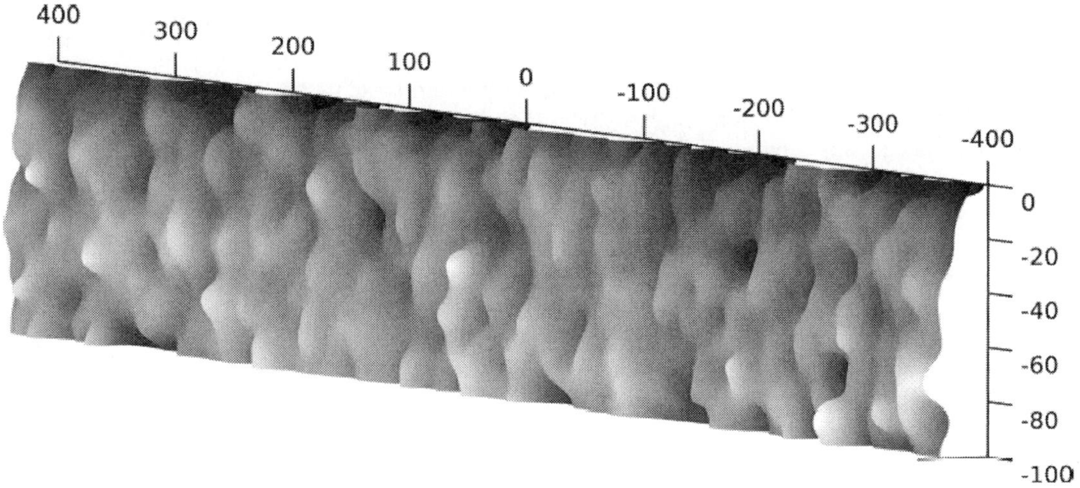

FIGURE 3. A three dimensional view of the boundary between exposed and unexposed resist. The surface is obtained from a simulated exposure of a 100 nm thick layer of CAR on an infinitely thick silicon substrate, assuming a Poisson distributed exposure dose of 40 $\mu C/cm^2$.

High Resolution EBI for Pattern Fidelity Monitor

Fei Wang, Brian Lee, Kevin Liu, Jack Jau

HMI, an ASML Company: 1762 Automation Parkway, San Jose, CA, 95131

INTRODUCTION

As design rules continue to shrink down to sub-10nm, multiple patterning technologies have been intensively implemented in order to transfer fine features from design to wafer. Pattern fidelity is becoming more and more important in determining process window and device yield. Traditionally, CD measurement by CD-SEM was used to characterize pattern fidelity, especially for one-dimensional pattern. However, due to limited throughput, CD-SEM could not meet the increasing requirement on the volume of measurements, both from research and development phase as well as from process monitoring stage. A high speed metrology method is needed for pattern fidelity characterization and monitoring. In this work, we will study the metrology capabilities of an advanced e-beam inspection (EBI) system, which can provide order of magnitude throughput improvement relative to CD-SEM technology. High speed metrology can be used for OPC model calibration and verification in research and development phase; it can also be used for pattern fidelity monitoring during production phase.

METROLOGY CAPABILITY OF ADVANCED EBI

Over the past decade, image resolution of e-beam inspection systems has continuously improved and image quality has become sufficient to perform high precision measurement. In this work, HMI's eP4, a 2 nm resolution e-beam inspection system, was used to evaluate the metrology capabilities and to study feasibility for pattern fidelity measurement.

Study of Measurement Sensitivity

Measurement sensitivity is a critical spec, which defines the minimal CD change a metrology system could robustly measure. In order to meet requirements for sub-10nm technology node, a metrology system needs to capture 0.25nm CD change. A special-designed wafer with known pitch CD change was used to quantify measurement sensitivity in the experiment.

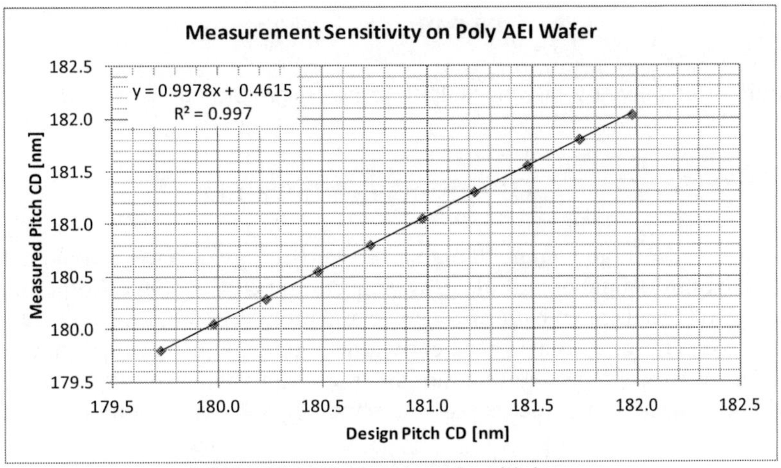

FIGURE 1. Measurement Sensitivity

Study of Contour Extraction on Complex 2D Pattern

To characterize a two-dimensional (2D) pattern, the information carried by simple cutline CD becomes limited. Edge placement (EP) information, which can be measured on contour, provides a solution for pattern fidelity study. In this work, we will study contour extraction from high resolution SEM image scanned by eP4 system. A basic work flow includes: image scan on eP4, align SEM image with design data, contour extraction, and edge placement measurement.

FIGURE 2. Contour Extraction from SEM Image

Cutline CD defined on contour was used to benchmark measurement results from eP4 against CD-SEM. High correlation can be achieved for both horizontal and vertical CD.

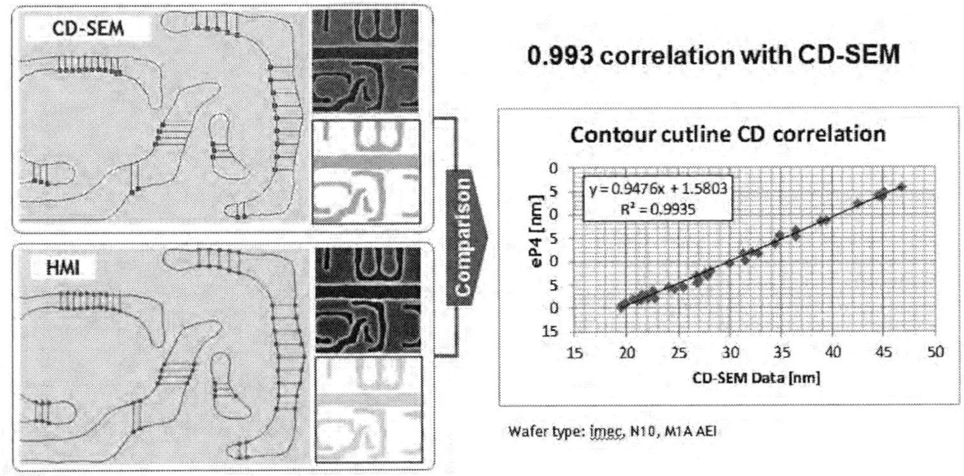

FIGURE 3. Benchmark with CD-SEM of cutline CD measurement on contour

APPLICATIONS

High speed metrology enabled by advanced EBI brings desirable capabilities for applications in R&D and production phase.

OPC Model Calibration

OPC model calibration needs large volume of high quality metrology data to get good model accuracy and shorten iteration cycles. Limited by throughput of metrology system, current OPC model calibration utilized tens of thousands of gauges, which can't provide sufficient pattern coverage and process condition coverage. Moreover, limited repeat measurements on the same pattern type can't efficiently reduce random process variation, which caused OPC mode wrongly compensate for the process variation but not the true mean value. To overcome the difficulties, high speed metrology system will be utilized to provide massive measurement data, in the volume of million gauges.

Pattern Fidelity Monitoring

Pattern fidelity monitoring (PFM) is very important element of ASML's holistic lithography. Defects library, which is observed by Lithography Manufacturability Check (LMC) prediction from OPC or wafer inspection result, provide high potential defects hotspots. With those hotspots and wafer topography and focus information which are observed by scanner and YieldStar, computation engine can predict high risk patterns and their wafer positions. High throughput EBI can measure pattern fidelity of these patterns then give feedback to user to judge pattern health. It's very helpful on yield enhancement.

SUMMARY

Foe leading edge technology, pattern fidelity is becoming an important factor limiting processing window and device yield. A high speed metrology provides capabilities of characterizing and monitoring large volume of critical patterns. New applications can be developed for OPC model calibration in R&D phase and pattern fidelity monitoring in production phase.

KEYWORDS

e-beam, massive measurement, pattern fidelity monitor, die to database, contour

In-line 3D AFM for Critical Dimension and Sidewall Roughness of Si Photonic Waveguide and Correlation with Its Propagation Loss

T.-G. Kim[1*], P. Verheyen[1], P. De Heyn[1], T. Vandeweyer[1], A. Miller[1], M. Pantouvaki[1], J. Van Campenhout[1], A.-J. Jo[2], S.-J. Cho[2], S.-I. Park[2]

[1]*Imec vzw., Kapeldreef 75, Leuven, B-3001, Belgium*
[2]*Park Systems, KANC 4F, Gwanggyo-ro 109 Suwon, 16229, South Korea*
**Tae-Gon.Kim@imec.be*

INTRODUCTION

Silicon photonics is considered to be a key potential technology to enable the scaling of bandwidth and power density beyond the bottleneck of electrical interconnects [1]. As an important building block of the Silicon Photonics circuitry, modulators are required to have low power consumption, high modulation speed, small footprint, large optical bandwidth and robust thermal sensitivity [2]. In this paper, we investigate the characterization of atomic resolution critical dimension of Si waveguide and its sidewall roughness using in-line 3D atomic force microscope (AFM) and correlate with their propagation loss properties in order to understand their correlation and to evaluate in-line 3D AFM capability for Si waveguide process monitoring.

SAMPLE AND MEASUREMENT TECHNIQUE

Strip Si waveguides, has a nominal width of 450 nm were manufactured on 300 mm wafers were used for this study and their height, sidewall slope and width at heights of 20, 50 and 70% from bottom as shown in Fig. 1.

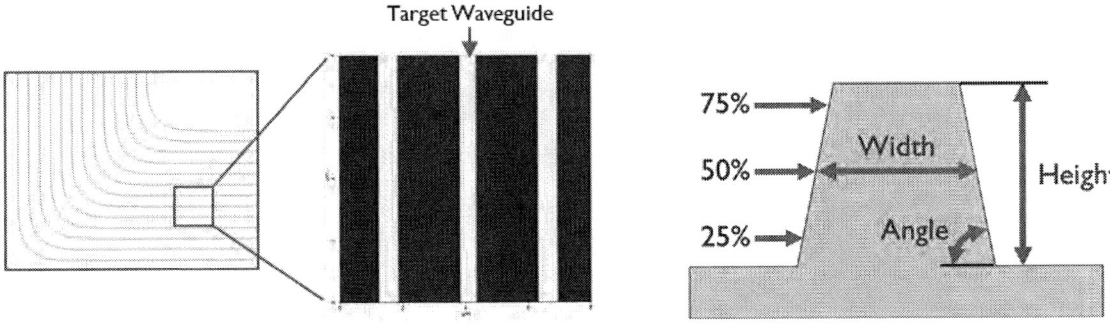

FIGURE 1. Schematic diagram of measured structure for the study

The measurements were done by in-line 3D atomic force microscope (NX-3DM, Park Systems). The AFM system is built based on a decoupled XY and Z scanning system which allows to tilt Z scanner independently by ±18 and ±39 degrees [3]. Tilted Z scanner design allows AFM probe to access the sidewall of structure to measure not only the sidewall roughness of structure, but also their critical dimension, sidewall slope and width at different height by stitching three measured data at three different angles for example, 0, -38, +38 degrees as shown in Fig. 2 [4]. Sidewall roughness measurement were performed on both pattern sidewalls at the tilted angle of ±38 degrees. AFM probe was scanned along the waveguide in order to get sufficient data acquisition for the sidewall roughness evaluation.

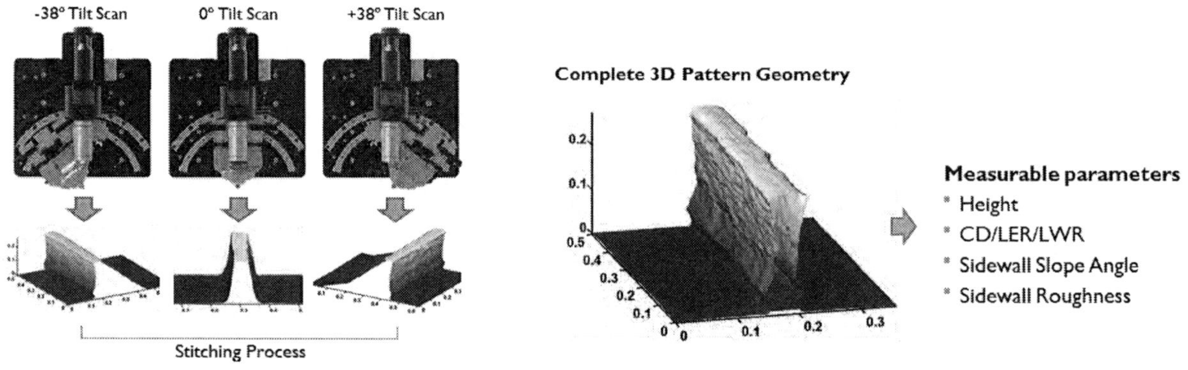

FIGURE 2. Schematic diagram of 3D AFM measurement method

RESULTS AND DISCUSSION

Critical Dimension and Sidewall Roughness of Si Waveguide

Pattern widths, heights and sidewall slopes were measured at different wafer location and their values were plotted at different figures as shown in Fig. 3, 4 and 5. Figure 3 shows that the pattern width in the wafer north has wider than at the wafer center and the pattern width at 75% height has smaller variation than at 25% height. On the other hand, height at the wafer center show slightly taller than at the wafer edge as shown in Fig. 4. As a results, the sidewall slope at the wafer edge shows wider than at the wafer center for both sidewalls as shown in Fig. 5. We observed some offset between left and right sidewalls and the amount of offset at the center is a bit wider than the other locations, but most of their values shows consistent across the wafer. The offset might be caused by hardware calibration issue, that is offset of the tilting angle of the Z scanner head or wafer bow.

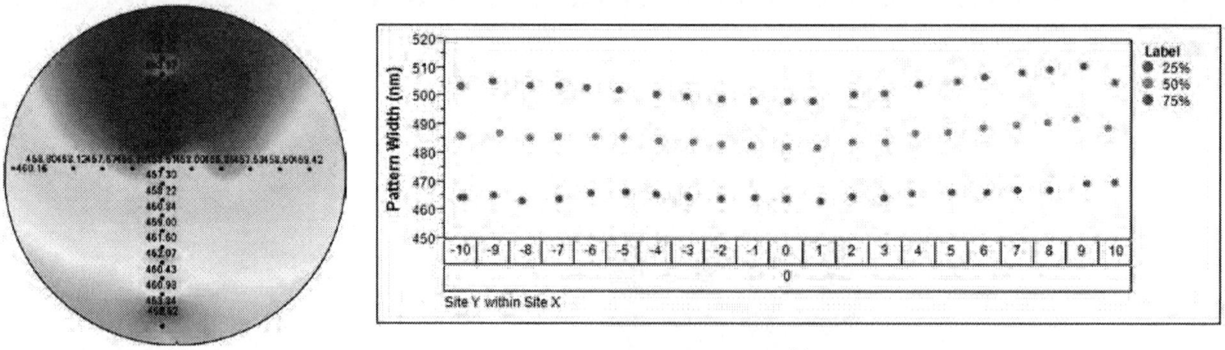

(a) (b)

FIGURE 3. (a) Wafer map of pattern width at heights of 75% from bottom and (b) their value changes, measured at heights of 25, 50 and 75% on Y axis of wafer

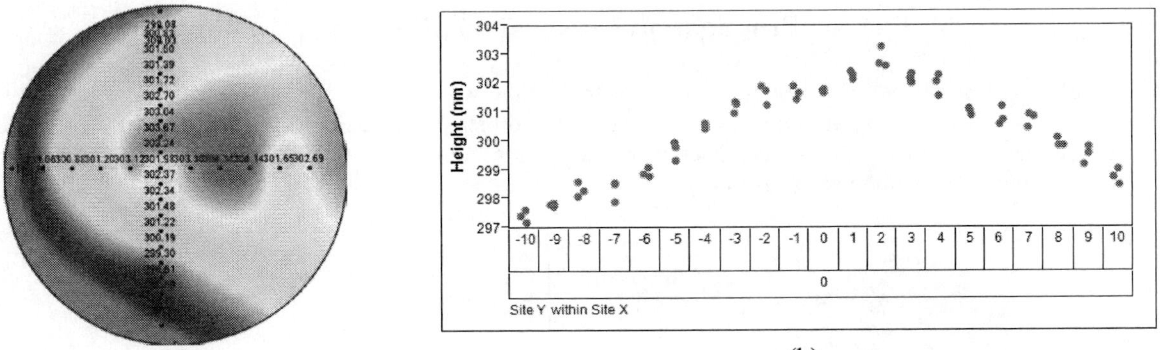

(a) (b)

FIGURE 4. (a) Wafer map of height and (b) their value change on Y axis of wafer

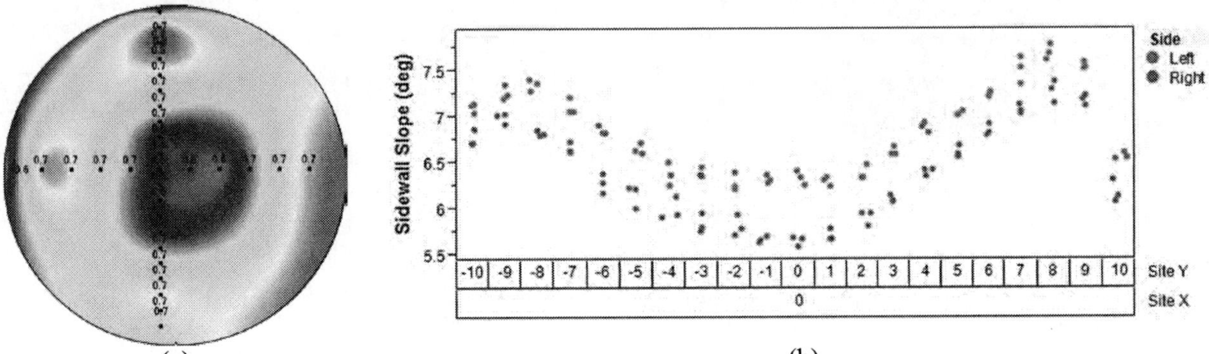

(a) (b)

FIGURE 5. (a) Wafer map of sidewall slope and (b) their value change on Y axis of wafer

The sidewall RMS roughnesses were measured at both sidewalls at different wafer locations. Figure 6 shows that the sidewall roughness at center is smaller than at edge no matter of pattern sides. The sidewall roughnesses of left and right sides show similar values at all locations except at the wafer edge. We presume that the offset might be caused by wafer bow and might not be caused by wrong head angle calibration because similar roughness values at the both sides. According to the results, the Z scanner head tilting measurement could provide atomic resolution sidewall roughness capabilities as well as critical dimension measurement in non-destructive way.

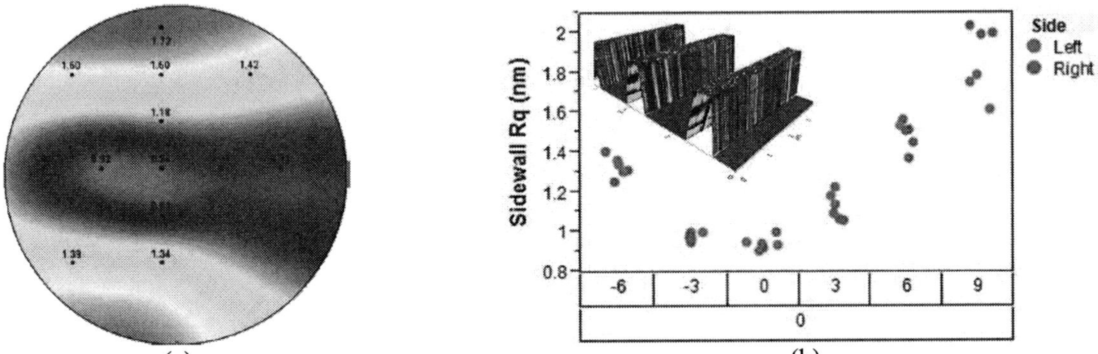

(a) (b)

FIGURE 6. (a) Wafer map of sidewall RMS roughness and (b) their value change on Y axis of wafer

Propagation Loss Correlation

The propagation loss of the Si waveguide was measured at the wavelength of 1550 nm and there results were shown in Fig. 7. When the propagation loss data is compared to dimension and sidewall roughness of waveguide, it shows that rougher sidewall roughness increases propagation loss and larger waveguide shows lower impact of roughness and lower propagation loss characteristics.

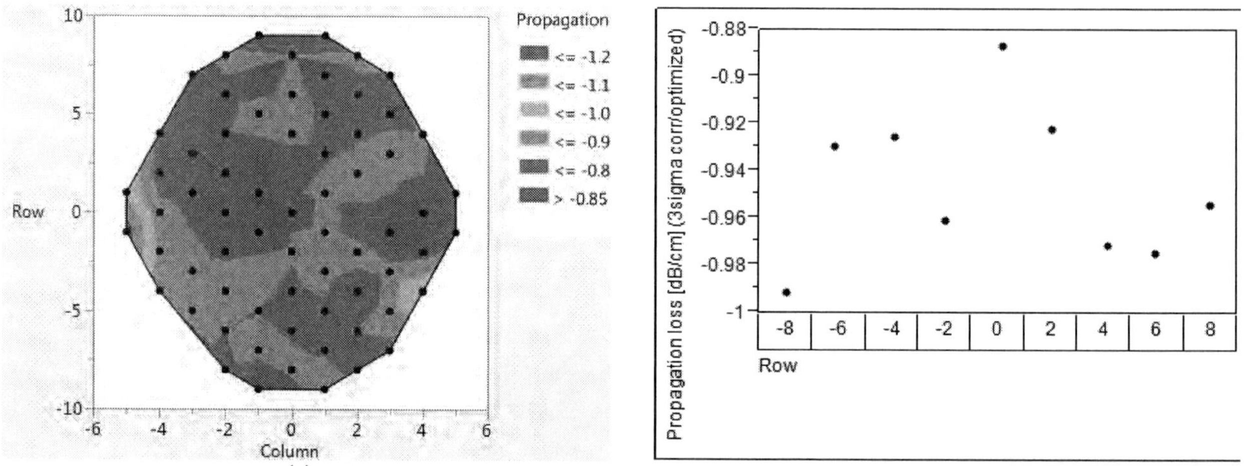

(a) (b)

FIGURE 7. (a) Wafer map of propagation loss, measured at the wavelength of 1550 nm and (b) their value change on Y axis of wafer

CONCLUSION

Atomic resolution height, width at different heights and sidewall slope angles and sidewall roughness of Si photonic waveguide can be characterized by using in-line 3D AFM and their values were correlated with their propagation loss properties. Dimension changes could make a correlation with sidewall roughness and rough sidewall increases the propagation loss of its waveguide. Larger waveguide shows lower impact of roughness and lower propagation loss characteristics. As a results, propagation loss could be controlled by controlling pattern dimension and in-line 3D AFM would be a good solution for their process monitoring.

ACKNOWEDGEMENTS

This work was supported in part by the Advanced Technology Center (ATC) Program (grant number 10045812) and Technology Innovation Program (grant number 10051701) funded by the Ministry of Trade, Industry and Energy to S.-J.C.

REFERENCES

1. D. A. B. Miller, *Proc. IEEE*, **97**, 7 , 1166-1185 (2009).
2. G. T. Reed, G. Machanovich, F. Y. Gardes and D. J. Thomson, Nat. Photon, **4**, 518-526 (2010).
2. J. Kwon, J. Hong, Y.-S. Kim, D.-Y. Lee, K. Lee, S. Lee, S.-I. Park, *Review of Scientific Instruments* **74**, 4378 (2003).
3. Y. Hua, C. Buenviaje-Coggins, Y. Lee, S.-I. Park, *Proc. SPIE* 8324, Metrology, Inspection, and Process Control for Microlithography XXVI, 83240I (2012).

KEYWORDS

Photonics, Si waveguide, AFM, 3D AFM, Sidewall roughness, Metrology, Critical Dimension, Propagation loss

Reference Materials to Enable Precise and Accurate Imaging with Electrical Scanning Probe Microscopes

J. J. Kopanski, L. You, J. Corona, and Y. S. Obeng

National Institute of Standards and Technology
Engineering Physics Division
100 Bureau Dr., stop 8120
Gaithersburg, MD 20899-8120

INTRODUCTION

Electrical scanning probe microscopes (eSPMs), such as the scanning Kelvin force microscope (SKFM), scanning capacitance microscope (SCM), or various scanning microwave microscopes (SMMs) are sensitive to the electric field between the sample and tip. Interpretation of measurements with these techniques can be confounded due to unknown tip shape and volume of interaction with the sample. Any two-terminal electrical measurement of electric field, capacitance, resistance, or inductance depends on the shape of the electrodes at each terminal. For simple one-dimensional metal-insulator capacitors, $C = \varepsilon_0\varepsilon_i A/t_i$ where C is capacitance, ε_o is the dielectric constant, ε_i the insulator dielectric constant, A is the device area, and t_i is the insulator thickness. For one-dimensional resistors, $R = \rho L/A$, where R is the resistance, ρ the material resistivity, and L the device length. Without knowing the device geometries, the material electrical properties (in these examples, ε_i and ρ) cannot be deduced regardless of how accurately the capacitance or resistance is measured. For complex electrode shapes varying in three dimensions, such as eSPMs, direct extraction of material properties is impossible without detailed information about the shape of the electrodes.

This work describes simulation, design and preliminary measurements of reference materials with precisely known geometries that will allow precisely calculable electric field gradients at a sub-micrometer scale to be generated. These structures have two intended uses: 1) as reference materials to determine the actual spatial resolution and accuracy of various eSPM techniques sensitive to electric field with different conductive tip technologies; and 2) as an electrical tip shape profiler. Electrical tip shape may vary significantly from physical tip shape. Knowledge of the electrical tip shape can then be fed back into image analysis software to improve the spatial resolution and accuracy of electric field measurements.

COMSOL MODEL OF ELECTRICAL SPMS

The COMSOL[1] MultiPhysics simulation software was used to simulate the electric field for candidate test structures. The COMSOL AC/DC Module combined with the RF Module allows simulation of high-frequency eSPMs such as SCM and SMM [1]. Many types of scanning probe microscope tips are integrated with conducting cantilevers and other conducting structures. For electrical measurements, all conducting parts of the probe assembly must be considered. We developed a complete COMSOL model of the SPM tip assembly to include the tip, the cantilever, the tilt of the cantilever with respect to the sample, and an arbitrary test structure (consisting of metal layers at defined bias voltages, ground planes, and dielectrics), Fig. 1. This model allows us to simulate the measured electrical potential between the defined test structure electrode and the defined probe electrode as a function of the terminal tip location.

[1]Certain commercial equipment, instruments, or materials are identified in this paper in order to adequately specify the experimental procedure. Such identification does not imply recommendation or endorsement by NIST, nor does it imply that the materials or equipment used are necessarily the best available for the purpose.

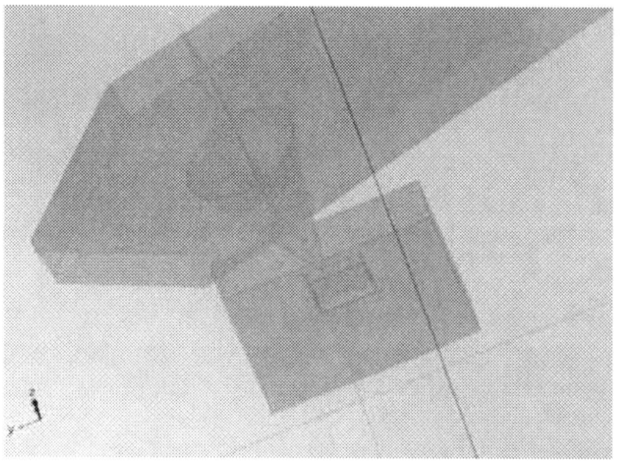

Figure 1: COMSOL generated image of the simulation geometry. A round base, conical tip (10 μm in height), the conductive cantilever, and a two-line test structure are shown.

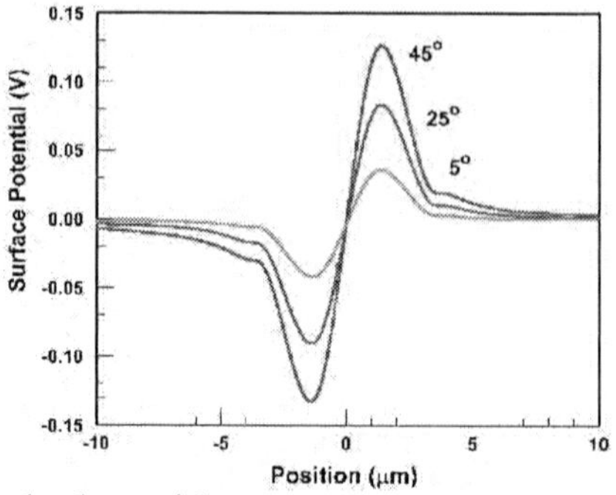

Figure 2: Simulated potential measurement across a test structure consisting of two lines, with the left line at negative potential and the right line at positive potential. The width of the response is related to the top cone width, while the slope of the response at the boundaries is inversely proportional to cone angle (5°, 25° or 45°).

A basic eSPM tip shape of a round cone is parameterized by the terminal tip radius, r_{tip}, the tip cone angle, Θ, the tip height (terminal tip to cantilever distance), h, and the top cone radius, r_{top}. For a perfect cone normal to the sample surface, the tip would be completely described by a single value of these four parameters (with $r_{top} = h \tan(\Theta)$). This model can be extended to simulate real eSPM tips that are not perfect cones (many are pyramidal), include some degree of asymmetry, and are used with their principle axis of symmetry tilted with respect to the sample surface. The physics of the tip-test structure interactions requires a dense mesh near the terminal tip, but also consideration of the long range contribution of all the parts of the test structure and the larger cantilever assemble. Many runs of the COMSOL model were require to determine a set of boundary conditions adequate to efficiently simulate the measurement while maintaining a realistic spatial domain and calculation time.

ELECTRICAL TIP PROFILER DESIGN STUDY

COMSOL simulations of the measured potential between a conical tip and simple test structures (generating an abrupt boundary in surface potential) revealed that both the cone angle and the top radius effect the measured surface potential in systematic ways, Fig. 2. The top radius broadens the measured response in direct proportion to the width of the tip as it crosses the boundary (Fig. 3a). Likewise, the slope of the response is inversely proportional to the cone angle at the point of maximum signal change (Fig. 3b). These results suggest a method of determining the tip parameters and asymmetry of an unknown tip from measurements on a test structure. If we consider the tip as an n-sided pyramid, with each side having an independent slope, then we can measure the width of the top of the pyramid and the slope of the side when that side cross the boundary in potential generated by the test structure. We suspect that most tips can be well characterized by four scans across the boundary with a 45° rotation between scans. While this approach will not be able to identify convex or concave tip sides, it will determine an equivalent tip shape (that is, the model tip parameters that produce a response equivalent to the tip under interrogation.) If the height of the tip is known by other methods, a complete equivalent tip shape can be determined and a figure of merit (measured field / actual field) for electrical profiling specified. The measured electrical tip parameters can then be used in a reverse model to deduce actual surface potential arising from structures whose geometry is not well known.

ELECTRIC FIELD MEASUREMENTS

The simulations have allowed us to design test structures capable of determining the electrical tip shape of eSPM tips. A simple biased line is sufficient if: 1) The line width is greater than the top radius of the tip, 2) regions beyond the active area are covered with a ground place, and 3) spacing of the active region to the ground plane is kept small. From a practical viewpoint a square test structure is preferable. With his geometry, the desired four angles of attack can be acquired from two images, one with the cantilever sides orthogonal to the directions of scan and a second with the sample-cantilever geometry rotated by 45°. A series of designed square test structures using this approach are shown in Fig. 4. As a preliminary test of our model, we imaged a wide aluminum line [2] with SKFM using a Point Probe Plus conducting eSPM tip. These tips have a symmetric orientation (left-to-right across the cantilever) and an asymmetric orientation (front-to-back along the cantilever). Figure 5 shows two line scans across the aluminum-to-oxide interface. When scanned with the symmetric tip profile parallel to the Al line, a symmetric contact potential difference response is obtained. When scanned with the asymmetric tip profile parallel to the Al line, an asymmetric response is obtained with the highly asymmetric side of the tip generating a kink in the CPD response as it passes over the Al-to-oxide interface.

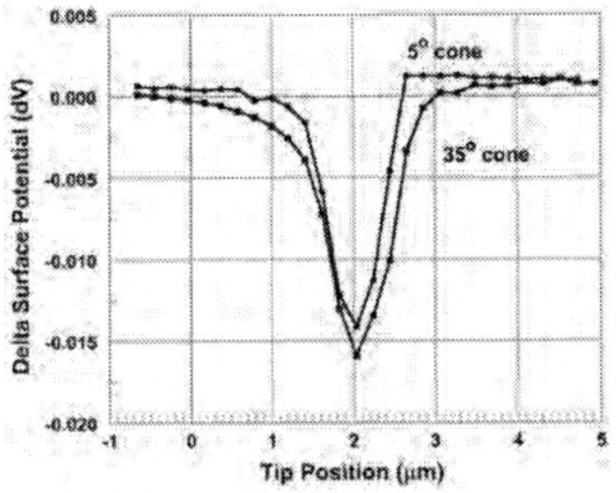

Figure 3a: Simulated potential measurement across a test structure consisting of a single biased line, surrounded by grounded planes on either side for conical tips with 5° and 35° cone angles. The broadening of the response provides a measurement of the top cone radius.

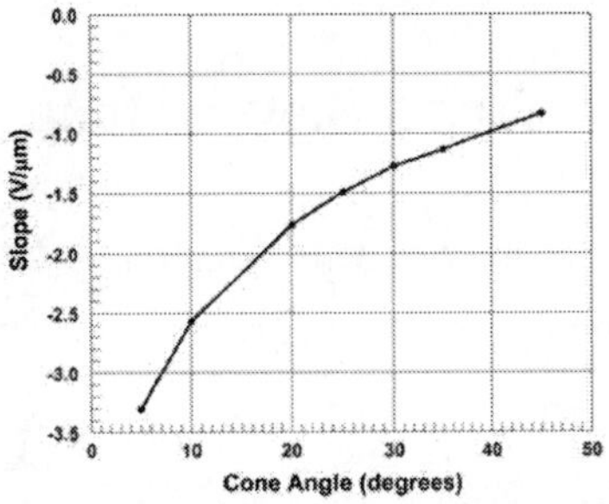

Figure 3b: Summary of the slopes at the boundary of simulated potential measurements across a test structure consisting of a single biased line, surrounded by grounded planes on either side for conical tips with cone angles of 5°, 10°, 20°, 30°, 35°, and 45°. Measured slope is proportional to cone angle.

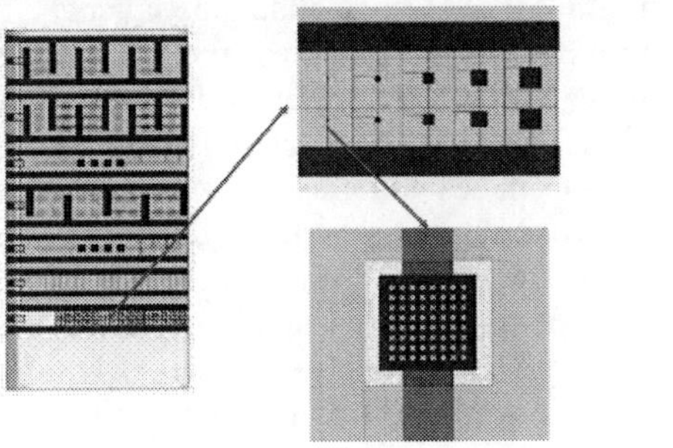

Figure 4: Section of test chip design with progressive enlargements showing the location of various sized 2-D electric tip profiler artefacts. Largest magnification shows buss structure, mid magnification shows different size 2D tip profilers, and highest magnification shows details of vias within a single profiler structure.

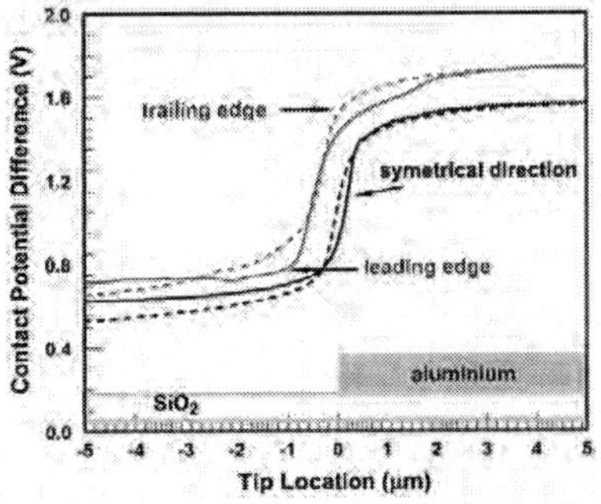

Figure 5: Measured surface potential across a biased metal line extracted from a SKFM image of a single metal line. The tip location data from the leading edge has been flipped and aligned with the trailing edge data for easier comparison. In the symmetric (purple lines) direction the tip (and the measured response) is broader. In the asymmetric direction, the response differs depending on whether the leading edge or trailing edge encounters the Al-to-SiO_2 boundary first. Symmetric data was offset for clarity.

CONCLUSIONS

In order to obtain accurate electric field based eSPM measurement results, or even to evaluate accuracy, it is necessary to have common calibration standards, ideally leading to SI traceability. COMSOL simulations show the electrical tip shape influences electric field based eSPM measurements in systematic ways. SKFM measurements with real tips on simple structures verify that such influences are real and can be experimentally measured. We have designed test structures that will amplify these effects and allow the tip electrical shape to be extracted.

REFERENCES

1. J. J. Kopanski, L. You, J. Li, J. Ahn, and Y. S. Obeng, ECS Transactions **72(2)**, pp. 131 -138 (May 2015).

2. L. You, J.J. Ahn, E, Hitz, J. Michelson, Y. Obeng, and J. J. Kopanski, Proc. 2015 IEEE Intl. Conf. Microelectronic Test Structures, pp. 235-239, Tempe, AZ (March 23-26, 2015).

KEYWORDS

Electric field measurement, scanning Kelvin force microscope, scanning probe tip profiler, reference materials

Ultra-thin ALD HfO_2 Growth Mechanism Studied By Atomic Force Microscope (AFM)

Charles C Wang, Steven Hung, Naomi Yoshida, Tatsuya E Sato, Yuri Uritsky

Applied Materials, Inc., M/S 0203, 3100 Bowers Ave, Santa Clara, CA 95054

INTRODUCTION

Due to the space constraints of device structures used in sub 1X technology node, IC manufacturing is challenged with the development of sub 10 Å ultra-thin films. One common concern about these ultra-thin films is their continuity on the substrate surface and morphological structures that can greatly affect their performance: a poorly grown and porous barrier cannot prevent the inter-diffusion of the materials; an etch stop layer with many pinholes will allow the etchants in the lithography process to attack the layers underneath it; and a high-k dielectric with many boundaries within the film will increase the leakage current through it. Due to the layer-by-layer deposition mechanism and the ability to provide good step coverage of nano-structures with high aspect ratios, atomic layer deposition (ALD) processes are largely being used for the deposition of conformal and continuous ultra-thin films in deep trenches and vias. However, depending on the interaction of the precursor and the substrate surface, this layer-by-layer growth model may not apply during the initial stage of the ALD film deposition and consequently the initial film growth can be of poor quality and with lots of defects [1]. In the case of ultra-thin film deposition, only a small number of ALD cycles are needed to reach the final thickness, therefore the morphology and properties of the ALD thin film can largely be controlled by its initial growth mechanism. Hence, the study of the film morphology and initial growth mechanism becomes extremely important for the successful development of these ultra-thin films. For this purpose, a high resolution AFM technique was developed for direct imaging of the surface morphology of ultra-thin films and its usefulness was demonstrated by the study of the growth of ALD HfO_2 on Si(100) substrate covered with a thin silicon oxide interface layer (IL). The reason that ALD HfO_2 was chosen was that, due importance, it has been the subjects of numerous studies and there seems to have a general consensus on its reaction mechanism [2, 3]. It is believed that it involves the reaction of the Hafnium precursor ligands with the –OH functional groups on the IL surface. Therefore, initial growth rate on a substrate rich in –OH on its surface is faster than that on a substrate deficient in –OH; an example of the former substrate is the ChemOx on silicon substrate after subjecting to SC1 clean [4] and an example of the latter is the HF cleaned silicon substrate, which has a hydrogen terminated surface [5].

AFM TECHNIQUE DEVELOPMENT

To achieve high resolution AFM imaging two criteria need to be met. The first one is that a robust scanning condition needs to be used to minimize the AFM tip damage during AFM imaging to ensure good quality images are obtained. To this end, a dynamic force AFM scanning technique operated in the attractive force regime was developed and documented elsewhere [6, 7]. The 2nd important criteria is that the AFM tip should have small enough radius of curvature (ROC) in order for it to resolve the fine surface feature for ultra-thin film morphology imaging. Typical commercially available AFM tips are with ROCs in the range of 10 nm to 5 nm. As illustrate in Figure 1(A), such kind of tips may not provide enough resolution to reveal the morphological details the surfaces of interest. Super sharp tips with ROCs in the range of 4 nm to 2 nm are also available commercially. However, attempts to use the commercially super sharp tips have not yielded satisfactory results. It is postulated that due to the high surface energy of the super sharp tip, it may have a limited shelf live - the apex of the tip can either curl up or become enlarged by ambient contaminants. In order to achieve desired high resolution, a proprietary technique was developed to sharpen the regular commercial AFM tips with ROCs in the 10 nm to 5 nm down to 3 nm or even better. As shown in Figure 1(B), the sharpened tip considerably improved the image quality from the same bare silicon substrate.

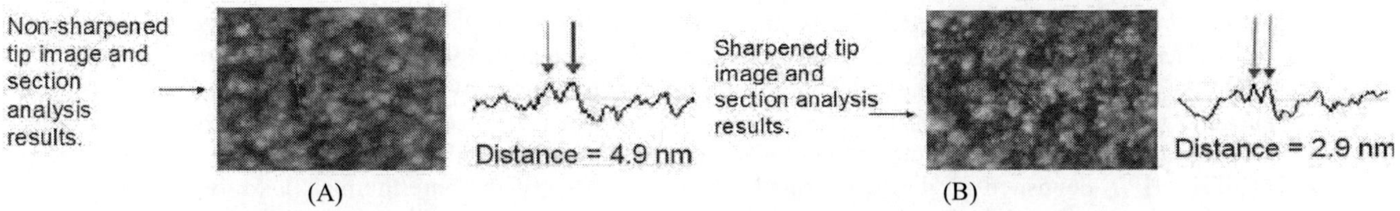

FIGURE 1. AFM images of a polished 300 mm silicon substrate: (A) image obtained using commercial AFM probe with ~ 300 kHz resonance frequency and force constants of ~ 40 N/m, the section view on the right (red line in the AFM image showing where the section was taken) showing a pair of features with 4.9 nm spacing was barely resolved; (B) image obtained using the same type but sharpened tip, the section view on the right showing that features with 2.9 nm spacing can be resolved.

APPLICATION – ULTRA-THIN ALD HFO_2 GROWTH MECHANISM STUDY

The same ALD process was used to deposit HfO_2 on five different silicon oxide IL layers. Three different thermal processes were used to grow the ILs on the silicon substrates. The first two ILs (IL1, IL2) were made with the same thermal process but were with two different thicknesses. Another two thermal processes were used to grow the 3rd IL (IL3) and the 4th IL (IL4). The 5th IL (IL5) were grown using the same thermal process as IL1 and IL2, but was treated with the SC1 solution and then cleaned with DI water before the HfO_2 deposition. Due to the great influence of the surface chemistry of the IL on the growth mechanism and properties of HfO_2 thin films, the emphasis of the IL effect in this study was warrantied [5]. The HfO_2 samples are grouped into 5 groups according to the IL layers used (see Table 1).

TABLE 1. Descriptions of the five ALD HfO_2 sample groups.

	Group 1	Group 2	Group 3	Group 4	Group 5
IL type/thickness^	IL1/10 Å	IL2/15 Å	IL3/15 Å	IL4/15 Å	IL5/15 Å
Oxide process*	T1	T1	T2	T3	T1
Surface treatment	none	none	none	none	SC1 (25°C)+

^ Thicknesses of the IL layers were verified by XPS measurements; * "T" stands for thermally grown oxide;
+ IL5 was IL1 dipped in SC1 (25°C) solution for a few seconds then rinsed clean with DI water.

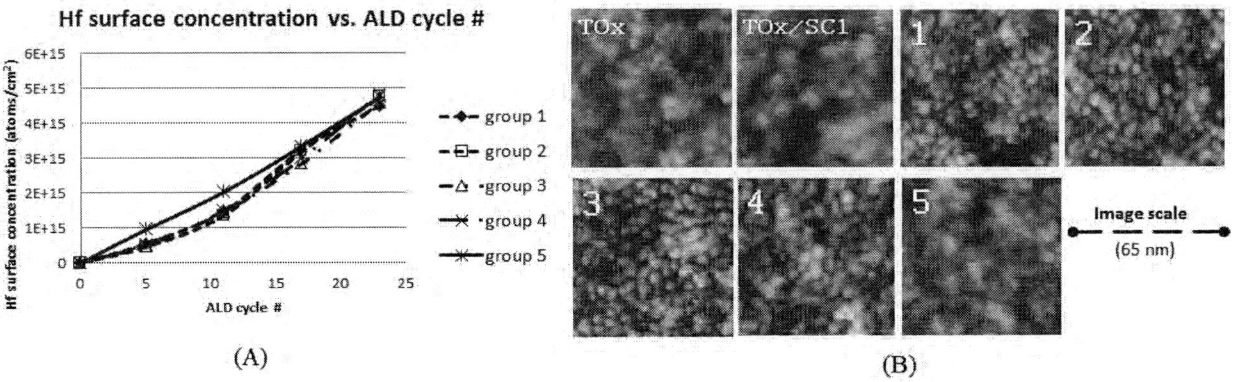

FIGURE 2. (A) Amounts of Hafnium deposited on ILs as a function of ALD cycles determined by RBS; (B) AFM images of IL surfaces with no HfO_2 deposition together with those deposited with 5 cycles of ALD HfO_2, the labels on the top left corner provide the sample descriptions: TOx indicates the image is from the bare IL1 with no HfO_2 deposition, TOx/SC1 indicates the bare SC1 treated IL1, the numbers indicate the images of 5 cycles of HfO_2 from the 5 different sample groups.

High resolution AFM imaging, Rutherford Backscattering Spectrometry (RBS) HfO_2 surface concentration measurements together with the electrical parameters (leakage current and equivalent silicon oxide thickness (EOT)) measurements were carried out. XPS measurements verified that compositions of all the Hafnium oxide films were close to the stoichiometric HfO_2. RBS results are summarized in Fig. 2(A). As can be seen, initially (5 ALD cycles) the growth rates of the HfO2 films deposited on thermally grown oxide were approximately half of that of the group 5 HfO_2 films, which were deposited on a thermally grown IL that was treated by the SC1 solution at room temperature prior to the HfO2 deposition. The growth rate of the group 5 HfO_2 films was constant with the ALD cycles as can be seen from the linear relationship between the HfO_2 surface concentrations and the ALD cycles. As the ALD cycles increased to 17, the surface concentrations of HfO_2 films on thermal oxides (groups 1 to 4) began to catch up with that of the group 5 film. At 23 cycles, all the groups became to have very close surface concentrations, in other words, the growth rates on thermal ILs were initially lower than but eventually surpassed that of the HfO_2 deposited on the SC1 treated IL. AFM image of the bare thermal IL (only IL1 image is shown here, because the other 3 types of thermally grown ILs have similar surface morphology) appears slightly rougher than the thermally grown IL treated with SC1 (IL5). The images of the 5 cycles of ALD HfO_2 on the four thermally grown IL show clearly a grape-like morphology, while that of the HfO_2 grown on SC1 treated IL (IL5) shows the same smooth and almost featureless surface morphology as the starting IL surface.

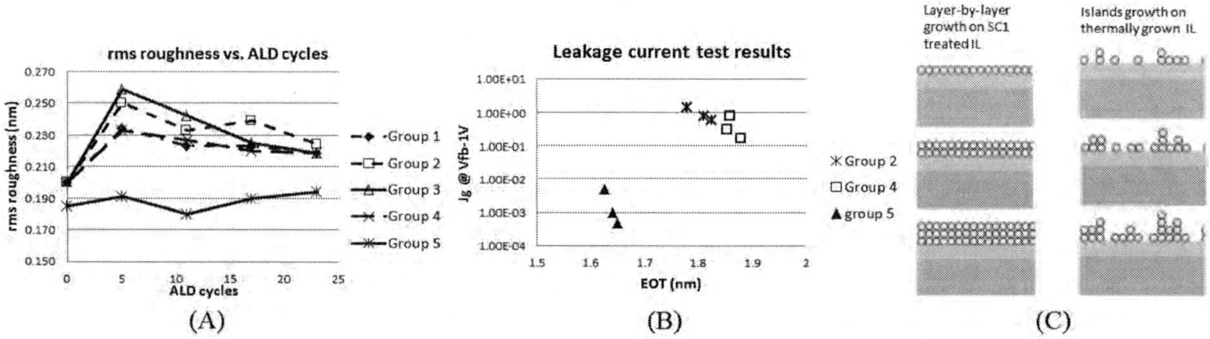

FIGURE 3. (A) rms roughness of the ALD HfO_2 grown on the ILs; (B) MOS capacitor leakage current (Jg) and EOT test data from 23 cycles of HfO_2 deposition on different types of ILs; (C) Growth model of ALD HfO_2 on different ILs

AFM roughness data (Fig.3(A)) shows that roughness of the HfO_2 on SC1 treated, thermally grown IL (IL5) stayed the same with increasing ALD cycles, while the roughness of HfO_2 grown on thermally grown ILs increased dramatically initially, then dropped gradually with increasing cycles. Electrical test data (Fig. 3(B)) shows that the HfO_2 on SC1 treated thermally grown IL, even with lower EOT, had at least 2 orders of magnitude lower leakage current than those deposited on thermally grown ILs. It is clear from all the results that the group 5 ALD HfO_2 followed a layer-by-layer growth model, while those deposited on thermally grown ILs followed islands growth model (Fig. 3(C)). The deposition rates of the HfO_2 on thermally grown IL surpassed that of the group 5 HfO_2 can be explained by the larger surface area resulted from the grape-like morphology in the initial stage of ALD film growth on the thermally grown ILs. The results also prove that the surfaces of the as deposited thermally grown ILs (IL1 to IL4) do not have as many –OH groups as that of the SC1 treated, thermally grown IL (IL5). This study clearly demonstrates the usefulness of the high resolution imaging method. This technique can be widely applied to the study of the ultra-thin film growth mechanisms and to tackle ultra-thin film continuity issues.

REFERENCES

1. R. L. Puurunen and W. Vandervorst, *J. Appl. Phys.* **96,** 7686-7695 (2004).
2. X. Liu *et al.*, *J. Electrochem. Soc.* **152**, G213-G219 (2005).
3. D. M. Hausmann, E. Kim, J. Becker and R. G. Gordon, *Chem. Mater* **14**, 4350-4358 (2002)
4. W. Kern and D. A. Puotinen: *RCA Rev.* **31,** 187-206 (1970)
5. J. C. Hackley, T. Gougousi and J. D. Demarre, *J. Appl. Phys.* **102,** 034101-1-034101-7 (2007).
6. C. C. Wang, Y. Pu, L. Fu, Y. Ma and Y. S. Uritsky, *AIP Conference Proceedings* **788**, 194-199 (2005)
7. B. Liu, C. C. Wang, P. Huang and Y. S. Uritsky, *Proc. of SPIE* **7729**, 77290O-1-77290O-11 (2010)

KEYWORDS

AFM, ultra-thin film, ALD, HfO2, morphology, thin film growth mechanism

Ferroelectricity in Atomic Layer Deposited $Hf_{1-x}Zr_xO_2$ Nanoscale Films: Characterization by Synchrotron Grazing Incidence X-ray Diffraction and Polarization Measurements

Sonal Dey[1], Kandabara Tapily[2], Steven Consiglio[2], Robert D. Clark[2], Cory S. Wajda[2], Gert J. Leusink[2], Arthur R. Woll[3], Pankaj Sharma[4], Suman Datta[4], and Alain C. Diebold[1]

[1]*Colleges of Nanoscale Science and Engineering, SUNY Polytechnic Institute, 257 Fuller Rd., Albany, NY 12203, USA.*
[2]*TEL Technology Center, America, LLC, 255 Fuller Rd., Suite 214, Albany, NY 12203, USA.*
[3]*Cornell High Energy Synchrotron Source (CHESS), Ithaca, NY 14853, USA.*
[4]*Department of Electrical Engineering, University of Notre Dame, Notre Dame, IN, 46556, USA*

Bulk hafnium oxide (HfO_2) is known to form a stable monoclinic phase at room temperature and pressure, tetragonal and cubic phases at elevated temperatures, and orthorhombic phase at higher pressures[1,2]. Nano-crystalline HfO_2, due to surface energy effects, is also known to form a stable tetragonal phase at lower temperatures[3]. The presence of a capping layer on nanoscale films could lead to a suppression of the monoclinic phase[4,5] along with a significant increase in the dielectric constant[6]. Doping the HfO_2 with other materials, such as Si, Zr, Al, La, Gd, Sr, and Y, has been shown to induce a ferroelectric orthorhombic phase in HfO_2[7].Theoretical studies by Materlik *et al.*,[8] predicts undoped HfO_2 to be ferroelectric when the grain sizes are less than 4 nm. In case of $Hf_{1-x}Zr_xO_2$ with x = 0.5, a slightly higher upper limit of the grain size (~ 5nm) is predicted[8] as a requirement to show ferroelectric properties. In addition, the transition to a ferroelectric phase is expected to happen under asymmetric stress conditions in both HfO_2[9] and ZrO_2 nanoscale films[10,11]. The well-developed process and integration schemes for HfO_2 as a gate dielectric material in semiconductor devices makes it an attractive candidate to study the formation and origin of the ferroelectric phase in detail. Ferroelectric HfO_2-based films have technological applications including ferroelectric field-effect-transistor (FeFET) memory[12] and low power consumption FETs based on the use of ferroelectric negative differential capacitance[13].

We have used atomic layer deposition (ALD) to prepare metal-insulator-semiconductor (MIS) structures. Three different thicknesses (3 nm, 5 nm, and 7 nm) of $Hf_{1-x}Zr_xO_2$ ($0 \leq x \leq 1$) were deposited by ALD on Si wafers with chemically grown oxide. The $Hf_{1-x}Zr_xO_2$ films were prepared with either a cycle ratio (CR) or a co-flow (CF) ALD method[14]. In the CR method, the hafnium and zirconium precursors were cycled one after another whereas in the CF approach, the precursors were injected simultaneously as was reported previously[14]. All films were capped with an ALD deposited TiN metal capping layer of 5 nm thickness. The metal oxide part of some of the thin film stacks were also subjected to post deposition annealing (PDA) before depositing the metal capping layer. Some of the stacks were also subjected to post metal deposition annealing (PMA) step. Synchrotron based grazing incidence x-ray diffraction (GI-XRD) was used to determine the structure of the films. By collecting GI-XRD data using a linear detector which could cover a wide range of the reciprocal space at one scan point, we were able to collect high quality GI-XRD data and grazing incidence d-spacing maps (GI-DSM) within a reasonable time frame[15,16]. The line-like features in the GI-DSMs at a particular d-spacing value indicate that there is no preferred orientation in the films[16]. All the films reported here showed a line-like feature in the GI-DSM (Figure 1), that is, the films are found to be nano-crystalline with a random grain orientation. The integrated XRD data over a certain angular range is plotted as intensity vs. d-spacing (plots in Figure 1) for phase identification purposes. The GI-XRD data in Figure 1 shows the presence of a non-monoclinic feature between $d \approx 2.95 - 3.05$ Å. The most prominent peak position for

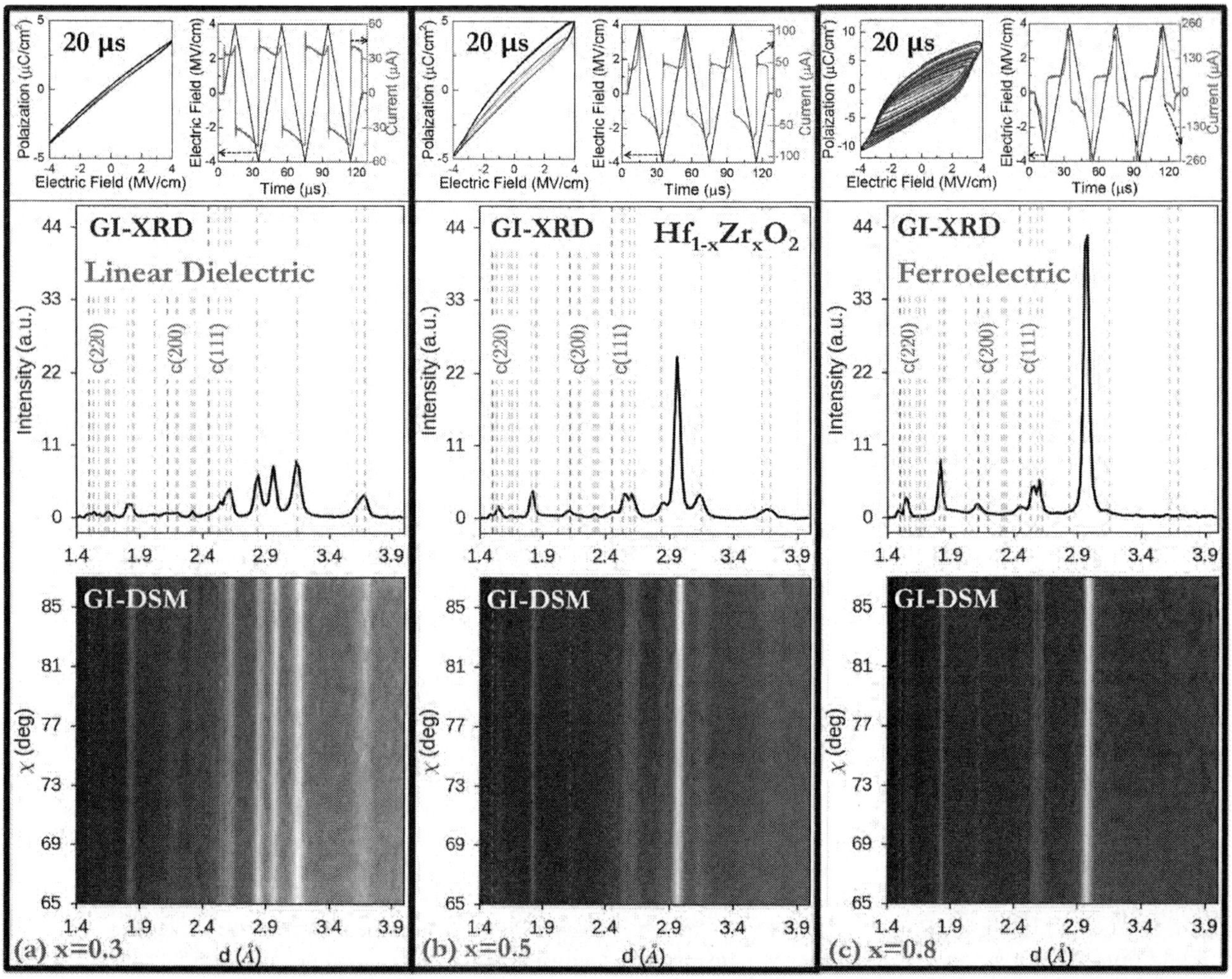

FIGURE 1. Figure showing the increases in non-monoclinic phase (by following the peak intensities between d ≈ 2.7 – 3.3 Å in the GI-XRD data) as the film stack goes from a largely linear dielectric to a ferroelectric (as measured in the electrical data) with increasing ZrO_2 content. The values of d-spacing labeled c(220), c(200), and c(111) correspond to cubic TiN (which are also shown by vertical dotted-lines on the GI-DSM) and the other unlabeled dashed lines correspond to values of d-spacing for monoclinic HfO_2. The label of 20 μs on the polarization vs. electrical field data refers to the pulse period and the accompanying transient response data is also shown.

this non-monoclinic feature matches well with some of the possible orthorhombic, tetragonal, and cubic space groups for HfO_2 and ZrO_2 based compounds. At the time of writing this abstract, the exact nature of the non-monoclinic feature in the GIDSM is still being investigated. With increasing ZrO_2 concentration, the monoclinic peaks appear to be suppressed whereas the intensity of the non-monoclinic peak increases. At the same time, the electrical measurements in Figure 1 show an increase in the polarization response of these films. Although the exact nature of the non-monoclinic phase could not be identified at this point, the presence of the non-monoclinic phase is found to be directly related to the observation of a close hysteresis loop in the polarization vs. electrical measurements (Figure. 1). We have also observed an increase in the non-monoclinic phase for films with PMA but without PDA as compared to the films without PMA but with PDA. Thus the post metal annealing seems to be favorable for stabilizing the non-monoclinic phase in these nanoscale films. We have also observed a decrease in the monoclinic phase with decreasing thickness which could be attributed to surface energy effects of the nano-grains[3,17]. The smallest thickness films (~3 nm) seemed to be in a higher symmetric phase (tetragonal or cubic). The CF process seems to increase the percentage of the non-monoclinic phase when both type of films with and without PDA and PMA are characterized. For the thickest film (~7 nm), the CF method was more effective in suppressing the monoclinic phase formation as compared to the CR method. The CF method could lead to a film with better intermixing of the HfO_2 and ZrO_2

components compared to the CR method which may be responsible for better suppression of the monoclinic phase formation in these films. Additional electrical characterization is currently underway along with physical characterization to gain a better understanding of the phase responsible for ferroelectric behavior in these films.

REFERENCES

[1] X. Zhao and D. Vanderbilt, Phys. Rev. B **65**, 233106 (2002).
[2] X. Zhao and D. Vanderbilt, Phys. Rev. B **65**, 75105 (2002).
[3] R.C. Garvie, J. Phys. Chem. **69**, 1238 (1965).
[4] D.H. Triyoso, P.J. Tobin, B.E.W. Jr, R. Gregory, and X.D. Wang, Appl. Phys. Lett. **89**, 132903 (2006).
[5] S. Consiglio, K. Tapily, R. D. Clark, T. Hasegawa, F. Amano, G. J. Leusink, J. Jordan-Sweet, R. Vasić, M. Medikonda, and A. C. Diebold, J. Vac. Sci. Technol. B **32**, 03S122 (2014).
[6] S. Migita, Y. Watanabe, H. Ota, H. Ito, Y. Kamimuta, T. Nabatame, and A. Toriumi, in *2008 Symp. VLSI Technol.* (2008), pp. 152–153.
[7] M.H. Park, Y.H. Lee, H.J. Kim, Y.J. Kim, T. Moon, K.D. Kim, J. Müller, A. Kersch, U. Schroeder, T. Mikolajick, and C.S. Hwang, Adv. Mater. **27**, 1811 (2015).
[8] R. Materlik, C. Künneth, and A. Kersch, J. Appl. Phys. **117**, 134109 (2015).
[9] T. Nishimura, L. Xu, S. Shibayama, T. Yajima, S. Migita, and A. Toriumi, Jpn. J. Appl. Phys. **55**, 08PB01 (2016).
[10] E.H. Kisi, C.J. Howard, and R.J. Hill, J. Am. Ceram. Soc. **72**, 1757 (1989).
[11] E.H. Kisi, J. Am. Ceram. Soc. **81**, 741 (1998).
[12] A. Chen, Solid State Electron. **125**, 25-38 (2016).
[13] M. Hoffman, M. Pešić, K. Chatterjee, A. I. Khan, S. Salahuddin, S. Slesazeck, U. Schroeder, and T. Mikolajick, Adv. Funct. Mater. doi:10.1002/adfm.201602869 (2016)
[14] S. Consiglio, K. Tapily, R.D. Clark, G. Nakamura, C.S. Wajda, and G.J. Leusink, J. Vac. Sci. Technol. A **31**, 01A115 (2013).
[15] S. Dey, K. Tapily, S. Consiglio, R.D. Clark, C.S. Wajda, G.J. Leusink, A.R. Woll, and A.C. Diebold, J. Appl. Phys. **120**, 125304 (2016).
[16] S. Dey, K. Tapily, S. Consiglio, K.-H. Yu, R.D. Clark, C.S. Wajda, G.J. Leusink, A.R. Woll, and A.C. Diebold, MRS Adv. **1**, 269 (2016).
[17] J.-H. Ahn and S.-H. Kwon, ACS Appl. Mater. Interfaces **7**, 15587 (2015).

ACKNOWLEDGEMENTS

S.D. thanks TEL Technology Center, America, LLC for funding support. This work is based upon research conducted at the Cornell High Energy Synchrotron Source (CHESS) which is supported by the National Science Foundation and the National Institutes of Health/National Institute of General Medical Sciences under NSF award DMR-1332208.

KEYWORDS

Ferroelectric, ALD, Hafnium Oxide, Zirconium Oxide, HfO2, ZrO2, Synchrotron, GI-XRD, GI-DSM

Helium Ion Microscopy For Si Materials With Less Mechanical Damage

Shinichi Ogawa[1)], Nobuaki Tarumi[2)], Tomohiko Iijima[1)], Yumiko Shimizu[2)], and Yuji Otsuka[2)]

1)Nanoelectronics Research Institute, National Institute of Advanced Industrial Science and Technology (AIST) West 7A, 16-1 Onogawa, Tsukuba, Ibaraki 305-8569 Japan
2)Toray Research Center Inc., 3-3-7 Sonoyama, Otsu, Shiga 520-8567 Japan.

A secondary electron (SE) mode imaging with less damage and less transformation of soft materials such as low dielectric materials and photo resist fine patterns for Si semiconductor processing has been reported using the helium ion microscopy (HIM) [1], and an observation of helium ion beam induced luminescence was reported [2]. However the damages themselves have not been well studied. In this study, after an observation of a SiO_2 film and Si substrate using the HIM at standard helium ion beam irradiation conditions for observations, the sample mechanical damages were characterized by a cross section transmission electron microscopy (TEM) - electron energy loss spectroscopy (EELS) method [3].

Helium ion beams of a few pA, in which a dose amount was 1E13/cm^2 - 7E14/cm^2 at 30 kV, was irradiated onto the SiO_2 film of 400 nm thick and Si-sub., and luminescence was studied at room temperature. And then those SiO_2 films and Si sub. were examined by the cross section TEM-EELS method. Furthermore, for Si-sub. samples with higher doses up to 2E16/cm^2 were characterized by the cross section TEM-EELS.

As shown in Fig. 1, in the HIM luminescence, shifts in wavelength of the spectra from the SiO_2 with different doses were not seen, while the intensity became higher with the higher dose, that might mean no damage, which was detectable by the luminescence, generated during these irradiation conditions. From the Si-sub., no luminescence was observed in the range of 250nm – 800nm because of its energy gap of 1.2 eV. Energy loss near edge structure (ELNES) and valence EELS (V-EELS) spectra at distances of 20, 100, 250, and 400 nm from the SiO_2 surface were shown in Fig. 2. ELNES and V-EELS give us information of electronic structure of conduction band, and defect states in the band gap, respectively. A project range of irradiated helium ions was 250 nm and it might be damaged the heaviest, if such damage was generated. Any difference in the spectra was not observed among all spectra of the ELNES and V-EELS. For a Si substrate, quite the same results have been obtained by ELNES (Fig. 3) without any detectable damage or bubble in the samples. It has been revealed that no damage occurred during the HIM observations of standard helium ion beam irradiations within a detection limit of the EELS.

High resolution characterization of Si single crystalline substrate after the higher doses up to 2E16/cm^2, which is needed for the HIM high resolution observation, is now under evaluation using the TEM, and results will be shown from a point of view of change in crystallinity of Si substrate by the TEM-EELS.

Collisions of helium ion beams of "sub-nm diameter" onto materials, SiO_2 and Si-sub. in this study, including damage generation in the materials have not been well understood, and further discussions will be required for further HIM applications with less damage.

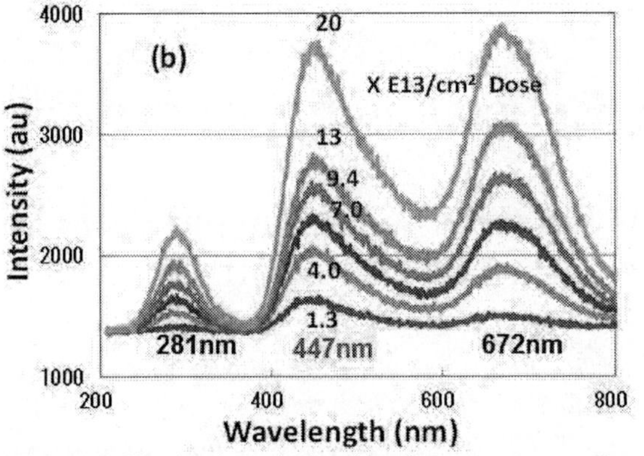

FIGURE 1. Helium ion beam induced luminescence spectra from SiO_2 at different ion doses of 1.3-20 E13/cm^2 at 30kV.

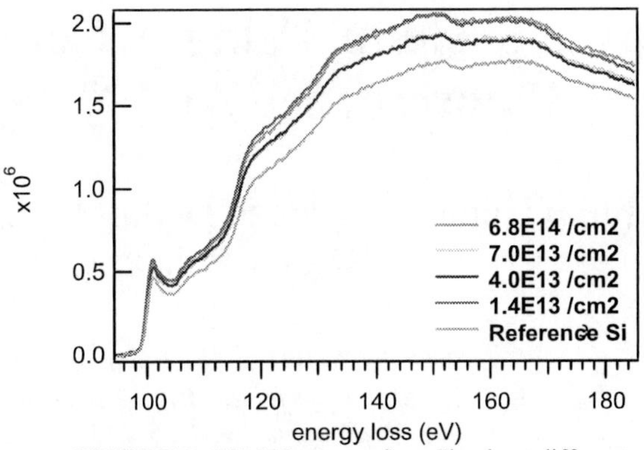

FIGURE 3. ELNES spectra from Si-sub. at different doses at 100 nm depth from the Si-sub. surface.

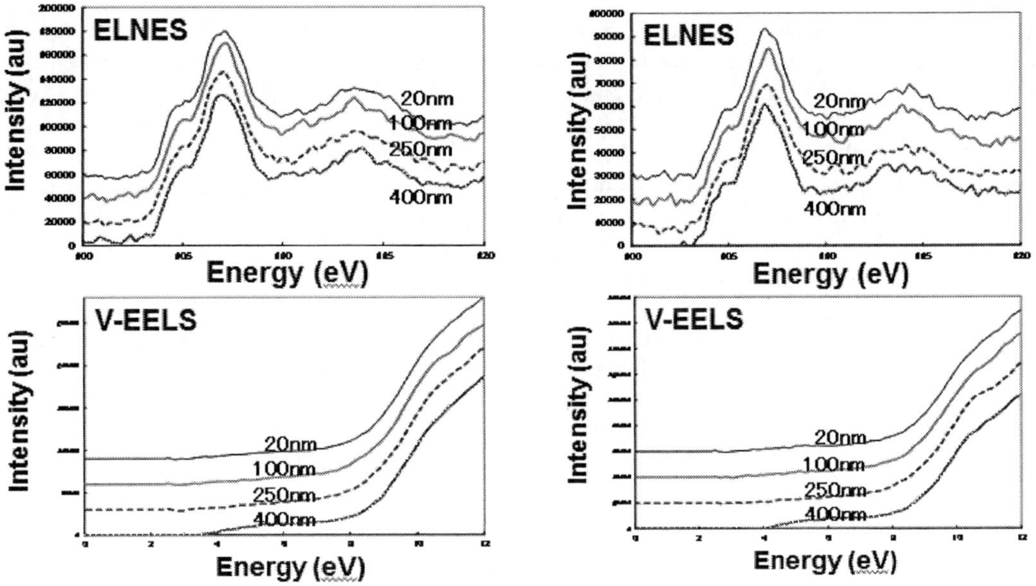

(a) (b)

FIGURE 2. TEM ELNES and V-EELS spectra at distances of 20, 100, 250, and 400 nm from the SiO_2 surface without helium ions beam irradiation (a) and with 2E14/cm^2 helium ions beam irradiation (b).

REFERENCES

1. S. Ogawa, et al, Jpn. J. Appl. Phys., 49 04DB12 (2010),
2. S. Ogawa, et al, Proc. of International Interconnect Technology Conference P1.31 (2011),
3. Y. Otsuka, et al, Jpn. J. Appl. Phys., 49 111501 (2010)

KEYWORDS

Helium ion microscopy, Si, SiO2, Damage, TEM, EELS, ELNES

Reference-free In-Depth Characterization Of Nanoscale Layer Systems Using A Combined Grazing Incidence X-ray Fluorescence And X-ray Reflectometry Approach

Philipp Hönicke[1], Blanka Detlefs[2], Janis Eilbracht[1], Yves Kayser[1], Uwe Mühle[3], Beatrix Pollakowski[1] and Burkhard Beckhoff[1]

1: Physikalisch-Technische Bundesanstalt (PTB), Abbestr. 2-12, 10587 Berlin, Germany
2: CEA-LETI/ESRF, 71 Avenue des Martyrs, 38043 Grenoble, France
3: Fraunhofer IKTS, Maria-Reiche-Str. 2, 01109 Dresden, Germany

INTRODUCTION

The accurate in-depth characterization of nanoscale layered systems is an essential topic for today's development in many fields of materials research. Especially nanolaminates, which are a stack of multiple thin layers, are technologically relevant for current and future electronic devices and other applications [1,2]. However, the metrological challenges to sufficiently characterize such systems with respect to their in-depth elemental distributions require a further development of the current analytical techniques.

The combined analysis using Grazing Incidence X-ray Fluorescence (GIXRF) and X-Ray Reflectometry (XRR) as proposed by de Boer [3] has already been shown to be capable of contributing to the in-depth analysis of nanoscaled materials. The general approach to model the experimental data using the density ρ, thickness d and roughness σ of each layer as independent parameters can quickly result in numerous degrees of freedom and thus unreliable results [4].

In this work, XRR is combined with reference-free GIXRF, providing access to mass depositions (ρd) of relevant elements. This allows for a significant reduction of the degrees of freedom and thus improves the characterization reliability of the combined XRR-GIXRF methodology. Employing the in-house built instrumentation [5] and radiometrically calibrated detectors at the laboratory of the PTB at the BESSY II synchrotron radiation facility, the reference-free XRR-GIXRF method is applied to the in-depth analysis of Al_2O_3-HfO_2 nanolaminate stacks [4].

COMBINED GIXRF-XRR METHODOLOGY

A combined GIXRF-XRR analysis is usually modeled by using thicknesses, densities and roughnesses of every layer as the optimization parameters [6,7]. Any remaining discrepancies, which result from deviations from box-like layers are then taken into account by adding mixture layers with additional modeling parameters. This usually provides a very good qualitative reproduction of the experimental data but this approach can result in a large number of free modeling parameters. A validation whether non-physical correlation effects between parameters hinder the reliability of the derived results is usually missing.

In this work, we show that this conventional GIXRF-XRR modeling can have a negative impact on the reliability of the obtained results. It will also be demonstrated how the reference-free X-ray fluorescence spectrometry methodologies of the PTB [8] can be used to address this critical issue. This reference-free approach allows deriving quantitative information for each element of interest without requiring any external calibration standard or reference sample.

The thereby derived elemental mass depositions, defined as the product of a materials density and thickness, can then be used to either validate [4] any conventionally modeled mass deposition results or they can serve as a direct input to the modeling of a combined reference-free GIXRF-XRR approach. As the mass deposition of each element is fixed through the reference-free quantification, this allows deriving the different layers thicknesses from their modeled densities or vice versa. Thus the degrees of freedom are significantly reduced.

EXPERIMENTAL

Thin Al_2O_3/HfO_2 nanolaminate layer stacks in different sequences with up to three layers and thicknesses in the nanometer range have been deposited on silicon wafers with a native oxide layer using atomic layer deposition (ALD). A description of the samples with respect to their layer sequence can be found in [4]. In addition, a part of the samples was annealed in N_2 atmosphere at two different temperatures to induce interface diffusion.

The experiments were carried out using PTB's in house built setup [5] for GIXRF and XRR at variable angles of incidence. The emitted fluorescence radiation is detected by means of a calibrated fluorescence detector. The samples were measured at two PTB beamlines [9,10] at the BESSY II electron storage ring. Two photon energies were used to optimize the excitation conditions for Al-K and Hf-L lines and the GIXRF and XRR measurement were conducted in parallel at both photon energies. The recorded fluorescence intensities for oxygen, aluminium and hafnium can directly be transferred into mass depositions by the reference-free XRF approach [8]. The used fundamental parameters are either derived from databases or from dedicated experiments at PTB [11,12].

RESULTS

Results Of The Conventional Modeling Approach

For the sake of comparison and to demonstrate the weaknesses of the conventional modeling strategy for XRR and combined GIXRF-XRR data, selected XRR measurements were first modeled using a conventional fitting approach (using the layer densities, thicknesses and roughnesses as the modeling parameters). In left side of Fig. 1, the resulting modeled curve is compared to the experimental data taken at 10 keV for an annealed sample with a Al_2O_3 / HfO_2 / Al_2O_3 layer sequence. The right hand side shows the corresponding calculation of the Hf GIXRF signal which corresponds to the layer stack modeled with the XRR data in comparison to the experimental curve.

Even though both calculations nicely reproduce the experimental data, significant deviations occur when comparing the resulting modeled mass depositions (product of thickness and density of each layer) to the quantified mass depositions. In the case of Al, the total modeled mass deposition is exceeding the quantified one by up to 62 % which indicates that the conventional modeling suffers from severe correlation effects due to the large number of free parameters.

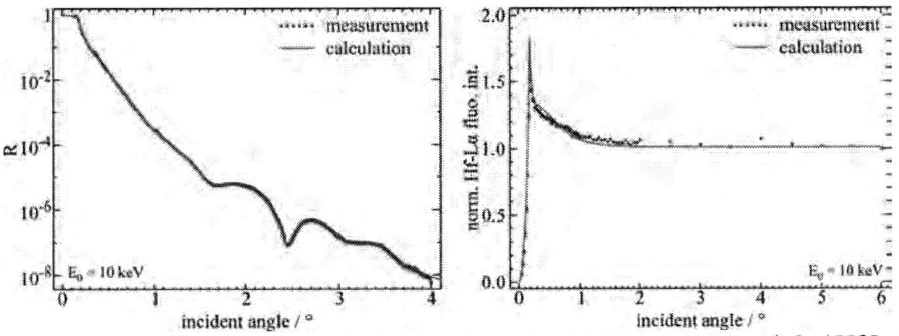

FIGURE 1. Comparison of conventionally modeled XRR of sample S4 800 °C (left side, Al_2O_3 / HfO_2 / Al_2O_3 layer sequence) and the resulting GIXRF signal (right side) for Hf in comparison to the experimental data.

Results Of Reference-Free GIXRF-XRR

The combined modeling for reference-free GIXRF-XRR calculates quantitative concentration depth profiles for every layer. These depth profiles can overlap at the interfaces depending on the respective intermixing coefficient. The concentration depth profiles are then used to calculate depth profiles for each optical constant at the respective photon energies. The intermixing regions are separated in thin sublayers in order to calculate both the resulting XRR as well as the X-ray standing wave fields (XSW). The derived intensity distribution within the XSW is then used in conjunction with the calculated concentration depth profiles and other relevant instrumental and fundamental parameters to calculate the angular fluorescence profiles for Al and Hf.

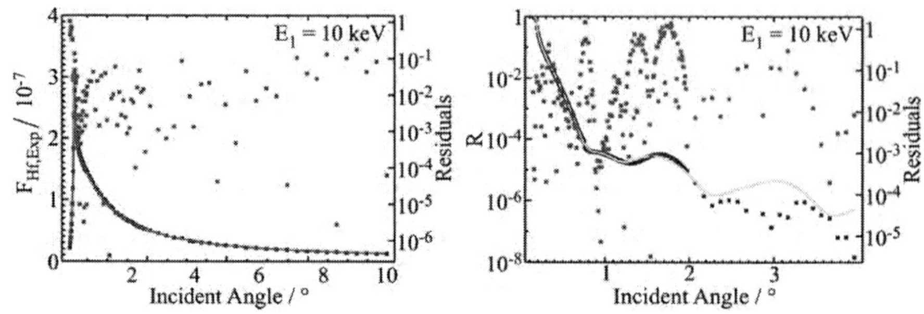

FIGURE 2. Results of the reference-free GIXRF-XRR (solid lines) for sample S5 (Al_2O_3 / HfO_2 / Al_2O_3 layer sequence) in comparison to the experimental data (black dots). The residuals are also plotted as violet dots.

In Figure 2, the modeling results of the combined reference-free GIXRF-XRR modeling routine for sample S5 are shown. The agreement between modeled and experimental data is good as all relevant features of the experiment are well reproduced. The deviations in the fluorescence data are the highest in the very low incident angle regions, where the uncertainties associated with the calculation of the solid angle of detection are the highest.

A similar procedure was applied to the samples S4 and the annealed S4 800 °C sample. The concentration depth profiles are shown in Figure 3, where the solid line depth profiles correspond to sample S4 and the dotted ones to sample S4 800°C. An increase of the interfacial intermixing for the annealed sample is clearly visible for all interfaces. The as-deposited sample shows no relevant intermixing, which is in line with the expectations of ALD processing [13]. The observed symmetric intermixing for the sample, which was annealed for 40 s at 800 °C is in-line with the findings in the work of Lan et al. [14] and also with the TEM image obtained on the non-aged region of sample S4 900 °C (see right hand side of Figure 3). Note also the increase of the modeled thickness of the SiO_2 layer on the annealed sample. The SiO_2 layer thickness in the TEM image in Figure 3 is in the same order as for the modeling results.

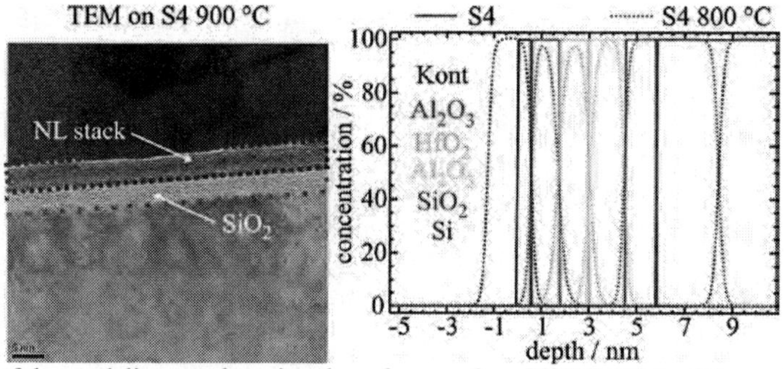

FIGURE 3. Comparison of the modeling results using the reference-free GIXRF-XRR of the as-deposited and the annealed sample S4. For comparison, also a TEM image is shown.

CONCLUSION

In summary, we show how a conventional modeling of XRR or combined GIXRF-XRR data, where the density, thickness and roughness of each layer serve as independent modeling parameters can suffer from correlation effects because of the large amount of free model parameters. In contrast, the reference-free GIXRF-XRR, where a direct access to the present mass depositions is enabled, allows for a reduction of the degrees of freedom. This leads to a more reliable interpretation of the experimental data compared to the conventional modeling approach.

ACKNOWLEDGMENT

This research was performed within the EMPIR projects '3DStack' and '3D MetChemIT'. The financial support of the EMPIR program is gratefully acknowledged.

REFERENCES

1. R.D. Clark, *Materials* **7**, 2913-2944 (2014).
2. J. Azadmanjiri, C.C. Berndt, J. Wang, A. Kapoor, V.K. Srivastava, C. Wen, *J. Mater. Chem. A* **2**, 3695-3708 (2014).
3. D. de Boer, W. van den Hoogenhof, *Spectrochim. Acta B* **46(10)**, 1323 (1991).
4. P. Hönicke, B. Detlefs, M. Müller , E. Darlatt, E. Nolot, H. Grampeix, B. Beckhoff, *Phys. Stat. Solidi A* **212(3)**, 523 (2015).
5. J. Lubeck, B. Beckhoff, R. Fliegauf, I. Holfelder, P. Hönicke, M. Müller, B. Pollakowski, F. Reinhardt, J. Weser, *Rev. Sci. Instrum.* **84**, 045106 (2013).
6. A.J.G. Leenaers, D.K.G. de Boer, *X-Ray Spectrom.* **26**, 115-121 (1997).
7. B. Caby, F. Brigidi, D. Ingerle, E. Nolot, G. Pepponi, C. Streli, L. Lutterotti, A. André, G. Rodriguez, P. Gergauda, M. Morales, D. Chateigner, *Spectrochim. Acta B* **113**, 132-137 (2015).
8. M. Müller, P. Hönicke, B. Detlefs, C. Fleischmann, Materials **7(4)**, 3147 (2014).
9. F. Senf, U. Flechsig, F. Eggenstein, W. Gudat, R. Klein, H. Rabus und G. Ulm, J. Synchrotron Rad. **5**, 780 (1998).
10. M. Krumrey and G. Ulm, Nucl. Instrum. Meth. A **467-468**, 1175 (2001).
11. P. Hönicke, M. Kolbe, M. Krumrey, R. Unterumsberger, B. Beckhoff, *Spectrochim. Acta B* **124**, 94 (2016).
12. P. Hönicke, M. Kolbe, M. Müller, M. Mantler, M. Krämer, B. Beckhoff, *Phys. Rev. Lett.* **113(16)**, 163001 (2014).
13. M. Knez, *Semicond. Sci. Technol.* **27**, 074001 (2012).
14. X. Lan, X. Ou, Y. Cao, S. Tang, C. Gong, B. Xu, Y. Xia, J. Yin, A. Li, F. Yan, Z. Liu, *J. Appl. Phys.* **114**, 044104 (2013).

KEYWORDS

GIXRF, XRR, quantitative depth profiling, nanolaminates

From Large Research Instruments To An Industrial Control: X-ray Photoelectron Spectroscopy Characterizations Of Advanced Technology Gate Stack.

L. Fauquier[1,2,3*], B. Pelissier[2,3], J. Avila[4], M.-C. Asensio[4], D. Le Cunff[1], D. Doloy[1], C. Beitia[2,5] and T. Baron[2,3]

[1] *STMicroelectronics, 850 Rue Jean Monnet, 38920 Crolles, France*
[2] *Univ. Grenoble Alpes, F-38000 Grenoble, France*
[3] *CNRS, LTM, MINATEC Campus, F-38054 Grenoble, France*
[4]*Synchrotron SOLEIL, Saint Aubin, BP 48 91192 Gif-sur-Yvette, France*
[5] *CEA, LETI, MINATEC Campus, F-38054 Grenoble, France*

INTRODUCTION

With the continuous downscaling of integrated circuit, metrology techniques have to face new challenges to support the development of advanced technology nodes. One of them is the composition determination of nanometer thick multilayers films. Since the 32nm node, the SiO_2 gate insulator has been replaced by the High-k Metal Gate (HKMG) film stack **[1]**. For the 14nm node, this HKMG stack is composed of a high-k HfON layer deposited on a SiON interfacial layer (IL), to prevent the formation of a bad quality SiO_2 layer between the HfON layer and the channel **[2]**. To reach the transistor specifications, thickness and composition of the different deposited layers must be precisely controlled. For the HKMG stack, the amount of nitrogen in the HfON and SiON layers is specifically critical. Nevertheless, for such thin layers, conventional optical metrology techniques, well suited for volume measurements, are difficult to implement. This is why X-ray based techniques, more suited for surface characterization, are increasingly used. X-Ray Photoelectron Spectroscopy (XPS) is a well-established method for the analysis of ultrathin films thanks to its surface sensitivity (<100Å) and its ability to chemically characterize samples **[3]**. Hence, in this work we present three different XPS systems, used to characterize the HfON/SiON HKMG stack: a Nano Angle Resolved Photoelectron Spectroscopy (Nano–ARPES) system using the SOLEIL synchrotron ANTARES beam line **[4]**, a Thermo Fisher Scientific laboratory pARXPS (parallel Angle Resolved XPS) system and an in-line NOVA system. The Nano-ARPES and the pARXPS, thanks to the angular information, can provide depth resolved information on the chemical state of near-surface layers **[5]**. The in-line NOVA system, optimized for an industrial use, can quickly and accurately provide ultra-thin film stack information. These three tools, based on the same technique, exhibit different way to implement it, whether it is resolution driven or used in a production purpose (Figure 1).

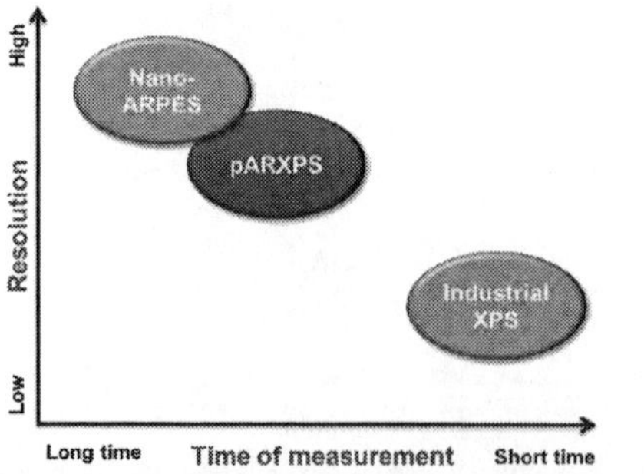

FIGURE 1. Comparison of the three XPS capabilities based on the resolution and the time of measurement.

Hence, these tools where used in conjunction to address the needs of industrial control by measuring the same HfON/SiON stack.

XPS ANALYSIS

The same industrial HfON/SiON HKMG gate stack developed for the 14nm node has been measured on the three different XPS equipment. First a comparison of the XPS spectra will be shown. Chemical profile reconstructions of this stack have also been done thanks to the angular information obtained with the Nano-ARPES and pARXPS. Finally, based on the results acquired with these two advanced laboratory tools, a robust in line XPS measurement program has been developed on the in-line NOVA system to control the process monitoring.

XPS Spectra Comparison

The Hf4f, Si2p, N1s and O1s spectra of the HfON/SiON industrial gate stack have been acquired on the three different XPS equipment. First, a comparison of these spectra have been done and will be presented (Figure 2). The main assets (resolution, sensibility, rapidity…) and the limits of each tool will be discussed, based on these results and the nature of the tool.

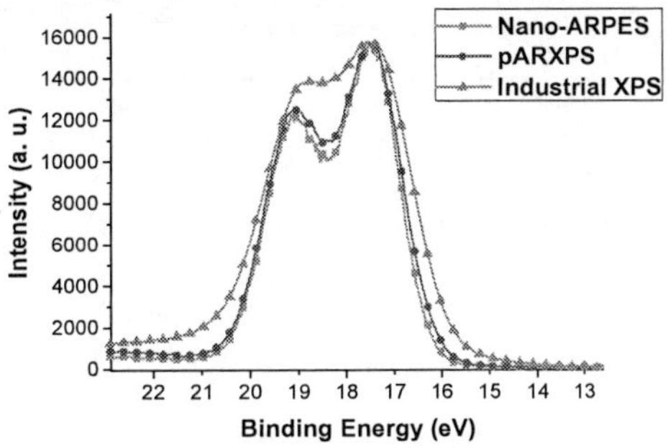

FIGURE 2. Comparison of Hf4f XPS spectra obtained by Nano–ARPES, pARXPS and industrial XPS.

XPS Chemical Profile Reconstruction

One key characteristic to investigate after the HfON/SiON deposition process is the nitrogen distribution in this stack. Thanks to angular information obtained by Nano–ARPES and pARXPS, chemical profile reconstructions can be performed, allowing us to precisely determine the nitrogen distribution. The chemical profile reconstructions have been extracted with the same software, developed by ThermoFisher, which is based on the maximum entropy concept. It is to be noted that a substantial work has been done on the Nano-ARPES data to be compared with the pARXPS.

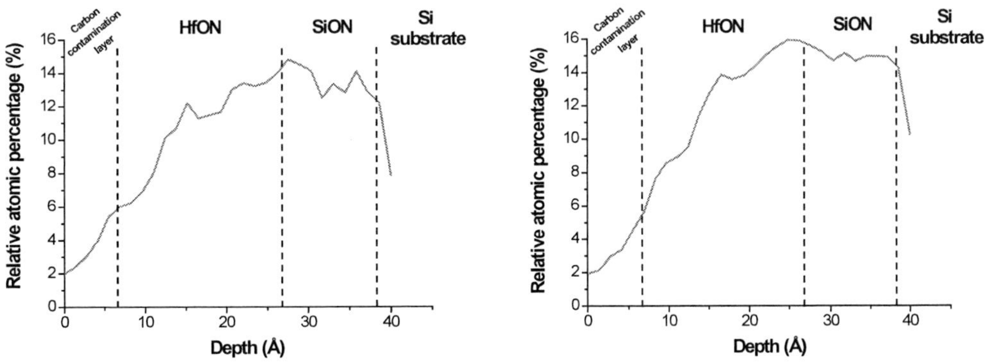

FIGURE 3. Nitrogen distribution obtained by chemical profile reconstruction from the pARXPS spectra (left) and the Nano–ARPES spectra (right).

The chemical profiles, and more specifically the nitrogen profiles, obtained by Nano-ARPES and pARXPS (Figure 3) will be compared. The profile differences and similarities will be discussed based on the XPS spectra. Although a significant difference in the tool characteristics, we will see that these two XPS give consistent results.

XPS For Industrial Control

From an industrial perspective, one of the key requirements to ensure the transistor integrity is to control very precisely the nitrogen percentage in the HfON and SiON layers. Based on the Nano–ARPES and pARXPS analysis, we will show that a robust in-line XPS measurement program has been developed (Figure 4). This program allows to measure simultaneously and independently the nitrogen amount in the HfON layer and in the SiON layer, as well as the HfON and the SiON thicknesses.

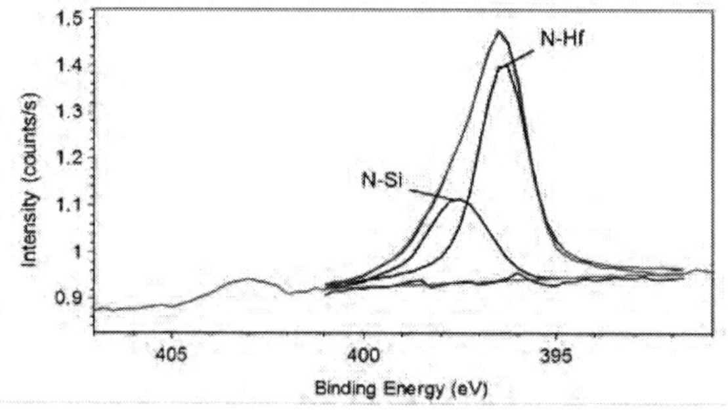

FIGURE 4. N1s XPS spectra obtained on the in-line XPS and the measurement program modeling.

We will show that this program has been tested on fifty wafers that have undergone the same process. The results will be presented and the accuracy of this measurement program discussed.

CONCLUSION

The capabilities of three different XPS equipment have been compared on a same HfON/SiON sample. These tools were then used in conjunction to develop a robust in line XPS measurement program to characterize the HKMG 14nm node gate stack. From the Nano-ARPES at the SOLEIL synchrotron, a large research instrument, to an in-line XPS system, XPS was found to be a very valuable metrology technique to characterize ultrathin layers.

ACKNOWLEDGEMENT

This work has been partially supported by the French Government program "Investissements d'Avenir" managed by the National Research Agency (ANR) under the contract number ANR-10-EQPX-33.

REFERENCES

1. ITRS, "Roadmap for Semiconductors - Process Integration, Devices, and Structures," 2013.
2. M. Cho, J. Park, H. B. Park, C. S. Hwang, J. Jeong, and K. S. Hyun, *Applied Physics Letters* **81,** 334 (2002).
3. J. F. Watts, J. Wolstenholme, "An introduction to surface analysis by XPS and AES", *Wiley-VCH* (2003).
4. J. Avila, I. Razado, S. Lorcy, B. Lagarde, J.L. Giorgetta, F. Polack and M. C. Asensio, *Journal of Physics: Conference Series* **425**, (2013).
5. M. Py, J. P. Barnes, R. Boujamaa, M. Gros-Jean, K. Nakajima, K. Kimura, C. Roukoss, B. Pelissier , N. Gambacorti , *J. Vac. Sci. Technol.* pp.B 29 (2011).

KEYWORDS

High-k Metal Gate (HKMG), Nano-ARPES, pARXPS, Profiling, XPS measurement program, 14nm node, Metrology.

XPS/ARXPS In Thin Film And Nanomaterial Process Control

Chris Moffitt[1], Simon Hutton[2], Sarah Coultas[2]

Kratos Analytical, Inc., 100 Red Schoolhouse Rd., A6, Chestnut Ridge, NY 10977 USA
Kratos Analytical, Ltd., Wharfside, Trafford Wharf Road, Manchester M171GP UK

INTRODUCTION

The measurement and understanding of interfacial properties and surface chemistry is a primary concern across a wide range of technological disciplines, and numerous examples of materials systems failures due to unexpected interface phenomena exist throughout industry. The characterization of surface properties become even more critical for nanomaterials, since by definition the entire volume of a nanostructure can lie within what was traditionally considered the surface region of a material, so that the bulk materials properties are no longer dominant and the surface properties manifest themselves as the prevalent factors governing the final nanomaterial properties.

Having the ability to measure the chemical bonding of elements in the first few atomic layers of a material is vital to interfacial development for all materials systems. X-ray photoelectron spectroscopy (XPS) provides signal from just these outer few atoms, and the information that can be retrieved from it can provide chemical bonding and elemental depth data, illuminating the layered chemistry of thin films, including film structures on nanoparticles, yielding core shell thickness and chemical information[1].

Nanoelectronics development will require careful measurement of this surface chemical information for process control. Similar to how angle resolved XPS (ARXPS) became critical in the tuning of the modern transistor gate oxide structure[2], ARXPS will have a large role to play for the wide range of materials for application in nanoelectronics, measuring the various films and process results.

ARXPS IN THIN FILM METROLOGY AND SURFACE CHEMISTRY

ARXPS has an enduring place in the semiconductor metrology arsenal, providing chemical and distribution information from the thinnest layers of materials. By simply changing the collection angle with respect to the surface normal of a thin film, ARXPS allows the data from atoms in the outer ~10nm of the material to be measured. Powerful mathematical models have been developed to reduce data to physical information, revealing depth and stoichiometric details. Figure 1 shows the modeling results from ARXPS data taken from a carbon based plasma polymer deposited on a silicon wafer with native passivation, revealing the film thickness and the detail of the material interface region.

Developing sample structures to test and validate measurements for different materials classes will be necessary to improve accuracy and detail for a particular materials system. Figure 2 shows the thickness sensitivity of modelling ARXPS data from planar film structures. It is through understanding the effects in traditional planar thin film systems that the basis for interpreting nanostructure results must be built. In this poster we will present the results from sets of samples from several different materials classes, showing how elemental properties and material densities affect the depth information, as the transport of electrons through covering materials governs the signal strength from buried elements. This effect will be displayed by comparison of buried layer depth sensitivity between light element films and denser films having constituents with higher atomic number.

Choices made for physical constants and for data fitting parameters can have significant effects on model results, and must be carefully assessed, and the effects of these choices will be seen in an example. The strength of the measured signal also gives rise to the effects of counting statistics on the modeling results, and the importance of these on the interpretation of deeper layer information will be discussed.

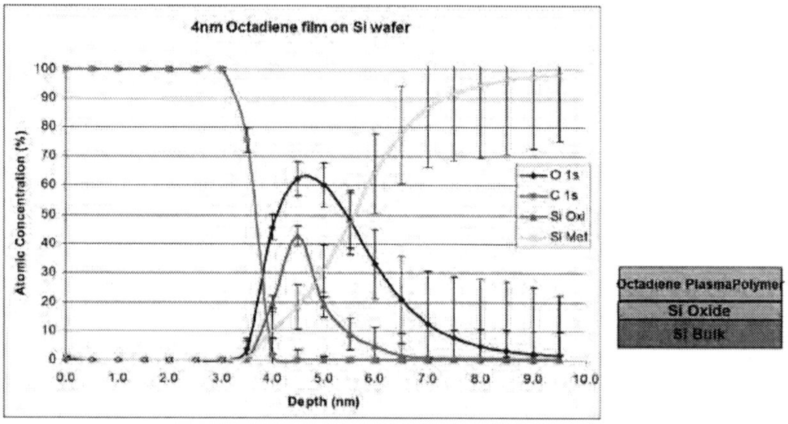

FIGURE 1. Maximum entropy modeling results from ARXPS data from an oxtadiene plasma polymer film deposited on the native oxide of a silicon wafer.

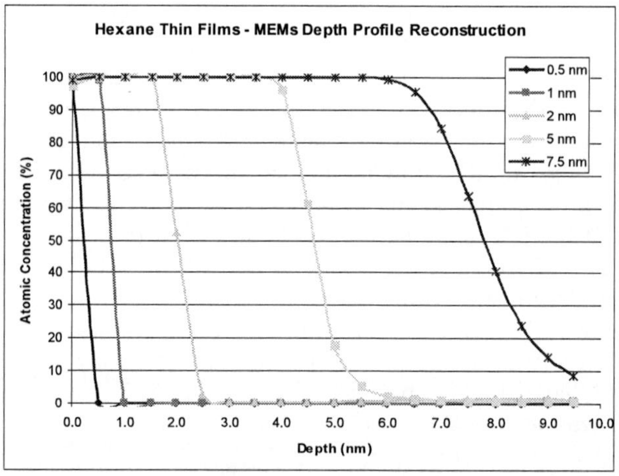

FIGURE 2. ARXPS/MEMs results from varying thicknesses of hexane plasma polymer films on silicon wafers, showing the thickness measurement capability inherent in the technique.

REFERENCES

1. A.G. Shard, *J. Phys. Chem. C*, 2012, **116** (31), pp 16806–16813.
2. C.R. Brundle, G. Conti, P. Mack, *J. Elec. Spec. and Rel. Phen.* **178**, 433-448 (2010).

KEYWORDS

XPS, ARXPS, Maximum Entropy Method

Evaluations of graphene to graphene contacts

Jia-Bin Chen and Dung-Ching Perng*

Institute of Microelectronics and Dept. of Electrical Engineering
National Cheng Kung University, One University Road, Tainan 701, Taiwan

INTRODUCTION

Graphene is a two-dimensional material that has a hexagonal lattice structure with carbon atoms tightly packed via strong sp^2 bonding. Graphene film has attracted tremendous interest in basic research and applications due to its remarkable physical, chemical, and electronic properties [1]. It is considered as an attractive photonic material [2,3], as well as an option for post-silicon electronics [4]. In addition, interconnect performance is a dominated factor in nano-era integrated circuits. The current density of copper limits the performance of interconnects with nano-scaled dimensions [5]. The development of future high-performance power-efficient technology demands thinner and more conductive interconnects with better current drivability. Recently, carbon-based materials have been considered as one of the most promising candidates for future interconnect technology. Carbon nanotube interconnect [6,7] and graphene structures [8] are in the early stages of development. Graphene nanoribbons have been proposed for use as an interconnect conducting material to transport large currents [9]. Furthermore, tri-layer graphene film (~1nm thick) has also been proposed as an excellent Cu diffusion barrier for Cu interconnect [10]. Graphene wires are capable of passing the same current with a much thinner wire thickness, which helps reduce parasitic capacitance and, hence, chip power consumption. It draws many research interests for graphene's application to integrated circuits.

Junctions between graphene and various metallic contacts have been extensively studied [11−14]. However, graphene to graphene contact is possible for graphene electronics and has not been thoroughly studied. Here, we investigated graphene to graphene contact resistance. Graphene stripes were fabricated using 900°C annealing of patterned C/Ni stacked thin films. We transferred one graphene stripe at a time on a SiO_2 coated silicon substrate such that the stripes crossed each other. The individual stripe's resistance and the graphene to graphene contact resistance were studied.

SAMPLE FABRICATION

A 400nm-thick Ni film was deposited on a SiO_2-coated Si substrate using e-beam evaporation method, followed by sputtering deposition of a ~5nm-thick carbon film. A shadow mask with 2 mm-wide lines was used to pattern the stacked Ni/C films. After deposition, the sample was then loaded into a rapid thermal annealing (RTA) chamber. The temperature of the RTA chamber was increased to 900°C. The dwell time of 900°C annealing was 10 min. To grow graphene film, the RTA chamber cooling rate was well-controlled at 10°C/s. Well know standard graphene-transfer method [15], spin coating PMMA, $FeCl_3$ etching of Ni film, and acetone removal of PMMA, was adopted to transfer the graphene film to another SiO_2-coated Si substrate. We repeat this transfer of graphene stripe to the same sample such that a cross-patterned graphene stripes was fabricated, as shown in Figure1. Line resistances of the 2 mm-wide graphene stripes and their graphene to graphene contact resistance were measured and investigated by TEM.

RESULTS AND DISCUSSION

Figure 2 shows the Raman analyses of the graphene stripes at (a) top graphene stripe, (b) bottom graphene stripe, and (c) crossed-area, in Fig 1. Three different points on each stripe and crossed-area were analysed. From the intensity ratio of 2D/G bands of Fig.2a and 2b, it indicates that each individual graphene stripe is likely bi-layer graphene sheet. Few graphene layers may be grown on the edge of the stripes because of non-uniform Ni film thickness, as can be observed by its darker colour from the optical image (OM) in Fig1b or Fig.3a. The high D band intensity indicates that defect density is high in these graphene sheets which is typical for graphene grown by solid carbon source on Ni film. The I_{2D}/I_G ratio of the crossed area indicates the cross-area consists of multi-layer graphene.

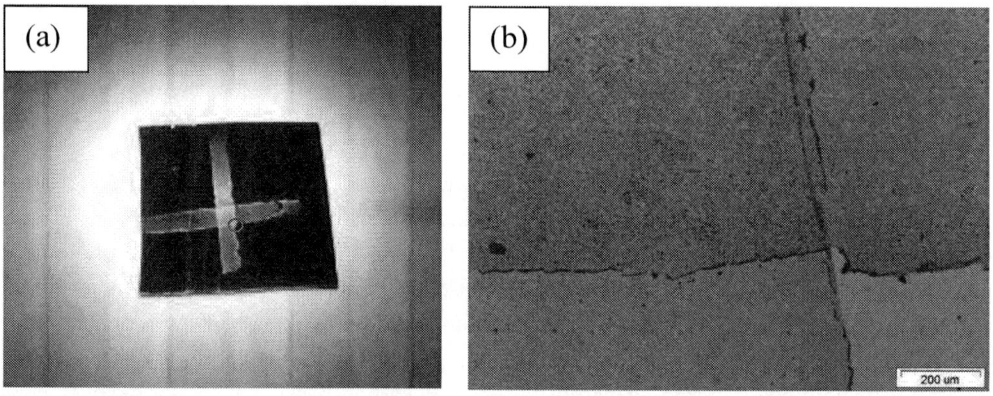

FIGURE 1. (a) Image of the cross-patterned 2-mm wide graphene stripes and (b) enlarged optical image of the red circle at the lower-right corner of the graphene cross in (a).

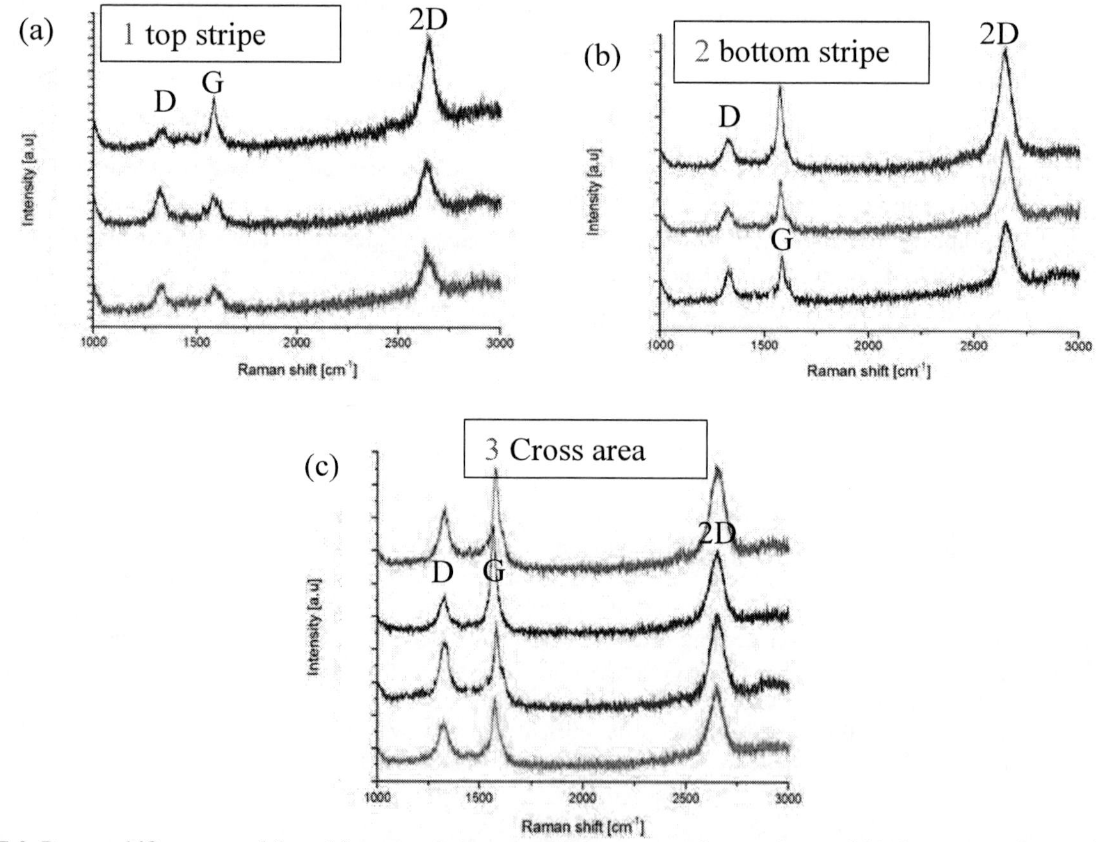

FIGURE 2. Raman shifts measured from (a) top graphene stripe (b) bottom graphene stripe, and (c) the two graphene stripes crossed or contact area.

Several double-crossed graphene stripes samples were fabricated for resistance measurements as shown in Figure 3a. More graphene layers were grown at the stripe's edge (brighter purple colour) due to more carbon out-diffuse at thinner Ni film during the RTA cooling step. The numerical numbers indicated in Fig 3a are the probe locations. We placed indium dots at the probe locations as the contact material. Graphene stripe's line resistance was measured first. Probing point 5 to point 2 can obtain contact A's contact resistance (after minus some stripe resistance), or simple using four-point probe technique. And probing point 1 or point 2 to point 4 can extract contact A and contact B contact resistances. The average resistance of these 2-mm-wide graphene stripes is 1.14 kΩ/sq. and the average graphene-graphene contact resistance is 4.12 kΩ. The studied graphene sheets are polycrystalline graphene which has much higher resistance than single-crystal graphene because a graphene grain boundary can contribute ~2.1kΩ to the resistance [16]. Figure 3b shows the cross-sectional TEM image of contact B as the red circle indicated in Fig.3a. It indicates that there are voids between the multi-layer graphene stripes. Average distance of the gap is ~1.04 nm. The extra resistance provided from the gap is yet to be studied. This gap, created by typical graphene transfer, has important implications for future studies and applications in graphene electronics.

FIGURE 3. (a) Double-crossed graphene stripes (the numerical numbers are the probe locations) and (b) TEM cross-sectional image of contact B at the red circle indicated in (a).

REFERENCES

1. K. Geim, K. S. Novoselov, *Nature*, **6**, 183, (2007).
2. F. Bonaccorso, Z. Sun, T. Hasan, and A.C. Ferrari, *Nat. Photonics*, **4**, 611, (2010).
3. N. Grigorenko, M. Polini, and K. S. Novoselov, *Nat. Photonics*, **6**, 749, (2012).
4. F. Schwierz, *Nat. Nanotechnol.*, **5**, 487, (2010).
5. J. Liang, S. Yeh, S. S. Wong, and H.-S. P. Wong, *Proc. IEEE IMW*, 2012, pp. 1–4.
6. B. Q. Wei, R. Vajtai, and P. M. Ajayan, *Appl. Phys. Lett.* 79, 1172 (2001).
7. G. F. Close, S. Yasuda, B. Paul, S. Fujita, and H. S. P. Wong, *Nano Lett.* 8, 706 (2008).
8. H. M. Wang, Y. H. Wu, Z. H. Ni, and Z. X. Shen, *Appl. Phys. Lett.* 92, 053504 (2008).
9. A. Naeemi and J. D. Meindl, *Proc. of IEEE IITC, IEEE Xplore* (2008), pp.183–185.
10. B.S. Nguyen, J.F. Lin and D.C. Perng, *Appl. Phys. Lett.* 104, 082105 (2014).
11. S. Russo, M.F.Craciun, M.Yamamoto, A.F. Morpurgo, S. Tarucha, *Physica E* 42, 677 (2010).
12. F. Xia, V. Perebeinos, Y.M Lin, Y. Wu and P. Avouris, *Nat. Nanotech.* 6, 179 (2011).
13. K. Nagashio, T. Nishimura, K. Kita, and A. Toriumi, *Appl. Phys. Lett.* 97, 143514 (2010).
14. A. Venugopal, L. Colombo, and E. M. Vogel, *Appl. Phys. Lett.* 96, 013512 (2010)
15. W. Regan et al., *Appl. Phys. Lett.* 96, 113102 (2010).
16. Q. Yu et al., *Nat. Mater.* 10, 443 (2011).

KEYWORDS

Graphene, Nickle, solid carbon source, graphene to graphene, contact resistance

Nanoscale Analysis of Conductance Switching Dynamics and Current Hysteresis in $(GeTe)_2$/ Sb_2Te_3 Superlattice Films Using Scanning Probe Methods

Leonid Bolotov, Tetsuya Tada, Yuta Saito, Noriyuki Miyata, and Junji Tominaga

Nanoelectronics Research Institute, Nat. Inst. of Advanced Industrial Science and Technology (AIST), 1-1-1 Higashi, Tsukuba, Ibaraki 305-8565, Japan
Phone: +81-29-861-3760, e-mail: bolotov.leonid@aist.go.jp

INTRODUCTION

New class of 2D materials such as chalcogenide superlattice (SL) films made of a stack of nanometer-thick layers of GeTe (ferroelectric) and Sb_2Te_3 (topological insulator), have been of practical interest for memory applications in order to realized multi-level and energy-efficient memory cells. In contrast to conventional alloy materials, crystalline chalcogenide superlattices offer new functionality owing to its magneto-electric and magneto-optic properties. [1,2]

While promising characteristics of phase-change SL-based memory cells have been reported in the literature, there remain challenges in superlattice's growth control and its integration with device fabrication processes.[3,4] Conventional techniques give area averaged film characteristics, which is insufficient for analysis of nanocrystalline SL films with a grain size less than 100 nm. Therefore, nanoscale characterization of electrical and structural transformations in the SL films under external voltage and current are highly desirable to predict the device characteristics. In order to optimize superlattice structure as a recording media and to evaluate spatial uniformity of electrical characteristics for designed superlattice films, we have developed an approach integrating multiple nanoscale methods of scanning probe microscopy. Here, we presented results on variations of switching voltage, current hysteresis, and stability of the ON-state for $[(GeTe)_2/ (Sb_2Te_3)]_4$ superlattice films as a function of position and film growth process. The presented nanoscale analysis methods are usable for evaluation a wide range of 2D systems for nanoelectronics.

SUPERLATTICE STRUCTURE AND NANOSCALE ANALYSIS METHODS

$[(GeTe)_n/ (Sb_2Te_3)_m]_k$ superlattice films (n, m, k are integer indexes) were grown on Si substrates by Ar ion sputtering technique. In this report, we focused on properties of SLs with n=2, m=1 and k=4 fabricated using single- (called A-sample) and two-step (called B-sample and C-sample) processes. The two-step process included the following: (1) the deposition of a 3-nm-thick Sb_2Te_3 seed layer at room temperature and subsequent annealing at 230°C, and (2) the deposition of 4 periods of GeTe(0.8 nm)/Sb_2Te_3(1 nm) by repeated sputtering of GeTe and Sb_2Te_3 alloy targets at a pressure of 0.5 Pa, with an Ar flow rate of 10 sccm, a power of 20 W and a substrate temperature of 230°C. Uniform crystal orientation and film thickness were confirmed by x-ray diffraction (XRD) and atomic force microscopy (AFM). [5,6]

To measure time evolution of electrical and structural responses of SLs caused by voltage pulses of different amplitudes, duration and injection power, we have been using a multimode scanning probe microscopy (MSPM) system [7] installed in an UHV chamber (~$8x10^{-8}$ Pa). Comparing with conventional scanning probe microscopy, the key advantages of our methods are (1) the non-contact operation eliminates mechanical stress on the SL film by the metal probe, which is applicable to soft films; (2) simultaneous acquiring current and interaction force signals allows adjustment of the probe-sample gap at a sub-nanometer level and fine tuning both injection current and local electric field; (3) continuous monitoring of time variations of SL film characteristics (electric current, electrostatic force and film morphology) in different measurements stages (WRITE, READ, ERASE phases).[8] We used voltage pulses and gap-voltage spectroscopy to vary local field strength and to observe the film distortion by the applied field. Additionally, to observe current hysteresis, series of current-voltage spectra were acquired by sweeping bias voltage in forward direction with a negative voltage increment and in backward direction with a positive voltage increment at a rate of 1...10 V/s, while the MSPM probe was immobilized in predefined position on the film.

RESULTS AND DISCUSSION

All measured SL films showed granular structure with an average grain size of 20-100 nm confirmed by transmission electron microscope (TEM) and x-ray diffraction (XRD). However, response of the SL films upon application of external voltage and injection current depends on details of the SL growth. Large degree of uniformity of structural and electrical characteristics were observed for the B- and C- samples. ON-OFF current ratio and current hysteresis were large for the A-sample. Observed position-dependent characteristics discussed below are essential for quantitative evaluation of the SL performance as a recording media of novel memory cells.

Conductance State Switching by Voltage Pulses

Figure 1(a - c) shows an example of conductance state switching upon voltage pulses with V_R=+2.1 V and different duration for the B-sample. Five high current spots appeared in a current map at a READ voltage of +1.0 V after applying 7 pulses in positions indicated by "x", while there were negligible structural changes in the constant-gap topograph in Fig. 1(a). Spots with a diameter less than 5 nm were created in positions marked by 'x'. Large probe-sample current in the spots reflects local transition of the SL grains from initial low conductance state (LCS) (~0.1 pA at +1.0 V) to high conductance state (HCS) (~300 pA at +1.0 V). It is the major feature of the fabricated SLs, where the transverse conductance is altered by internal structural change caused by applied voltage and current.

Dynamics of the conductance state transition shown in Fig. 1(c) revealed important details of the OFF-ON switching behavior which varied from grain to grain. Two grains out of 7 did not change its LCS as seen at 12 and 22 s. The spot at 30 s was created after significant incubation time. We note that there 3 kinds of SL grain response: (1) 50% - easy switching grains (such as at 15 and 18 s); (2) ~8% - no switching below 3 V (at 12 and 22 s); (3) ~42% - switching after long incubation time (at 30 s). Using different WRITE voltage amplitude (V_W) and duration (Δt), we obtained a minimum switching threshold voltage of 1.6 V and a switching power of ~20 pW for the B-sample.

The observed incubation time and the SL expansion emphasis the role of electric field strength in the observed switching.[8] We employed the gap-voltage spectroscopy to vary local field strength and to assess these parameters critical for memory devices. All observed transitions were accompanied with an SL expansion of ~0.3 nm (~3%) such as seen at 33.5 s, the characteristic feature of the crystal-to-crystal structural phase transition for the B-sample. We observed the SL expansion of ~10% for the A-sample, and contraction by ~10 % for the C-sample, which suggests significant deviation from the perfect SL structure. The observed expansion of the SL films under external electric film resembles that of a ferroelectric crystal caused by displacement of constituent atoms.

Position-Dependent Current Hysteresis

To address spatial uniformity of ON-OFF switching, we measured I-V spectra along the SL film and obtained position variation of HCS-LCS current difference. Figure 1(d, e) shows typical analysis steps for the B-sample. The spectra showed hysteresis loop at V > 0 V, where the current in forward direction was ~2 orders of magnitude larger than that in backward direction. At +2.8 V the SL experienced a transition from LCS to HCS, similar to that observed under voltage pulse in Fig.1(c), and we see a gradual transient increase in the current at +2.8 ... +2.5 V. Small current loop appeared at -2.7 V, and the SL changed back to the LCS, resulting in small current value at V > 0 V in the backward sweep direction. The presence of current loops is a characteristic feature of bi-polar switching of the SLs.

To make quantitative comparison, I-V spectra were measured along the SL film with a spacing of 2 nm, and the HCS-LCS current difference was obtained at +1.2 V as a function of position as illustrated in Fig. 1(e). Majority of I-V spectra showed hysteresis loops with a current difference of 10 – 100 pA. In areas between 5 and 15 nm, and 200 and 215 nm small current values (~50-100 fA) were observed, i.e. no hysteresis loops. A *steady HCS current* was observed in an area between 180 and 198 nm, demonstrating a HCS/LCS ratio of ~2000. In contrast, the area between 25 and 50 nm showed different values of current in different traces corresponding to *unsteady HCS current* such as that observed at 8 - 10 s in Fig.1(c). Statistical analysis confirms existence of 3 kinds of grain areas in both cases: pulse excitation and I-V spectroscopy.

The results demonstrate that integrated SPM-based analysis methods are useful approach for evaluation of nanocrystalline chalcogenide superlattice films and other 2D composite films at nanoscale.

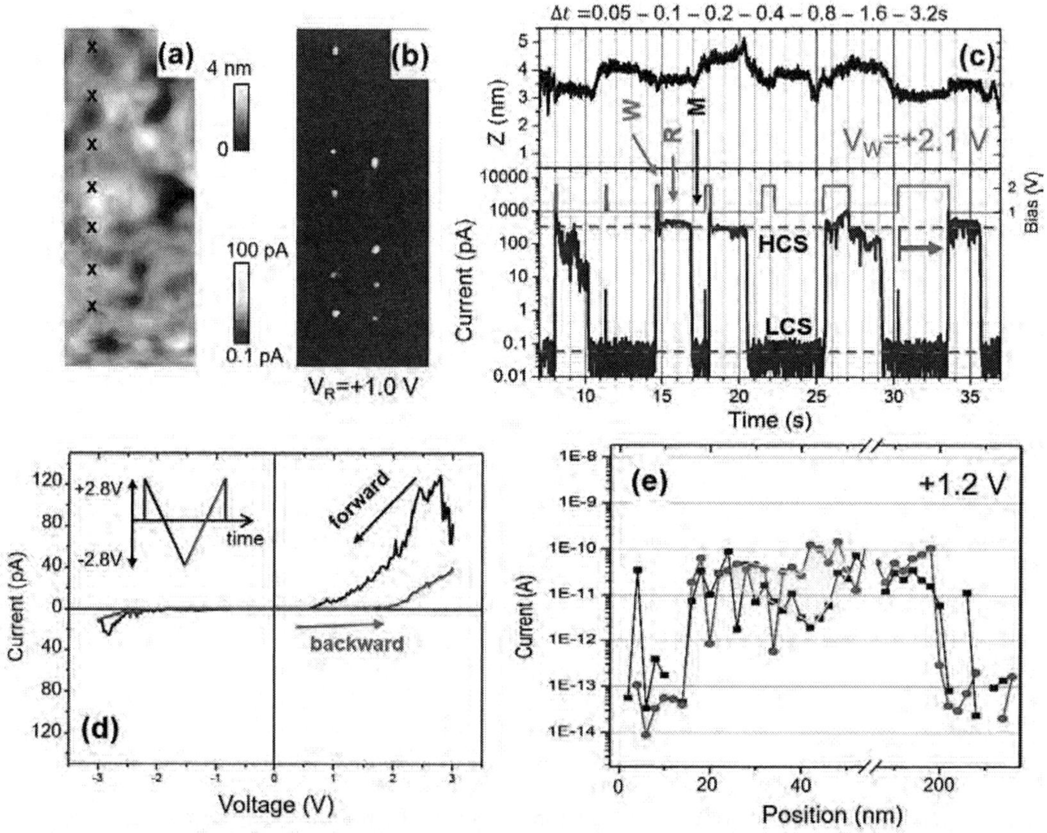

FIGURE 1. (a) constant-gap topograph (60x150 nm) and (b) corresponding current map at V_R after series of 14 voltage pulses. (c) time evolution of current (blue) and the gap (black) in response to voltage pulses (magenta) with WRITE (W), READ (R) and move (M) phases for positions marked by 'x'. (d) I-V spectrum with a hysteresis loop. Insert shows applied voltage vs. time. (e) Variation of HCS-LCS current difference at +1.2 V as a function of position for two traces.

REFERENCES

1. J. Tominaga et al., *Appl. Phys. Lett.* **99**, 152105 (2011)
2. D. Bang et al., *Sci. Rep.* **4,** 5727 (2014)
3. T. Ohyanagi et al., *Jpn. J. Appl. Phys.* **52,** 05FF01 (2013)
4. T. Egami et al., *Jpn. J. Appl. Phys.* **53,** 04ED02 (2014)
5. Y. Saito et al., *AIP Adv.* **6**(4), 045220 (2016)
6. Y. Saito et al., *Phys. Status Solidi* B **252**, 2151-2158 (2015)
7. L. Bolotov et al., *Jpn J Appl Phys* **54**, 04EK02 (2016)
8. L. Bolotov et al., *Sci Reports* **6**, 33223 (2016)

KEYWORDS

Chalcogenide superlattice, conductance switching, current hysteresis, scanning probe methods

Thermal cooling of high-power electronics using SiO_2 nanoparticle packings

Anil Yuksel[1], Eric J. Tervo[2], Baratunde A. Cola[2], Paul S. Ho[1], Michael Cullinan[1]
Zhuomin M. Zhang[2], Jayathi Murthy[3]

[1] *Department of Mechanical Engineering, The University of Texas at Austin, Austin, TX, 78712, USA*
[2] *GeorgeW. Woodruff School of Mechanical Engineering, Georgia Institute of Technology, Atlanta, GA, 30332, USA*
[3] *Henry Samueli School of Engineering and Applied Science, University of Califormia, Los Angeles, CA 90095, USA*

INTRODUCTION

Nanoparticles have received considerable interest due to their unique plasmonic behavior under the illumination of light and due to their potential for enhanced near-field radiative heat transport. Indeed, when metallic nanoparticles dispersed in solvents, they exhibit very strong and localized surface plasmon resonance (SPR) peaks which is caused by the collective free electron oscillations in the metal. Polar dielectric nanoparticle exhibit similar behavior due to their ability to support surface phonon polaritons (SPhPs), but these have not been explored in much detail. It has also been shown that at the nanoscale van der Waals forces are the dominant force between particles, which results in particle agglomeration and creates particle clustering[1]. Hence, the synthesis of nanoparticle packings with useful and tunable properties has been very important for the development of novel optical, thermal and electrical properties of nanostructures. In this study, we analyze the optical-thermal properties of SiO_2 nanoparticle packings by using Fourier transform infrared attenuated total reflection (FTIR-ATR) measurements on 10-20 nm and 60-70 nm diameter packed SiO2 nanoparticle beds as well as a bulk SiO_2 film. The reflectivity of the nanoparticle packing exhibits a resonant-like minimum at a certain wavelength, which is affected by the type of nanoparticle, size and interparticle spacing as well as the heat transport direction. We demonstrate the SPhP resonance in nanoparticle packing with theory and a computational method. The results show that SiO_2 nanoparticle packings exhibit a unique optical-thermal behavior and can be a very good candidate for next generation thermal cooling applications such as high heat dissipation electronics, LEDs, data centers, or other high heat flux applications.

FTIR Measurement

FTIR-ATR measurements on 10-20 nm and 60-70 nm diameter packed SiO_2 nanoparticle beds have been performed to observe the SPhPs resonance. It has been seen that SPhP resonance frequency, ω_{SPhP}, in the frequency range between the transverse optical (TO) and longitudinal optical (LO) phonon frequencies. The FTIR-ATR spectroscopy analysis of the nanoparticle bed shows minima in the spectrum which agrees with the expected [2-4] values of the second resonance ω_{TO} for SiO_2. Moreover, SPhPs resonance is consistent with the Fröhlich resonance frequency[5] and recent experiments[6].

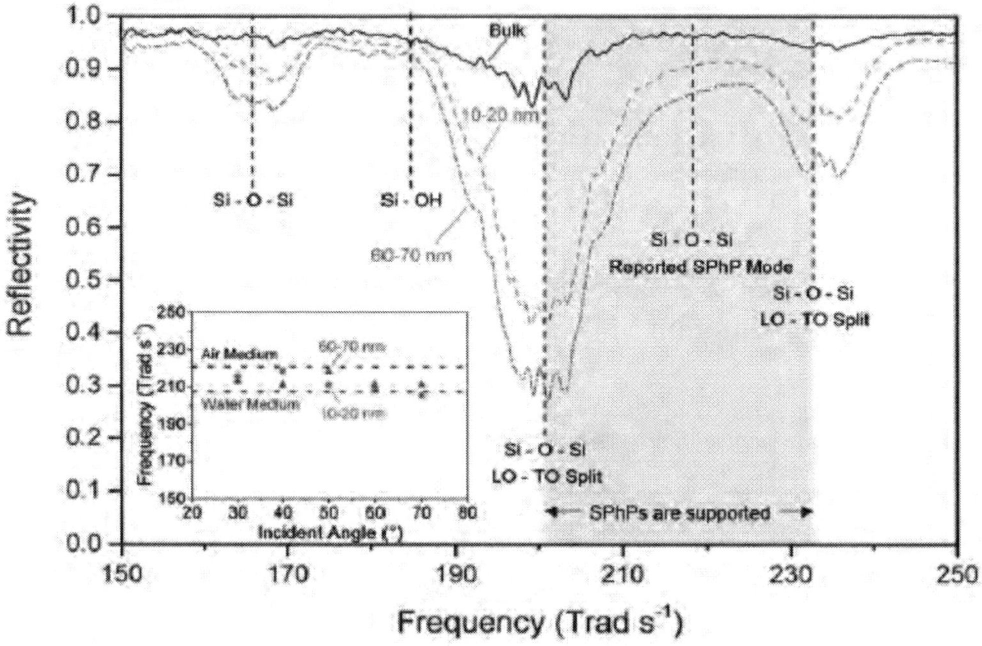

FIGURE 1. FTIR-ATR reflectivity of packed beds of 10-20 nm and 60-70 nm diameter SiO_2 nanoparticles, and a bulk SiO_2 film, vs frequency at an incident angle 70°

COMPUTATIONAL METHOD

Finite difference frequency domain has been applied to observe the reflection spectrum of the 10-20 nm and 60-70 nm diameter packed SiO_2 nanoparticle beds, and a bulk SiO_2 film by solving Maxwell's equations. Moreover, the SiO_2 nanoparticles were surrounded with air which the refractive index is taken as n=1 and dielectric function of the SiO_2 nanoparticle based on Lorentz model[7], $\varepsilon(\omega) = \varepsilon_\infty[1 + \frac{(\omega_{LO}^2 - \omega_{TO}^2)}{(\omega_{TO}^2 - \omega^2 - i\omega\Gamma)}]$, where ε_∞ is relative permittivity at infinite frequency, ω_{TO} and ω_{LO} are the resonance frequencies of transverse optical and longitudinal optical phonons, respectively, and Γ is the scattering rate. These properties for SiO_2 are also obtained from experimental data[7] ($\varepsilon_\infty = 2$, second resonance $\Gamma = 8.92$ Trads^{-1}, $\omega_{TO}=207$ Trads^{-1}, and $\omega_{LO}=234$ Trads^{-1}). We also analyze the scattering efficiency vs frequency of the SiO_2 nanoparticle bed by calculating the surface integration of the scattered Poynting vector.

CONCLUSION

Our analysis shows that nanoscale phenomenon affects the bulk material characteristics and exhibits unique optical-thermal behavior which demonstrates the possibility of a new class materials useful for better heat dissipation from high-power electronics, LEDs, data centers and other high heat flux applications. As surface phonon polariton coupling plays a key role within the nanoparticle packing and could be capable of enhancing thermal transport over macroscopic length scales. Additionally, tunability of heat transport could be achieved because the SPhPs may be sensitive to external electric fields. The enhanced heat transport by SPhPs could be activated by thermal self-emission[9] or propagating infrared light[10, 11]. It is also concluded that heat transport at discrete resonance frequencies could enable novel methods for controlling heat flow in packed nanoparticle beds.

REFERENCES

1. Yuksel A, Cullinan M. Modeling of nanoparticle agglomeration and powder bed formation in microscale selective laser sintering systems. Additive Manufacturing. 2016 Jul 18.
2. R. M. Almeida, Phys. Rev. B: Condens. Matter Mater. Phys., 1992, 45, 161–170.
3. G. Borstel, H. J. Falge and A. Otto, Solid-State Physics, Springer, Berlin, Heidelberg, 1974, vol. 74, ch. 2, pp. 107–148.
4. W.-T. Liu and Y. R. Shen, Phys. Rev. B: Condens. Matter Mater. Phys., 2008, 78, 024302.
5. H. Frohlich, Theory of Dielectrics, Clarendon Press,. 1949
6.. A. Babuty, K. Joulain, P.-O. Chapuis, J.-J. Greffet and Y. De Wilde, Phys. Rev. Lett., 2013, 110, 146103
7. Z. M. Zhang, Nano/microscale heat transfer, McGraw-Hill New York, 2007.
8. E. D. Palik, Handbook of Optical Constants of Solids, Academic Press, Orlando, 1985.
9. J.-J. Greffet, R. Carminati, K. Joulain, J.-P. Mulet, S. Mainguy and Y. Chen, Nature, 2002, 416, 61–64.
10. S. Dai, Z. Fei, Q. Ma, A. S. Rodin, M. Wagner, A. S. McLeod, M. K. Liu, W. Gannett, W. Regan, K. Watanabe, T. Taniguchi, M.Thiemens, G. Dominguez,A. H. C. Neto,A.Zettl, F.Keilmann, P. Jarillo-Herrero, M. M. Fogler and D. N. Basov, Science, 2014, 343, 1125–1129.
11. Yuksel, A., Cullinan, M., Murthy, J., 2016. "The Effect of Nanoparticle Clustering on Optoelectronic Property", Solid Freeform Fabrication Symposium, August 8-10, Austin, TX, USA
12. Tervo, E. J., O. S. Adewuyi, J. S. Hammonds, and B. A. Cola. "High thermal conductivity in polaritonic SiO_2 nanoparticle beds." Materials Horizons (2016).

KEYWORDS

Nanoscale Heat transport, Next generation thermal cooling, Nanoparticle packing

An Electromechanical Spectroscopy for Determining the Atomic-Configuration of Single-Molecule Devices

Joshua Hihath

Department of Electrical and Computer Engineering, One Shields Ave., Davis, CA 95616

INTRODUCTION

With the continued miniaturization of electronic devices, the conventional paradigm for manufacturing electronic devices is reaching physical limitations. Because of this looming boundary, a variety of novel, nanoelectronic platforms are being explored, each of which requires the development of new characterization techniques that can provide insights into their function.. Molecules represent a unique class of electronic materials that could potentially extend the development of nanoelectronics. These devices are inherently quantum mechanical in nature, can be designed and constructed with atomic-level precision, are natural 1-dimensional (1D) transport devices, and possess unique opportunities for device fabrication, function, and implementation.[1]

The development of molecular-scale electronic devices has made considerable progress over the last decade, and single-molecule transistors, diodes, and wires have all been demonstrated.[2-4] Despite this remarkable progress, the agreement between theoretically predicted conductance values and those measured experimentally remains limited. One of the primary reasons for these discrepancies lies in the difficulty of experimentally determining the contact geometry and configuration of a molecule when bound between two electrodes. In this report, we describe the development of a novel electromechanical spectroscopic ("alpha" spectroscopy) tool that is capable of determining the most probable binding and contact configurations for a molecular junction at room temperature in solution.[5] These results provide insight into the complex configuration of single-molecule devices that can be used to further improve the agreement between theory and experiment.

DEVELOPMENT OF ALPHA SPECTROSCOPY

In order to extract detailed configuration information from single-molecule devices, it must first be possible to reliably make contact to, and measure the electrical conductance of, a single-molecule. To achieve this, we employ the STM-break junction (STM-BJ) technique (Fig. 1a).[6-8] The measurements proceed by bringing the atomically sharp STM tip down into contact with the substrate electrode which is covered with the molecules of interest. Once contact is made between the two electrodes, the tip is withdrawn until the current reaches the resolution of the current amplifier. If no molecules are bound between the electrodes, then this withdraw process results in an exponential decay in the current. However, if molecules bind between the two electrodes, then steps occur in the current vs. distance trace (Fig. 1d). By repeating this process thousands of times it is possible to perform a statistical analysis of the results to determine the most probable conductance of a single-molecule junction for a given molecular species (Fig. 1e).[6]

To provide insights into the most probable electrode and molecular configuration for a single molecule junction, we have developed an electromechanical, derivative-spectroscopic technique that measures the change in conductance due to a change in electrode separation (dG/dz), which we call "*alpha spectroscopy*".[5,9] This technique is based on the use of a high-frequency, sinusoidal, mechanical modulation (~2 kHz) applied to the tip electrode.[5,9] Since the current is exponentially dependent on the separation between the two electrodes in a tunneling gap, a small perturbation of the tip-surface distance causes a large modulation in the current at that frequency. In general, the current can be described by a simple tunneling barrier with $I = VG_C e^{-\beta\zeta}$, where V is the applied bias, G_C is the contact resistance, β is the characteristic exponential decay constant that describes how the resistance of the junction changes with length,[10] and ζ is the length of the tunneling gap.

Alpha is measured by applying a periodic mechanical modulation to the electrodes of the form $z = z_0 + z_{AC}\cos(\omega t)$. This variation results in a modulation of the tunneling gap of $\zeta = \zeta_0 + \Delta\zeta\cos(\omega t)$. Performing a Taylor expansion of the tunneling current equation around ζ_0 yields the total measured current as: $I \cong I_{DC} + \left.\frac{dI_{DC}}{d\zeta}\right|_{\zeta_0} \Delta\zeta \cos\omega t + \cdots$, where the first term is the usual current $I_{DC}(\zeta_0)$, and the second term is the high frequency current response to the mechanical perturbation (I_{AC}). By normalizing I_{AC} by I_{DC} and z_{AC}, we obtain a value which we call α.[9] If the electrodes can be regarded as perfectly stiff, then $z_{AC} = \Delta\zeta$, $I_{DC}(\zeta_0) = VG_Ce^{-\beta\zeta_0}$, and $\left.\frac{dI_{DC}}{d\zeta}\right|_{\zeta_0} = -\beta VG_Ce^{-\beta\zeta_0}$, which results in $|\alpha| = \beta$. However, this is not typically the case when a molecule is bound to both electrodes because not all of the modulation applied to z will result in a change in the tunneling distance, i.e. $z_{AC} \neq \Delta\zeta$. This difference occurs because the actual change in ζ induced by z_{AC} depends on the elasticity of the electrodes and the molecule, properties which depend on the structural details of the junction configuration. This means that the α-value is a function of the particular mechanical stiffness of the junction at a given elongation z.

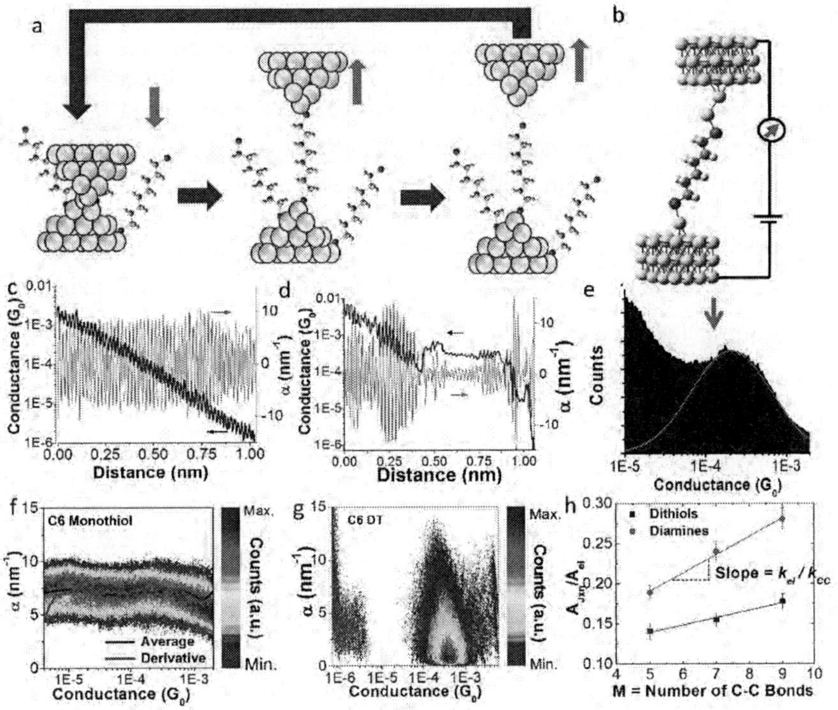

FIGURE 1. Single-molecule conductance and α-measurements. a) In an STM-break junction the current is measured as the tip is brought into/out of contact with the surface. b) Schematic of the configuration of a hexanedithiol junction derived from the alpha measurements. c) In combined conductance/α measurements a high frequency mechanical oscillation is applied to the tip. The current at the modulation frequency is constant when no molecules bind. d) When a molecule binds there is a step in the current decay trace, and a clear reduction in the high-frequency amplitude. e) 1-dimensional (1D) conductance histogram obtained from break junction measurements. f) 2D histogram showing the most likely conductance and α-value for a non-binding hexanemonothiol molecule and g) a binding hexanedithiol. h) By changing the length of the molecule the ratio of the molecular to electrode spring constants can be determined and the most likely binding configuration extracted (b).

RESULTS AND DISCUSSION

When alpha measurements are performed on a hexane-monothiol self-assembled monolayer (SAM) it is not possible to create a single-molecule junction because the molecule cannot bind to both electrodes. In this case, the obtained current vs. distance traces are exponential decays, the slope of these traces gives β, and the amplitude of the high frequency component is constant (Fig 1c, gray). By adding the amplitude of the high frequency current component as a function of the conductance for thousands of traces, it is possible to obtain a 2-dimensional (2D) histogram showing both the most probable conductance and α-values (Fig. 1f). Additionally, by averaging the current vs. distance traces from the same experiments and taking the derivative, it is possible to obtain a β-value for the system (Fig. 1f, green curve). By following this procedure it is apparent that $\alpha = \beta$ when no molecule binds between the tip and the surface. Alternatively, when a hexanedithiol SAM is used instead, clear steps occur in the current vs. distance trace, and the high frequency response is no longer constant over the entire trace (Fig. 1d).

Figure 1g shows a 2D histogram obtained for hexanedithiol. The total counts can be projected onto the axes to obtain 1-dimensional (1D) histograms and determine the most probable conductance and α-values. To determine β for the alkanedithiols, we repeated these measurements for octanedithiol and decanedithiol, and plotted the conductance values of these molecules in a semi-log plot vs. length. The slope of this plot gives the β-value for this family, which was determined to be 7.1 ± 0.8 nm^{-1}, and is consistent with previous single-molecule measurements on the alkanedithiols[11-13]. These results demonstrate that α and β are not equivalent in the case of a single-molecule junction, and in fact the α-values are only on the order of ~15% of the β-value.

The discrepancy between α and β is likely due to the fact that much of the applied modulation is absorbed by the gold contacts, and as such the effective amplitude across the molecular system is much smaller than the applied value. In this case, the actual amplitude of the mechanical perturbation that is applied to the molecular junction, $\Delta\zeta$, can be extracted directly from the relation $\alpha/\beta = \Delta\zeta/z_{AC}$, where the junction is defined as everything between the apex Au atoms.

To examine this hypothesis we applied a mechanical model to the molecular device. The system can be divided into three different springs, k_{el}, k_{ME}, and k_{CB}, which are obtained from the spring constants of the molecular device's chemical bonds. The first spring represents the combination of both Au electrodes, $k_{el}^{-1} = k_{el1}^{-1} + k_{el2}^{-1}$, note that k_{el1} and k_{el2} may each include several Au-Au bonds, and they are not necessarily equal. The second spring represents the two molecule-electrode contact regions; and the third spring represents the carbon backbone, $k_{CB} = \frac{k_{CC}}{M}$, where k_{CC} is the carbon-carbon bond spring constant and M is the number of bonds in the backbone. Furthermore, since we know both the applied amplitude (z_{AC}) and the junction amplitude ($\Delta\zeta = \zeta_{ME} + \zeta_{CB}$), where ζ_{ME} and ζ_{CB} represent the modulation across the molecule-electrode contact and the carbon backbone, we can also extract the amplitude applied to the Au contact region (ζ_{el}). It is then straightforward to demonstrate the relation: $\frac{\Delta\zeta}{\zeta_{el}} = \frac{k_{el}}{k_{CC}}M + \frac{k_{el}}{k_{ME}}$ Thus, by plotting the ratio of the extracted junction and contact modulation amplitudes for each molecule as a function of the number of C-C bonds in the system, it is possible to determine the ratio of the contact spring constant to a single C-C bond spring constant (k_{el}/k_{CC}) from the slope, and the ratio of the contact spring constant to the molecule-electrode coupling spring constant (k_{el}/k_{ME}) from the ordinate intercept (Fig. 1h). Because the Au-Au and C-C bond spring constants are known, it is possible to determine the most-probable binding configuration and contact geometry of a single-molecule junction from this information. As is shown in Fig. 1b, for the thiols, this analysis yielded a configuration where the thiols are bound in a top-site on both ends, and the Au apex atoms are likely bound in a bridge-site on one electrode and a top-site on the other.

Thus, this novel electromechanical spectroscopic technique is capable of providing atomic-level configuration information about how a single-molecule junction forms, and its most probable configuration.

REFERENCES

1. Tao, N. J. *Nature Nanotech.* **2006**, *1*, 173.
2. Reed, M. A.; Tour, J. M. *Scientific American* **2000**, *282*, 86.
3. Song, H.; Reed, M. A.; Lee, T. *Adv. Mater.* **2011**, *23*, 1583.
4. Mantooth, B. A.; Weiss, P. S. *Proc. IEEE* **2003**, *91*, 1785.
5. Rascon-Ramos, H.; Artes, J. M.; Li, Y.; Hihath, J. *Nature Materials* **2015**, *14*, 517.
6. Xu, B.; Tao, N. J. *Science* **2003**, *301*, 1221.
7. Venkataraman, L.*et al. Nano Lett.* **2006**, *6*, 458.
8. Jang, S. Y.; Reddy, P.; Majumdar, A.; Segalman, R. A. *Nano Lett.* **2006** *6*, 2362.
9. Xia, J. L.; Diez-Perez, I.; Tao, N. J. *Nano Lett.* **2008**, *8*, 1960.
10. Magoga, M.; Joachim, C. *Phys. Rev. B* **1998**, *57*, 1820.
11. Li, C.*et al. J. Am. Chem. Soc.* **2008**, *130*, 318.
12. Li, X.*et al. J. Am. Chem. Soc.* **2006**, *128*, 2135.
13. Haiss, W.*et al. J. Phys. Chem. C* **2009**, *113*, 5823.

KEYWORDS

Single-Molecule Devices, Molecular Electronics, Scanning Tunneling Microscopy (STM), Break Junction, Nanoelectronics

Pico-Second Laser and Broad Argon Beam Tools For Characterization Of Advanced Packages And Devices

Y.Wang, C. Spence, S. Coyle, J. Hunt and M. Hassel Shearer

Gatan Inc. 5794 W. Las Positas Blvd. Pleasanton Ca. 94588 USA

INTRODUCTION

"More than Moore" is emerging as the solution to continue the process development means for increasing speed and functionality of semiconductor devices. More than Moore relies on stacking multiple devices within an advanced package. Although this methodology relaxes the pressure to find new materials and extremely expensive production tools used by process engineering it may increase the difficulty confronting failure analysis engineers.

3D stacked chips may challenge failure analysis tools to keep pace. A key tool set for Failure Analysis Engineers has been and will continue to be electron microscopy. Electron microscopy imaging and analytical techniques have been pushed to 0.1 nm resolution over planar areas as large as 10 microns x 10 microns with device geometries shrinking to the 5 nm to 10 nm design rules. More than Moore maintains the same planar resolution but with advanced packages and stacked chips, cross sectional depths increase to perhaps as large as a few millimeters. Until now, mechanical polishing or FIB have been the tools of choice for preparing cross sections for Failure Analysis. Mechanical preparation may be limited due to the increased fragility of the silicon devices thinned to less than 50 microns within these advanced packages. Ga FIB's and even the newer Plasma FIB's may not have sufficient milling speed necessary to expose regions within a large thick package or cross sectional surfaces of stacked chips.

An alternative or compliment to FIB is proposed as a precision pulsed laser tool designed specifically for this application. These tools have "milling rates" orders of magnitude faster than FIB's. This higher speed can be utilized to cross section whole packages or open boxes completely through thick packages or cut out large cubes from a package for subsequent precision polishing with an FIB or broad ion beam tool.

In this paper, a workflow consisting of a pico-second laser tool (microPREP ™) followed by a Broad Argon Ion Beam tool is proposed as a solution for extremely large area preparation with surfaces suitable for SEM, TEM, FIB based analysis directly after polishing in the broad argon beam tool.

Results will be presented discussing the advantages of this workflow in terms of speed, size and quality of the surface for electron microscopy analysis. Anticipated enhancements that will come as the technique is developed will be discussed.

CROSS SECTIONING A PACKAGED STACKED CHIP SET

As an example of the proposed workflow of pico-second pulsed laser plus broad argon ion beam tool for SEM analysis is shown in the following slides. A commercially available 3 chip package was selected.

Figure 1 shows a low magnification SEM cross section of a Samsung K4AAG045WD-4CRB packaged chip set sectioned with the microPREP™ followed by cross section polishing in the Ilion$^+$ II a broad argon ion beam tool. The processing time for this specific workflow consists of 15 minutes to cross section completely through the length 9 mm and thickness 0.76 mm of the packaged part using the laser tool microPREP™. This was followed by 1 hr. in the Ilion$^+$ II. The polished cross sectional area is approximately 2 mm wide and through the complete thickness of the chip. The broad argon beam removes the Heat Affected Zone damage created by the ablation process of the laser tool. Estimates of the damaged layer are approximately 1 micron. Figure 2 presents the details of the cross-sectional package with a higher magnification micrograph.

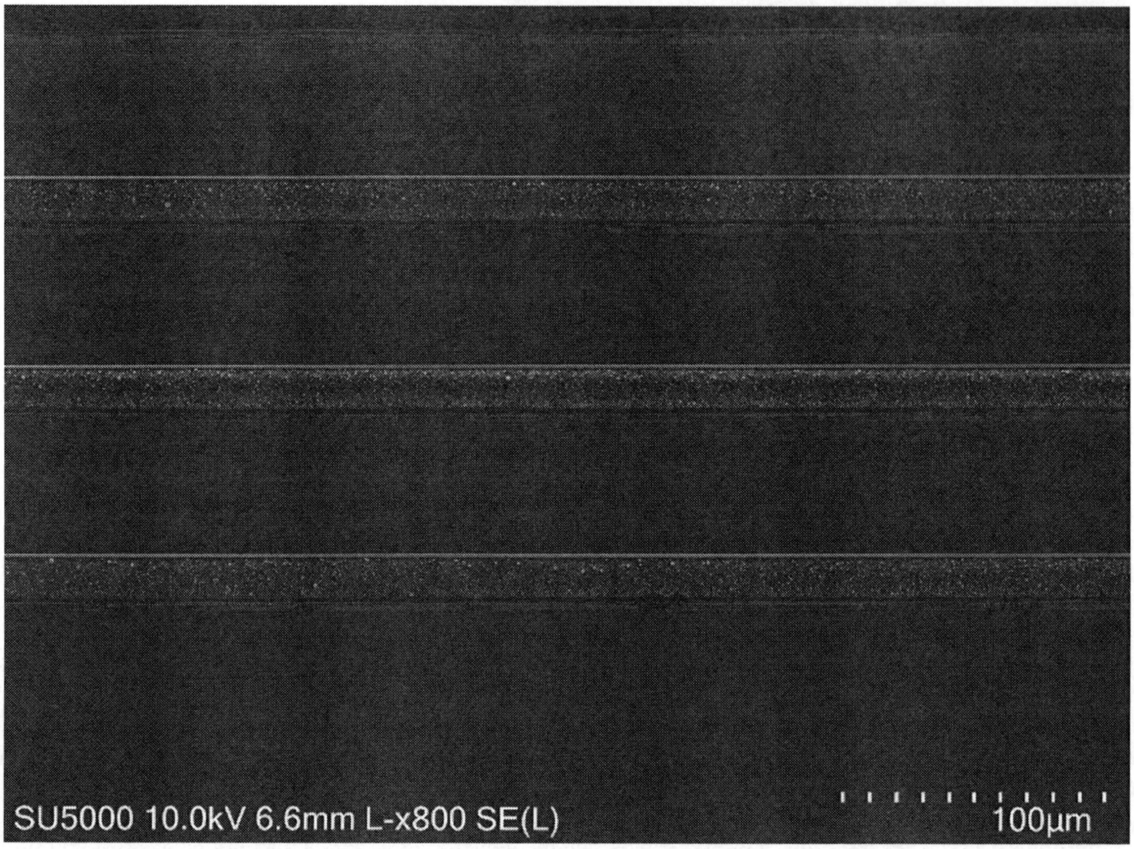

FIGURE 1. SEM Image of Cross Section of stacked chips prepared with a pico–second laser tool and Broad Argon ion Beam Tool

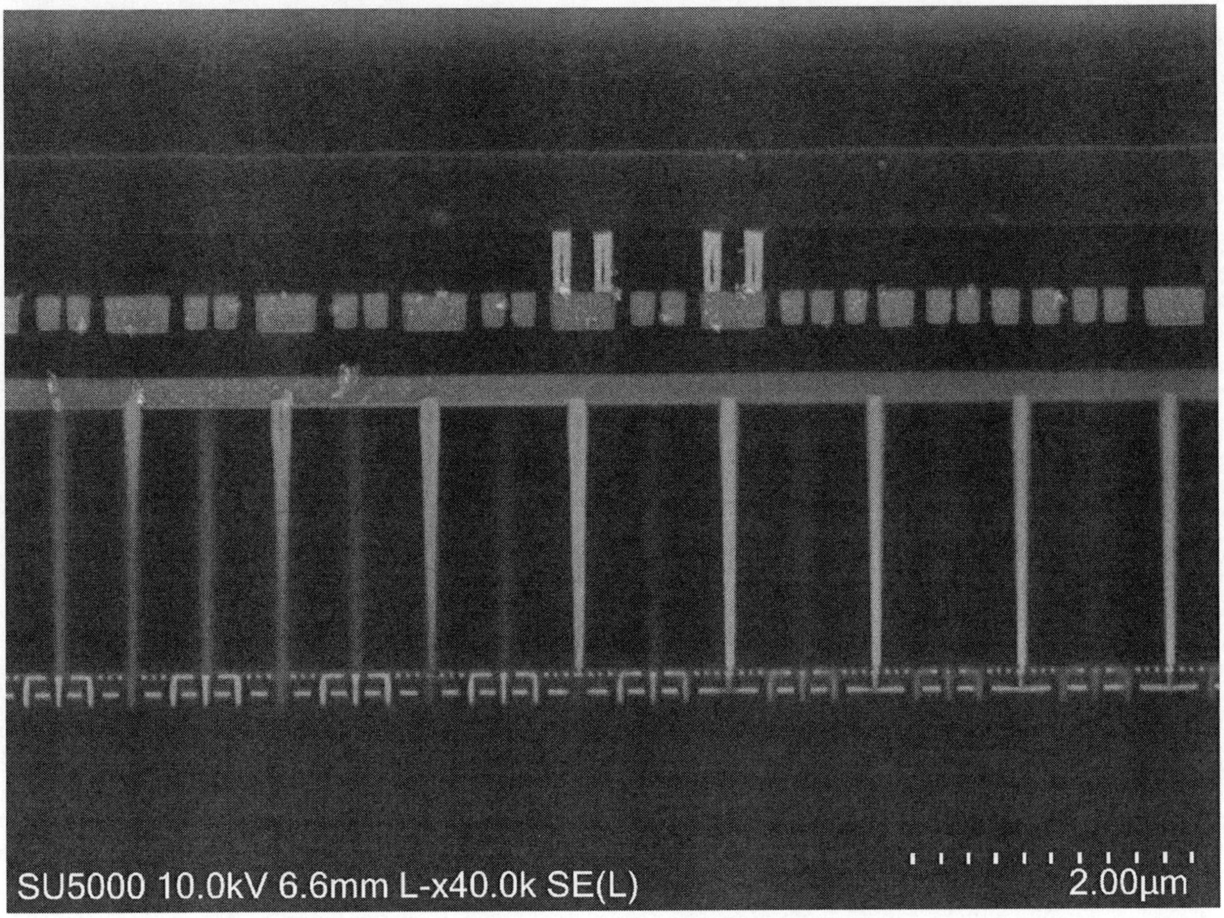

FIGURE 2. Higher Magnification of Cross Section of stacked chips Prepared with a pico-second laser tool and Broad Argon ION Beam Tool.

KEYWORDS

Pico-second pulsed laser ablation, broad argon beam tool, advanced packages, 3D stacked chips

Examination of Advanced Technologies in Characterization, Diagnostics, and Verification at Different Stages in the Manufacturing Lifecycle of Packaged IC Devices

Colin Ritchie[1], Scott West[1], Stuart Neches[1], Eiji Kato[2] and Masaichi Hashimoto[2]

1 Advantest America, Inc., 3061 Zanker Rd., San Jose, CA 95134 USA
2 Advantest Corporation, 2-48 Matsubara, Kamiayashi, Aoba-ku, Sendai 989-3124 Japan
stuart.neches@advantest.com

INTRODUCTION

This talk explores 3 applications of technology in the evaluation of IC packaged devices and systems at different stages in the manufacturing process: in the areas of Characterization, Diagnostics, and Verification/Reliability testing. Nondestructive mold thickness metrology using terahertz (THz) wave is examined for characterization of the moldcap and related manufacturing process in real time. Ultra-high resolution pulsed TDR/TDT is explored for diagnostic forensic analysis of packaged devices including complex 2.5D and 3D packages and interconnects. At the system level, verification of reliability and endurance is studied with respect to thermal consistency using Flash Memory-based SSD (Solid State Drive) SLT (System Level Test).

NONDESTRUCTIVE MOLD THICKNESS METROLOGY FOR CHARACTERIZATION OF MOBLE DEVICES

The worldwide embrace of smartphones, advanced consumer electronics and intelligent devices for Internet of Things (IoT) is driving demand for smaller, more highly integrated semiconductor devices. Semiconductor packages for this market must be compact, thin, and tough enough to stand up to mobile or ubiquitous usage. Accordingly, optimization and balancing of package durability and thinness is increasingly important. However, existing metrology methods are not suitable for volume production lines because it is slow, laborious and destructive and manufacturers have had to test package thickness by inspecting samples late in the production process. This makes it difficult to assess the quality of the entire production run and problematic to trace the root cause of any defects in the final product.

Advantest's TS9000 Mold Thickness Analysis (MTA) system (Fig. 1a) solves these issues by enabling non-destructive, rapid, repeatable, and highly accurate ($\pm$ 3 μm) measurements of mold thickness, without the limitations of existing measurement methods. Furthermore, it can be deployed at various points in the assembly and packaging process, even immediately after the curing process, enabling earlier detection of production issues and improving product quality and yield. These innovative inspection capabilities are expected to contribute significantly to product quality improvement amid the trends towards smaller package sizes and its higher integration. As shown in Fig. 1b, in the mold thickness measurement with this system, pulsed THz waves with sub-picosecond duration are send to device under test (DUT) by THz emitter and specularly reflected waves are acquired by THz detector [1]. Since THz pulses are partially reflected at the package top surface, and partially at the mold-die interface due to Fresnel reflection, the reflected waveform from the DUT would have echo pulses like Fig.1c. The time difference between the echoes (Δt) and refractive index of the mold give mold thickness by calculation.

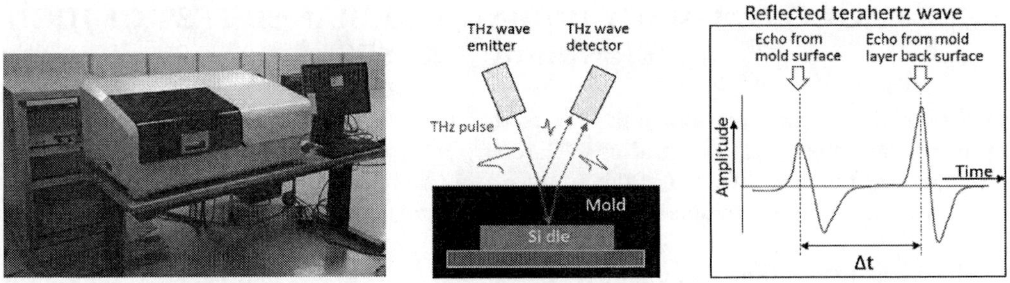

(a) (b) (c)

FIGURE 1. (a) TS9000 measurement unit (left side on the table), rack mounted controller electronics and optics (right), and controller PC (right side on the table). (b) Measurement configuration of overlaid mold thickness with pulsed THz radiation. (c) Example of reflected time-domain waveform from the DUT.

HIGH RESOLUTION TIME DOMAIN REFLECTOMETRY TOOL FOR DIAGNOSTIC FAILURE INVESTIGATION IN STATE-OF-THE-ART PACKAGES

Electronic device circuit fault analysis like open, short and high impedance is commonly performed by oscilloscope time domain reflectometry (TDR). However, as devices grow smaller and more highly integrated, the ability to locate fault with extreme spatial precision has become increasingly more important. The typical fault locations in cutting-edge semiconductor packages are shown in Fig. 2a. But existing TDR instruments have limited resolution, as the rise time of the TDR signal is facing the difficulty in shortening it much further, and this would be the risk that conventional TDR technologies will be inadequate to handle the requirements of fault isolation (FI) for highly integrated devices immediate future. Advantest's terahertz technology addresses these concerns and meets the need for high-resolution and long-haul measurement by utilizing ultrashort and low-jitter pulse generation and sampling technology.

Advantest's TS9000 TDR system relies on Advantest's market-proven TDR measurement technology to pinpoint and map circuit defects. The system is composed of an electro-optic sampling (EOS) system, TDR transceiver and a probe (Fig. 2b). The transceiver generates and detects ultrashort pulse signal with the help of femtosecond laser sources, of which the laser pulse width is less than <50 fs. Furthermore, an excellent jitter performance less than 30 fs is achieved [2]. The solution delivers circuit failure analysis with an extremely high spatial resolution of less than 5 μm, and a maximum measurement range of 300 mm, including for internal circuitry used in through-silicon via (TSVs) and interposers.

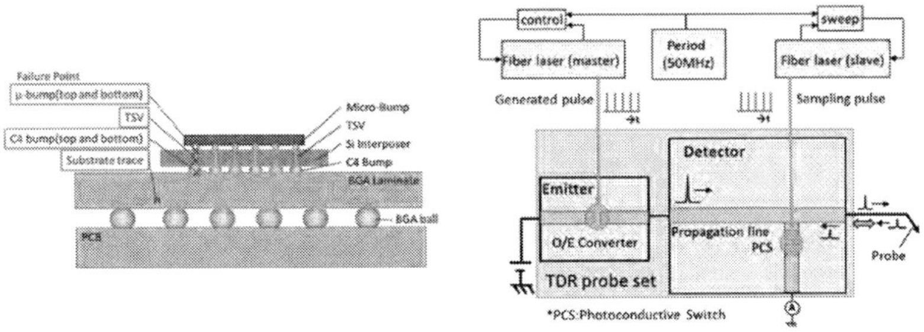

(a) (b)

FIGURE 2. (a) Advanced semiconductor package structure and typical fault locations. (b) Schematic diagram of the developed TDR transceiver and TDR signal generation/sampling scheme.

THERMAL CONSISTENCY DURING ENDURANCE TEST OF HIGH PERFORMANCE SSDS

The demand for high performance storage has driven the emergence and rapid growth of solid state drives with volumes to exceed 300M units by 2019, according to Gartner. Of key concern to SSD end-users in both the enterprise and consumer markets is the rated lifespan of the product. To ensure this lifespan, endurance testing must be done with long test times and elevated temperature to ensure statistical probably that the drives will meet lifespan specifications.

As SSD performance moves from that of single late 6G SATA devices to multi-lane 8G NVMe and 12G SAS, power consumption of drives increases to 25W or more per device. This creates a challenge during endurance testing to maintain thermal consistency across high power drives tested in parallel to achieve statistical confidence, while maintaining a low cost of test.

The MPT3000ENV test platform addresses this issue across a variety of SSD form factors and protocols, by maintaining a +/-5C thermal consistency for up to 256 25W devices, using a dual chamber design. Reliability demonstration testing (RDT) is performed, testing a sample of devices over time to prove that they will last over the desired lifetime. Typically, devices are run for 1,000 hours at high temperatures, simulating 3 months of constant read/write operations. Challenges in achieving the needed thermal consistency are explored.

REFERENCES

1. Wataru Momose et al., *Results in Pharm. Sci.* **2**, 29-37 (2012).
2. M. Hashimoto et al., 2016 IEEE 23rd Int. Symp. Phys. Failure Anal. Integ. Circ. (IPFA), Singapore, 238-240 (2016).

KEYWORDS

Molding, non-destructive, thickness, terahertz wave, time domain reflectometry, endurance test, thermal consistency

Characterization And Control Of The Surface Of The Topological Insulator Bi_2Se_3

Avery J. Green[1], Lee A. Walsh[2], Wesley Nolting[1], Carl A. Ventrice Jr.[1], Chris L. Hinkle[2], and Alain C. Diebold[1]

1.College of Nanoscale Science and Engineering, SUNY Polytechnic Institute
257 Fuller Road, Albany, NY 12203
2. Materials Science and Engineering, University of Texas at Dallas
800 W. Campbell Road, Richardson, TX 75080

INTRODUCTION

The semiconductor industry is searching for a replacement for CMOS based nanoelectronics. The unusual electronic properties at the surface of TI's results in dissipationless spin transport making TI's ideal candidates for spintronic devices. Among these materials, Bi_2Se_3 has been dubbed by some as the "ideal TI," due to its relatively simple surface band structure, which includes a single gamma-point Dirac cone, and a relatively large (~0.3 eV) band gap[1]. The two common methods of fabricating Bi_2Se_3 samples are scotch tape exfoliation and molecular beam epitaxy (MBE). Each method faces a variety of obstacles that need to be overcome before Bi_2Se_3 can be used for routine device fabrication. The obstacles include ambient oxidation, Se evaporation and vacancy formation, surface band bending, and polycrystallinity.[2,3] Previous work has shown that the Bi_2Se_3 surface oxidizes within hours of exposure to air, compromising both ex-situ production and electrical characterization.[4] In a collaboration with Prof. Chris Hinkle and Dr. Lee Walsh of UT-Dallas, we have recently begun to study Se-capped, MBE-grown Bi_2Se_3 in order to develop a method of protecting the surface. The Bi_2Se_3 surface may be affected by annealing during the Se decapping process. Temperature-programmed desorption (TPD) observation of Se desorption and scanning tunneling spectroscopy (STS) measurement of the Fermi energy are both being used to select the annealing rate and maximum temperature that result in the most appropriate surface for device fabrication. Preliminary TPD results have shown that bulk Se desorbs from the surface near 80 °C with an activation energy near 5 eV. Preliminary STS of uncapped samples has shown Fermi energies largely between 33 and 52 meV above the Dirac point of Bi_2Se_3 surface states. The median shift of 40 meV of the Fermi energy indicates a slight n-doping. In the future, we wish to complete these TPD and STS experiments and create larger STS data sets, in which the Fermi energy can be mapped over large surface areas.

We gratefully acknowledge Profs. Vince Labella and Carl Ventrice of CNSE for their help in TPD and STS measurements, and Prof. Chris Hinkle and Dr. Lee Walsh of UT-Dallas for their MBE-grown Bi_2Se_3 samples. We also thank Profs. Ken Burch of Boston College, Bradley Thiel of CNSE, and Robert Hull of RPI for discussions on TI's and related research techniques. Finally, we'd like to thank Dr. George Orji of NIST for help with AFM calibration.

TEMPERATURE PROGRAMMED DESORPTION

TPD is a technique in which a sample is heated at a predetermined rate controlled by software, and the desorption of surface species is monitored with a mass spectrometer. Se-capped Bi_2Se_3 samples in ultra high vacuum (UHV) (10^{-10} torr) were heated at a constant rate of 10 °C/min by a radiative tungsten filament, and the partial pressures of residual gasses, including H_2, H_2O, CO, CO_2, ^{78}Se and ^{80}Se, were measured with a Hiden™ quadrupole mass spectrometer. Final heating temperatures of 110 °C and 150 °C were used, and seen to have completely desorbed the bulk surface Se, as confirmed with scanning electron microscopy and atomic force microscopy. Once the heating temperature was reached, the sample remained at that temperature for 2 minutes before it was cooled.

The sensitivity of the mass spectrometer in secondary electron multiplier mode allows the time and temperature-dependent Se desorbate gas partial pressure to be analyzed, and activation energy to be extracted. Though the Se cap may be treated as a bulk, coverage was sample and location-dependent, thus causing some inconsistencies in desorption pressures. Nevertheless, a Polyani-Wigner equation of order 0 fit the early portion of the data with low error. Upon taking the slope of a linear fit to $\ln(p)$ vs. $1/T$, activation energies between 4 and 6 eV were extracted.

SCANNING TUNNELING SPECTROSCOPY

The main interest of this work is to measure the Fermi energy of Bi_2Se_3, after being capped by Se, and subsequently decapped. STS, with an appropriately chosen energy range and probe tip, provides a method of measuring the electronic properties of a sample at its surface. Using a lock-in amplifier, dI/dV spectra at various locations on each sample were acquired. This signal may be approximated as proportional to the local density of states (LDOS) at the sample surface. Since the Dirac point in the surface band structure of Bi_2Se_3 is ideally near the Fermi level, and is a minimum in the energy-dependent LDOS spectrum, any shifting between the Dirac point energy and the Fermi energy was easily observable. In one sample, several hundred spectra were measured, and many of these spectra had minima shifted ~-40 meV from the Fermi energy. This indicates that some n-doping occurred during or after Se desorption. This has been attributed in literature to Se vacancy formation. The mechanism(s) of doping in these capped and heated Bi_2Se_3 requires further investigation.

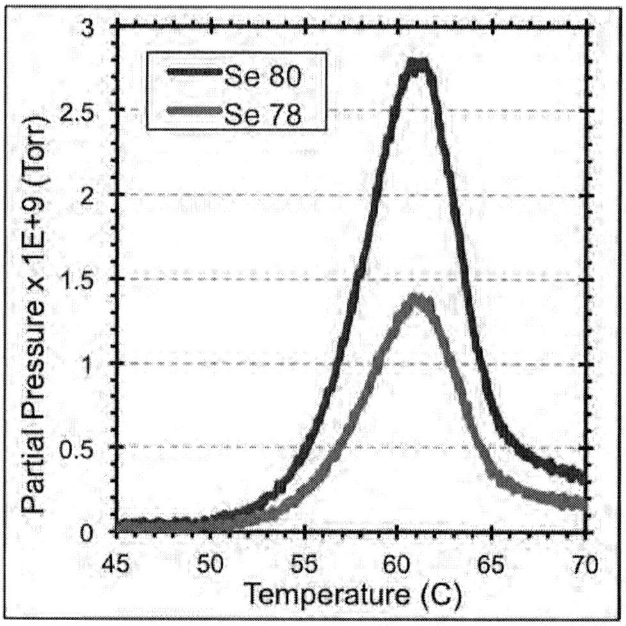

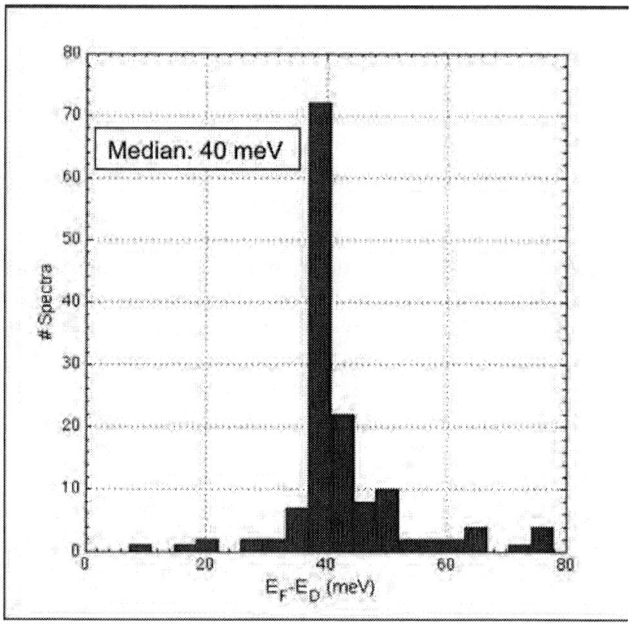

LEFT. Desorption pressure as a function of temperature in a TPD experiment. Both isotopes ^{78}Se and ^{80}Se in the capping layer are removed simultaneously, and at expected ratios. The portion of this curve before the inflection point of partial pressure increase is used to calculate E_a. For this sample $E_a = 4.02$ eV. **RIGHT.** Histogram of the difference between E_F and the Dirac point energy, E_D. Each spectrum was taken at a different point on the same sample surface. The median difference of this sample is 40 meV, and that value is seen frequently in measurements at various areas on the sample surface.

REFERENCES

[1] H. Zhang, C.-X. Liu, X.-L. Qi, X. Dai, Z. Fang, and S.-C. Zhang, Nat. Phys. **5**, 438 (2009).
[2] D. Kong, J.J. Cha, K. Lai, H. Peng, J.G. Analytis, S. Meister, Y. Chen, H.-J. Zhang, I.R. Fisher, Z.-X. Shen, and Y. Cui, ACS Nano **5**, 4698 (2011).
[3] H.M. Benia, C. Lin, K. Kern, and C.R. Ast, Phys. Rev. Lett. **107**, 177602 (2011).
[4] A.J. Green, S. Dey, Y.Q. An, B. O'Brien, S. O'Mullane, B. Thiel, and A.C. Diebold, J. Vac. Sci. Technol. A **34**, 061403 (2016).

KEYWORDS

Topological insulator, spintronics, temperature-programmed desorption, scanning tunneling spectroscopy

Strain and Composition Monitoring in Various (Si)Ge Fin Structures Using In-Line HRXRD

Andreas Schulze[1,*], Roger Loo[1], Liesbeth Witters[1], Hans Mertens[1], Nadine Collaert[1], Naoto Horiguchi[1], Matthew Wormington[2], Paul Ryan[3], Wilfried Vandervorst[1,4] and Matty Caymax[1]

(1) imec, Kapeldreef 75, 3001 Leuven, Belgium
(2) Bruker Semiconductor Division, 112 Robin Hill Road, Santa Barbara, CA 93117, USA
(3) Bruker Semiconductor Division, Belmont Business Park, Durham, DH1 1TW, UK
(4) KU Leuven, Dept. of Physics and Astronomy, Celestijnenlaan 200D, 3001 Leuven, Belgium
(*) Andreas.Schulze@imec.be

INTRODUCTION

The performance of heterogeneous 3D transistor structures critically depends on the composition and strain state of the buffer, channel and source/drain regions. In this paper we use an in-line high resolution X-ray diffraction (HRXRD) tool to study in detail the composition and strain state in various (Si)Ge based fin structures down to 16nm in width. We fabricated fin structures of identical dimensions into arrays which we analyzed using an in-line Bruker JVX7300LM tool. This tool is equipped with a micro-beam X-ray source as well as an automated pattern recognition system to facilitate measurements on patterned wafers. In order to analyze the fins' composition and strain state as well as modifications to the latter parameters during fabrication, we collected ω-2Θ scans using a conventional (0D) scintillation detector and reciprocal space maps (RSMs) using a linear (1D) detector. In this study we will discuss the relevance of such measurements using three representative examples: First we will analyze the composition and strain state of relaxed SiGe fins selectively grown in an STI matrix where we found an anisotropic in-plane relaxation as the fin width was reduced and significantly reduced relaxation in the direction along the fin. Secondly, we will analyze the strain state of etched Ge fins in the two in-plane directions and show that ω-2Θ scans can provide quick feedback on (undesired) relaxation occurring during fabrication. Finally, we will demonstrate the value of HRXRD for the analysis of complex multilayer fin structures which are relevant for horizontal nanowire FETs that are candidates to replace FinFETs below the 7 nm node.

SELECTIVELY GROWN SIGE/GE FINS (STI FIRST)

The selective epitaxial growth of a relaxed SiGe buffer and a thin strained Ge channel into narrow trenches defined by an STI matrix is considered as a CMOS-friendly integration scheme for the fabrication of compressively strained Ge channels [1]. It is important to note that the composition and relaxation of the SiGe buffer in such structures are crucial as both determine the lattice mismatch and hence the strain state of the Ge channel layer grown on top. Acquiring asymmetric RSMs parallel and perpendicular to the fins (Figure 1a/b) allowed us to determine the lattice parameters of the SiGe in all three spatial directions. We will show that the in-plane strain state of the selectively grown SiGe buffer is anisotropic in nature for narrower fins. Indeed, a significantly reduced relaxation in the direction along the fin becomes apparent (Figure 1c). This observation was verified using nano-beam electron diffraction and is explained based on the reduced probability for dislocation half-loops to evolve in trenches narrower than a few times the critical radius. Moreover, we will discuss a methodology for the determination of the composition in case of an anisotropic in-plane strain state which differs from the procedure commonly used for blanket layers. The obtained composition values are in excellent agreement with results from TEM-EDX.

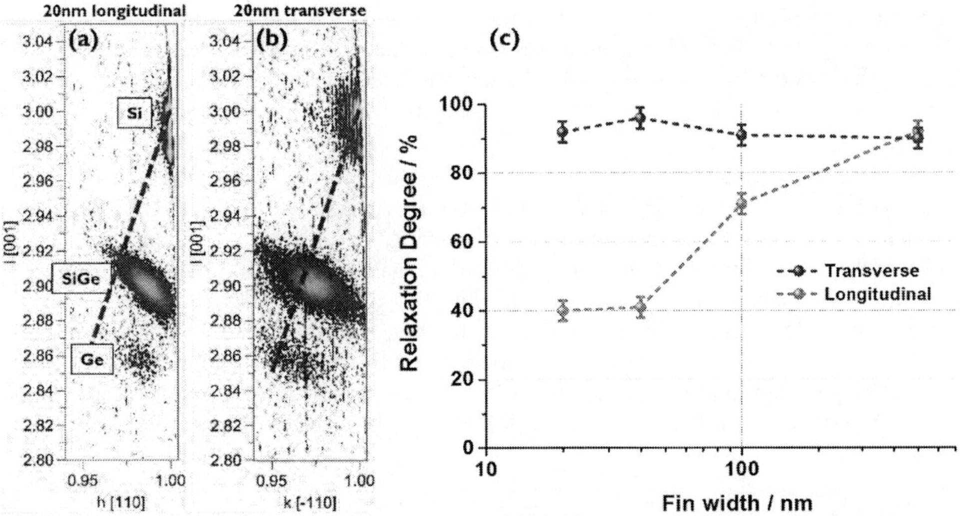

FIGURE 1. Reciprocal space maps (RSM) in the vicinity of the asymmetric (113) reflection acquired on 20nm wide fins in **(a)** longitudinal and **(b)** transverse direction. The dashed line in the RSMs connects the Si substrate peak with the center of the Ewald sphere and corresponds to the line of full relaxation. **(c)** Relaxation of the SiGe buffer in the transverse and longitudinal direction for different fin widths.

SIGE/GE FINS ETCHED FROM A BLANKET LAYER STACK (STI LAST)

An alternative fabrication scheme is based on the blanket growth of a thin strained Ge layer on top of a blanket SiGe strain relaxed buffer (SRB) followed by a reactive ion etch (RIE) process [2]. Such an SRB is typically a few micrometers thick in which the Ge content is stepwise increased to allow for a reasonably low density of threading dislocations at the top of the buffer. Asymmetric RSMs measured parallel and perpendicular to the fins immediately after fin patterning are shown in Figure 2a/b. From the longitudinal RSM (Figure 2a) it becomes apparent that the Ge channel layer has the same in-plane lattice parameter as the underlying SiGe, i.e. remains fully strained with respect to the latter during fin formation. This is essential information as strain along the fin (i.e. along the channel) leads to enhanced carrier mobility and is therefore beneficial for the device performance. From the transverse RSM we learn that the Ge exhibits a larger in-plane lattice parameter (smaller K value). This can be explained by elastic relaxation of the narrow Ge fin into the free space generated by the patterning process adjacent to the fin. The mosaicity of the plastically relaxed SiGe buffer leads to a substantial peak broadening in both the H and K direction. For this reason, satellite peaks caused by diffraction from the fin grating are not well defined.

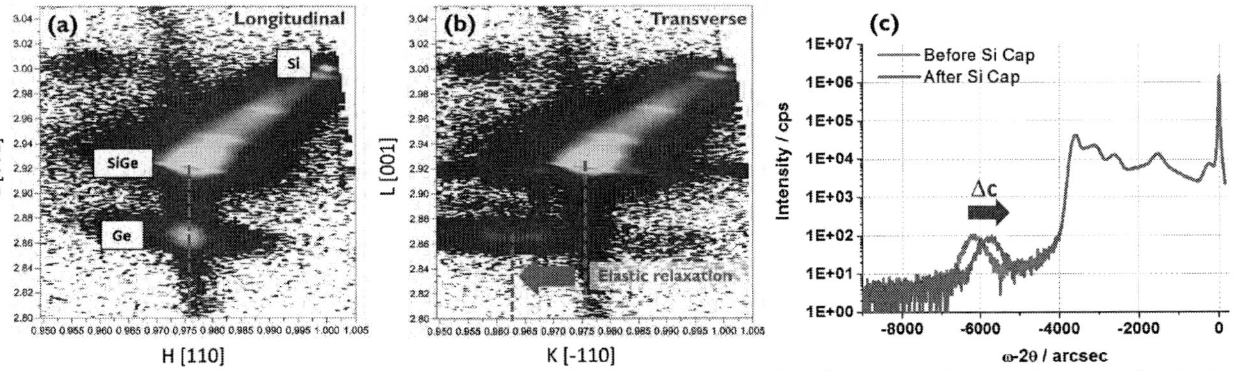

FIGURE 2. Reciprocal space maps (RSM) in the vicinity of the asymmetric (113) reflection acquired on 16nm wide fins in **(a)** longitudinal and **(b)** transverse direction. **(c)** (004) ω-2Θ scan reflecting the out-of-plane lattice parameter c measured before and after the conformal deposition of a Si cap around the Ge fin.

Since the acquisition of RSMs can be time consuming, we use symmetric ω-2Θ scans to quickly monitor whether a certain process step is modifying the strain state (or composition) of the structure. Using this approach we could e.g. detect an unexpected shift of the Ge peak after the growth of a Si cap layer (Figure 2c). By means of a longitudinal RSM (not shown) we could verify that the strain along the fin is maintained. In a separate experiment we skipped the Si cap growth but still exposed the wafer to the same thermal budget. In this case no peak shift could be observed for which reason a thermally induced relaxation of the fin could be excluded. As such, we believe that the observed peak shift (i.e. relaxation) is caused by the tensile strained Si cap or by Ge reflow.

MULTILAYER FINS ETCHED FROM A BLANKET LAYER STACK (STI LAST)

Horizontal nanowire (NW) FETs are the most promising candidate to replace FinFETs beyond the 7nm technology node as they offer improved electrostatics due to a gate-all-around (GAA) architecture. A multilayer fin fabricated by a top-down approach represents the basic building block for such an advanced device. The horizontal NWs are obtained by selectively etching out one of the materials from the layer stack [3]. In Figure 3a/b, asymmetric RSMs of Si/$Si_{0.75}Ge_{0.25}$/Si/ $Si_{0.75}Ge_{0.25}$/Si multilayer fins in the longitudinal and transverse direction are shown. The layer fringes representing the multilayer stack are nicely resolved and, moreover, are aligned with respect to the Si substrate in the *H* direction (Figure 3a). Hence, the stack is coherent with respect to the substrate in the longitudinal direction.

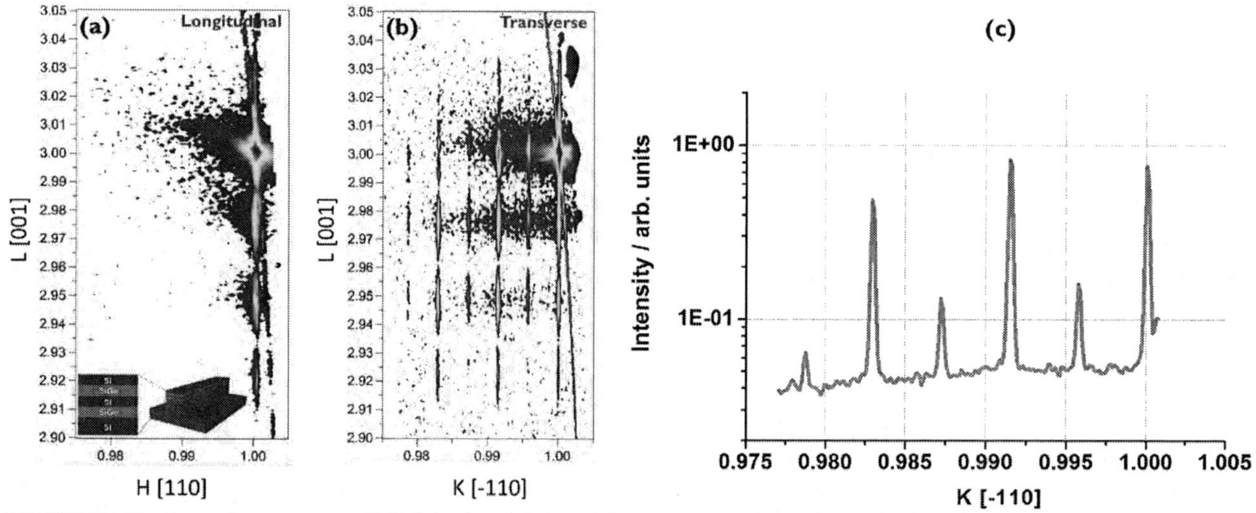

FIGURE 3. Reciprocal space maps (RSM) in the vicinity of the asymmetric (113) reflection acquired on 16nm wide fins in **(a)** longitudinal and **(b)** transverse direction. **(c)** Section along *K* through the transverse RSM averaged over L (2.91 - 2.99).

As the multilayers have been grown pseudomorphically, peak broadening due to mosaicity in the *H* and *K* directions is very limited. For this reason it is possible to resolve distinct 1st and 2nd order grating rods in the transverse RSM (Figure 3b/c). A section along *K* averaged over *L* through the transverse RSM is shown in Figure 3c. Diffraction from a fin array which acts as a grating placed on top of the Si wafer causes the 1st order grating rods in the *K* direction. We will show that the distance between the first order rods can be directly translated into fin pitch. The lower intensity second order grating rods confirm the presence of pitch walking which is induced by the multiple patterning process. The layer peak envelope is positioned at $K<1$ which proves the elastic relaxation of the stack in the perpendicular direction due to the limited fin width.

REFERENCES

[1] R. Loo *et al.*, *ISTDM,* pp. 19–20, 2014 (DOI:10.1109/ISTDM.2014.6874674).
[2] J. Mitard *et al.*, VLSI Technology, 2014 (DOI:10.1109/VLSIT.2014.6894391).
[3] H. Mertens *et al.*, VLSI Technology, 2016 (DOI:10.1109/VLSIT.2016.7573416).

KEYWORDS

Strain Metrology, HRXRD, Heteroepitaxy, FinFET, SiGe, Nanowire

In-Line Stress/Contamination Metrology for Advanced Semiconductor Device using micro Raman and Photoluminescence Measurements

Jae Hyun Kim[1,2], Chang Whan Lee[1], Hyung Won Yoo[1], Byoung Ho Lee[1], Seung Min Han[2]

[1]*SK hynix, Inc., Icheon-si, Gyeonggi-do, 467-701, Korea*
[2]*Korea Advanced Institute of Science and Technology, Daejeon, 305-701, Korea*
E-mail : jhyun.kim@sk.com

INTRODUCTION

As silicon complementary metal-oxide-semiconductor (CMOS) devices are continuously scaled, and approach 10 nm,1) the industry is facing difficult technical challenges, not only in process development, but also in quality control and/or device reliability. Occasionally, in volume device manufacturing, a large number of particles may be generated on cleaned Si wafers. Non-contact, optical characterization methods were proposed as practical in-line metrology for advanced semiconductor device using micro Raman and photoluminescence measurements

OPTICAL CHARACTERIZATION OF SILICON SURROUNDED BY CU THROUGH SILICON VIAS

Silicon surrounded by Cu TSVs goes through stress changes and is prone to Cu contamination during various process steps. For successful implementation of 3-D (three dimensional) device integration using Cu through-silicon-vias (TSVs), management of Si stress and prevention of Cu contamination in Si are two important factors. Non-contact, optical characterization methods were proposed as practical stress and Cu contamination monitoring techniques. Local stress of Si surrounded by Cu TSVs was characterized as a function of TSV diameter using micro-Raman spectroscopy at various thermal cycling process steps. Localized Cu contamination in Si was monitored by room temperature photoluminescence (RTPL) measurements..

SILICON SURFACE QUALITY VARIATION USING ROOM TEMPERATURE PHOTOLUMINESCENCE MEASUREMENTS

Conventional chemical analysis techniques for contamination are generally not able to distinguish between Si wafers with good and poor particle performance. No suspicious chemicals and elements were detected from any wafers regardless of characterization techniques. Surface photovoltage (SPV) measurement barely showed the differences between wafers with good and poor particle performance. Multiwavelength room temperature photoluminescence (RTPL) showed significant differences in intensity between them, indicating the presence of surface quality variations.

Higher failure rates have been observed for devices at or adjacent to areas that have been measured or inspected by in-line x-ray during front end of line process steps. The failures are thought to be related to SiO2/Si interface damage caused by x-ray radiation during routine in-line element, material, and process monitoring. This issue has been noted for highly advanced integrated semiconductor devices following device failure mode analysis of Si wafers, with and without x-ray in-line inspection. A multiwavelength room-temperature photoluminescence (RTPL) study was performed to identify the presence of such damage in x-ray irradiated wafers from various types of x-ray inspection steps during devicefabrication processes. It was found that x-ray radiation as low as 16 keV induced damage at the SiO2/Si interface. The damage to the SiO2/Si interface was successfully observed by using multiwavelength RTPL wafer mapping.

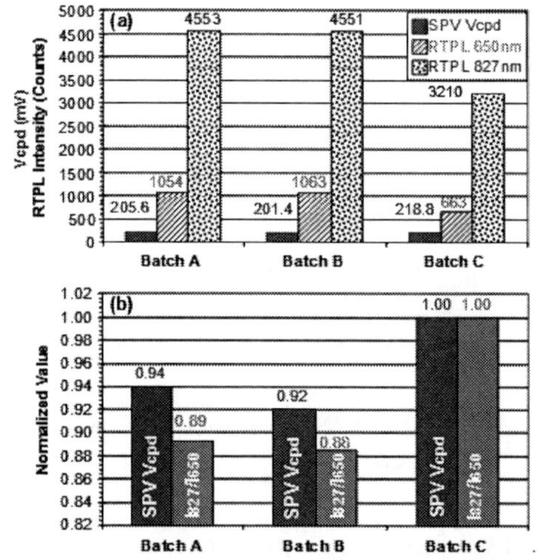

FIGURE 1. (a) average and (b) normalized Vcpd and RTPL intensity ratio (I827/I650) values over three 5-wafer batches (3 batches: 15 new blanket Si wafers).

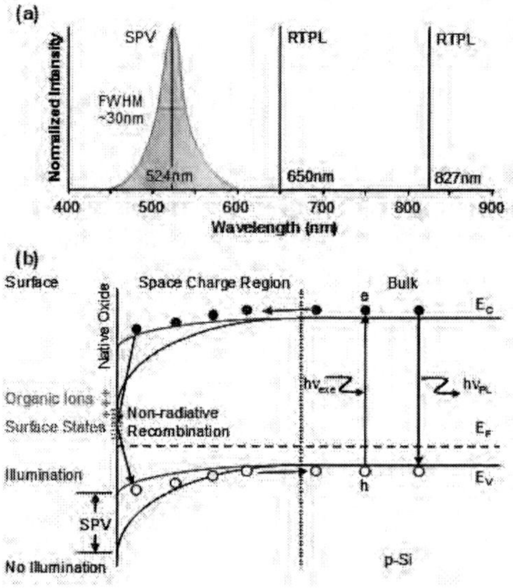

FIGURE 2. (a) spectral distribution of excitation light source and (b) schematic illustration of band bending with and without illumination.

REFERENCES

1. J. H. Kim, J. Y. Park, C. H. Lee, Y. J. Yoon, J. S. Yoo, T. Ishigaki, K. Kang, and W. S. Yoo, *J. Vac. Sci. Technol. B*, **34**, 041208 (2016).
2. W. S. Yoo, B. G. Kim, S. W. Jin, T. Ishigaki, and K. Kang, *ECS J. Solid State Sci. Technol.*, **3** (11), N142 (2014).
3. W. S. Yoo, J. H. Kim and S. M. Han, *J. Micro/Nanolithography, MEMS, and MOEMS*, **13** (1), 011205 (2014).
4. J.K. Kim, S.M. Han and Woo Sik Yoo, *ECS J. Solid State Sci. Technol.*, **5**(7), 438 (2016)

KEYWORDS

TSV Cu, Raman, Photoluminescence, SPV

Advanced Defect Classification By Optical Metrology

Jacques van der Donck, Esther Kramer, Peter van der Walle, Wouter Mulckhuyse, Jacqueline van Veldhoven, Loek Nijsten, Felipe Bernal Arango, Helma Spruit and Diederik Maas

TNO Stieltjesweg 1, 2628 CK, Delft, The Netherlands

INTRODUCTION

Particle defects are important contributors to yield loss in semi-conductor manufacturing. Therefore these have to be minimized and, when they do occur, to be characterized in order to determine their root cause. Figure 1 outlines the general approach for advanced defect classification in three consecutive steps; detection, review and classification. For defect detection, TNO has developed the RapidNano (RNx), a fast and sensitive dark-field microscope [1]. For each sample, the lower detection limit (LDL) is verified by an analysis of the speckle signal strength. The detected defects are ranked on their importance, thereafter the defects of interests (DoI) are redetected and reviewed. In previous work, the redetection process (R.2) has been simplified by the production of a stealth fiducial marker system [2]. The RapidNano 3 particle inspection system is capable of detecting 42 nm Latex Sphere Equivalent (LSE) particles on XXX-flat Silicon wafers. This sensitivity is achieved by illuminating the sample from multiple angles. In detection-mode (RN3.1) the signal from all these angles is added. In review-mode (RN3.9), signals from all nine individual arms are analyzed to derive the shape, material and size of deep sub-wavelength defects. Via this analysis we are able to classify and select defects for further analysis on much slower metrology tools. This paper will show such ADC results for a programmed defect wafer, using other review tools (SEM, AFM) to verify the ADC as made using the RN3.9.

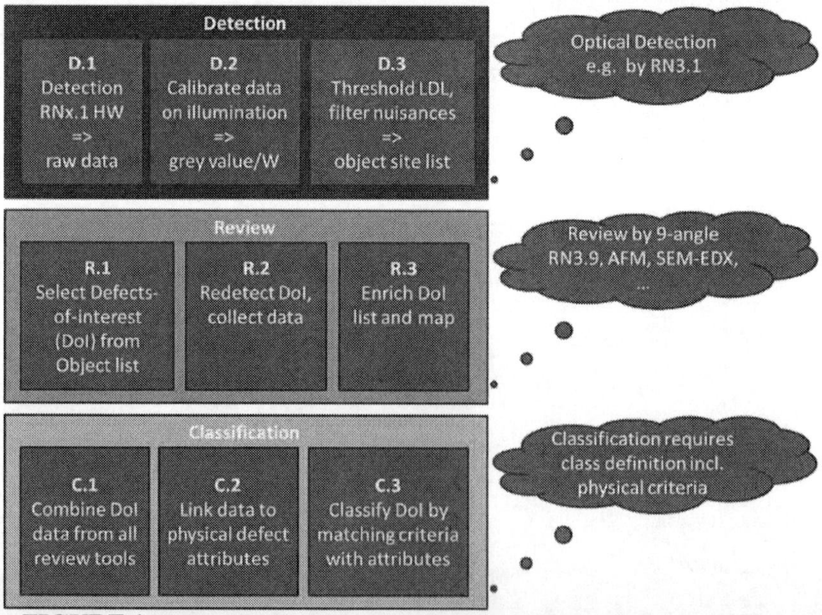

FIGURE 1 Process flow for the detection, review and classification of defects.

MATERIALS AND METHODS

The TNO RapidNano3 is a particle detection tool based on dark field imaging. Defects on a flat surface are illuminated under an angle and a dark field image is acquired. By stepping from field to field a pre-defined surface area can be imaged. The RapidNano3 is equipped with a 9 azimuth illumination for reduction of the speckle contribution in the dark field background image [3]. The RapidNano 3 can be operated in 2 modes. In RN3.1 mode the contribution of all azimuth angles are collected in one image. In RN3.9 mode nine dark field images are collected; one for each illumination azimuth angle.

Using electron beam lithography and a standard lift off technology, a 2D pattern is applied to an XXX-flat Si wafer blank. The 2D pattern comprises both large "stealth" markers for easy navigation in optical, atomic force, scanning probe and scanning electron beam microscopes (OM, AFM, SPM, respectively SEM) as well as a sub-micron set of programmed defects with a rectangular or elliptical shape at aspect ratios ranging from 1:1 to 1:5.

Using the WAVE3D module in the COMSOL modeling SW package, the optical scattering of sub-wavelength-sized metallic scatter centers (i.e. the programmed defects) into the NA of the RN3 microscope objective is calculated for the illumination conditions in the RN3.1 and RN3.9. **FIGURE** 2 shows the geometry definition in COMSOL as well as the 3D scattering profile for illumination by a collimated p-polarized 532 nm laser beam at 60° zenith.

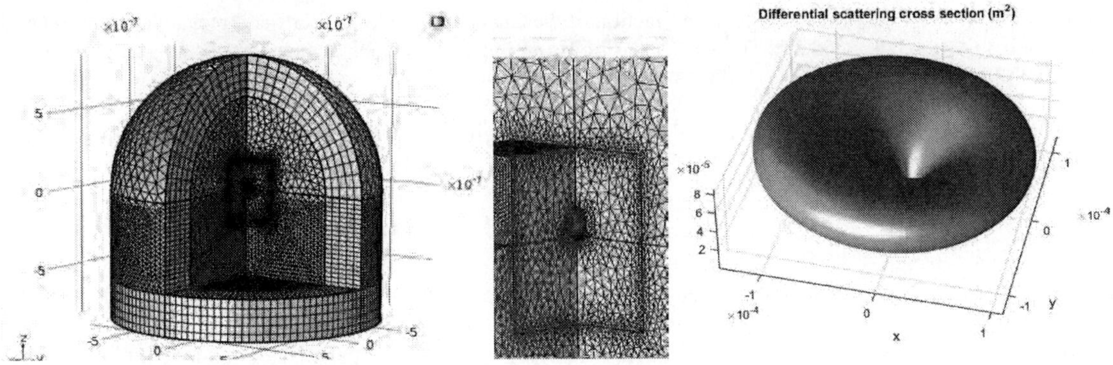

FIGURE 2 Geometry (left and center panel) and scattering profile (right panel) for typical RN3.9 illumination conditions. The differential scattering cross section calculated with an analytical model for a 30nm-radius spherical Au particle on a Si substrate illuminated by p-polarized light coming in at a 60° angle.

RESULTS

FIGURE 3 and 4 show RN3.9 data on the scatter signal into the nine individual illumination angles as recorded from a set of rectangular and elliptical programmed defects with aspect ratios between 1:1 and 1:5 and dimensions between 50 nm and several micron. The scatter intensity follows the theoretical prediction by Bobbert-Vlieger [5].

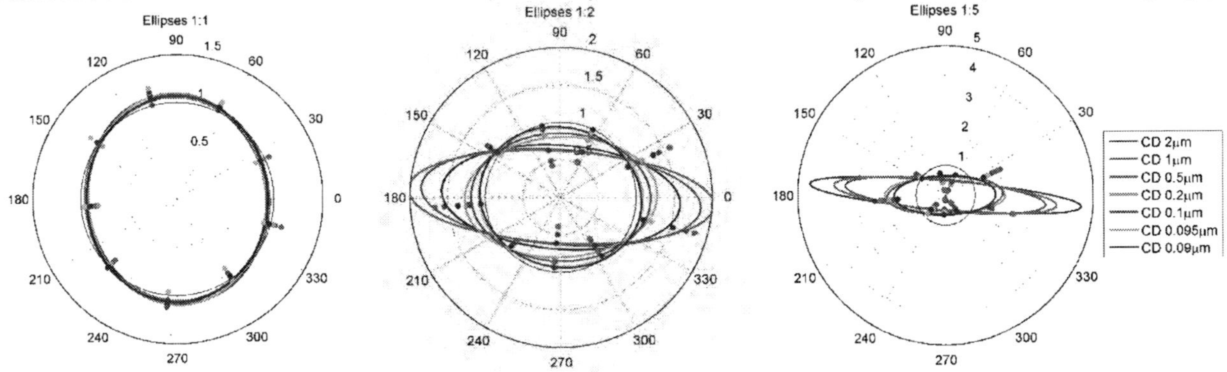

FIGURE 3 Polar plot of the normalized Rapid Nano signal for nine separate detection angles from Au ellipses of varying aspect ratios, for a number of different critical dimensions.

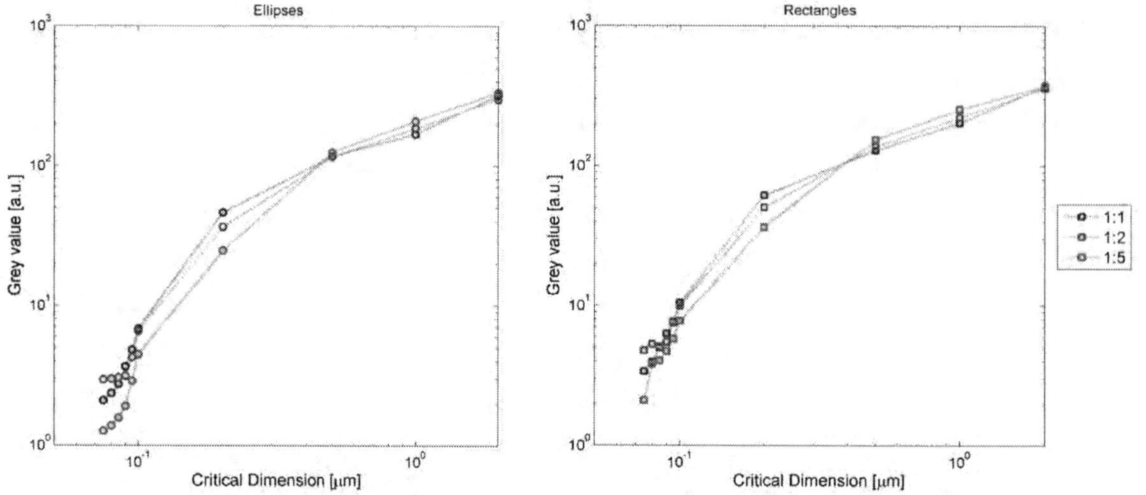

FIGURE 4 The measured RN3.9 signal strength as a function of the CD of a gold ellipse (left) and rectangle (right) on a Si wafer for aspect ratio's (AR) 1:1, 1:2 and 1:5. The AR has a significant impact on the DF scatter signal strength, leading to an under-estimation of the CD for defects <500 nm when using an "equivalent" sphere.

From FIGURE 4 we concluded that the critical dimension of high-AR defects is easily underestimated if only the total scatter signal is used.

CONCLUSIONS AND OUTLOOK

A workflow has been defined for Advanced Defect Review for (programmed) defects on blank wafers using multi-angle dark field optical metrology. The workflow was executed on a programmed defect wafer. Furthermore, a numerical model was developed to predict the scattering behavior of defects of various size, shape and material. This model was validated by comparison of the results for a scattering problem with an analytical solution, i.e. a spherical geometry, with those of the numerical model for identical settings. Both models showed the same scattering behavior. This model will be used to simulate the Rapid Nano signal for defects of arbitrary shapes and materials. By analyzing the simulated signals an algorithm will be developed to extract information on defect size, shape and material from measured Rapid Nano signals which can be matched to classification criteria.

A first experiment showed that it is indeed possible to extract more information from the multi-azimuth signal such as the aspect ratio and orientation of the defect.

REFERENCES

1. Van der Donck, J., Bussink, P., Fritz, E. and van der Walle, P. (2016) 'RapidNano: Towards 20nm Particle Detection on EUV Mask Blanks', MRS Advances, 1(31), pp. 2225–2236. doi: 10.1557/adv.2016.299.
2. Bouwens, M.A.J., Maas, D.J., van der Donck, J.C.J., Alkemade, P.F.A., van der Walle, P., Enhancing re-detection efficacy of defects on blank wafers using stealth fiducial markers, Microelectronic Engineering, 153 (2016) 48-54.
3. van der Walle, P., Hannemann, S., van Eijk, D., Mulckhuyse, W., van der Donck J.C.J., "Implementation of background scattering variance reduction on the Rapid Nano particle scanner", SPIE Proc. 9048 (2014)
4. COMSOL Multiphysics, v5.0, 3D wave optics module
5. Bobbert, P.A. and Vlieger, J., *Light scattering by a sphere on a substrate*, Physica **137A** (1986) 209-242

KEYWORDS

Dark field inspection, particle detection, scatterometry

Automatic Detection Of Dislocations In Strained SiGe With HCl Etch And Brightfield Inspection

Ralf Buengener, Jody Fronheiser

TD Research, GLOBALFOUNDRIES USA, 257 Fuller Road, Albany, NY 12203, USA, ralf.buengener@globalfoundries.com

Carol Boye, Shravan Matham, Stephen Bedell, James Demarest

IBM Research, 257 Fuller Road, Albany, NY 12203, USA

INTRODUCTION

One method to increase the carrier mobility in 3D transistors (FinFETs) is the use of strained SiGe as channel material. A common integration method for SiGe channel materials is to pattern trenches and selectively grow SiGe before fin formation [1]. However, a major concern regarding the introduction of SiGe channel materials is crystalline defectivity from incoming processes either from etch or in situ H_2 prebakes before epitaxial growth. These could have a detrimental impact on device performance and yield. Interfacial defects at the bottom of the cavity will cause threading dislocations that can extend up to the surface, relieving strain and degrading the electrical performance. These defects can be too small to be detected prior to film deposition.

One way to monitor defects has been a highlighting etch with SECCO [2] or more recently HCl [3,4,5]. Due to preferential etching, pits are created around the locations where the dislocations reach the surface while the rest of the surface is etched at a much slower rate. These pits can be counted manually under a microscope. This technique can only sample small areas and is not applicable to patterned structures with actual fins.

In this study we improved the method by using brightfield defect inspection (BFI) to detect and count etch pits automatically over a larger area across the wafer. To correlate with SECCO etching we established an HCl etch process utilizing a SiGe/Si epitaxial stack before defect highlighting.

METHOD DEVELOPMENT

As a first step we developed a method for highlighting and automatic detection on unpatterned pads. This method was implemented on planar pads on every chip of a wafer. Vaporous HCl etching was performed in the same chamber used for the epitaxial growth of the SiGe and Si films. The same wafer could be etched again and again to investigate how the defects develop over increasing etch times. A brightfield inspection tool was used for the detection of the etch pits. Defect counts were recorded before and after epi growth as well as after each etch step. Several defects were reviewed in a scanning electron microscope (SEM) after each inspection.

Fig. 1 shows how the defect count developed while etching the same wafer several times. T1 means the wafer went through the etch step once, T2 twice, etc. It was expected that the defect count would increase with increasing time as the holes around each defect grew big enough to be detected and then level out once every defect was found. However, we observed that the defect count kept increasing for longer etch times. Between T2 and T3 the curve bends and counts increase at a faster rate.

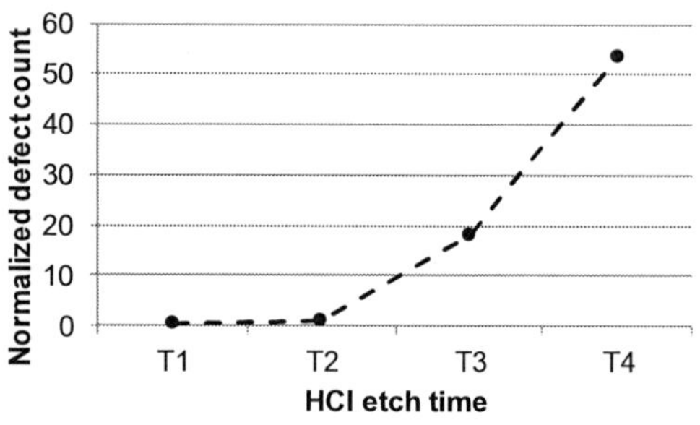

FIGURE 1. Defect count vs. HCl etch time

Almost all detected defects were the expected etch pits. Fig. 2 shows SEM images of the same defect at T1, T2, T3 and T4, respectively. At T1 a square pit is visible as well as three smaller pits left of it. The square shape is caused by preferential etching across the lattice planes once the etch process is seeded by the dislocation. The dislocation itself is invisible by inline SEM inspection. At T2, all pits are fully highlighted and have started to merge. No new pits have appeared, indicating that all initial defects are now highlighted. At T3 new defects appear, at T4 they have grown. These new defects may have been seeded by foreign material or by material lifted off elsewhere on the wafer and redeposited here.

This is one of several examples that show that all initially present defects are highlighted at T2 and new undesired defects appear at longer times. Therefore we set the etch time to be T2 but in a single etch step.

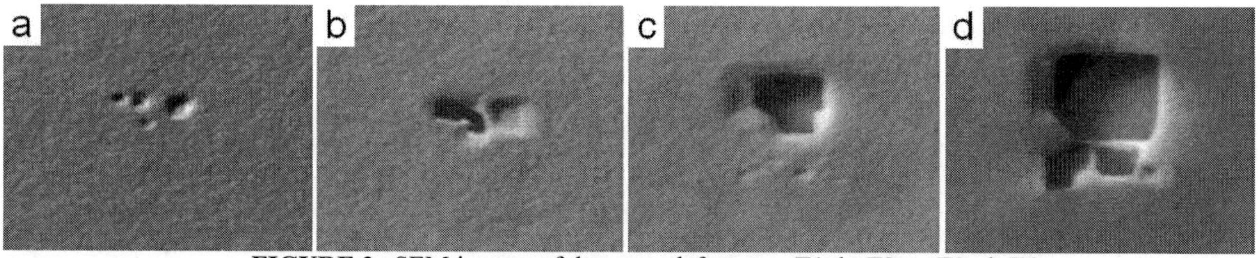

FIGURE 2. SEM images of the same defect at a: T1, b: T2, c: T3, d: T4.

RESULTS

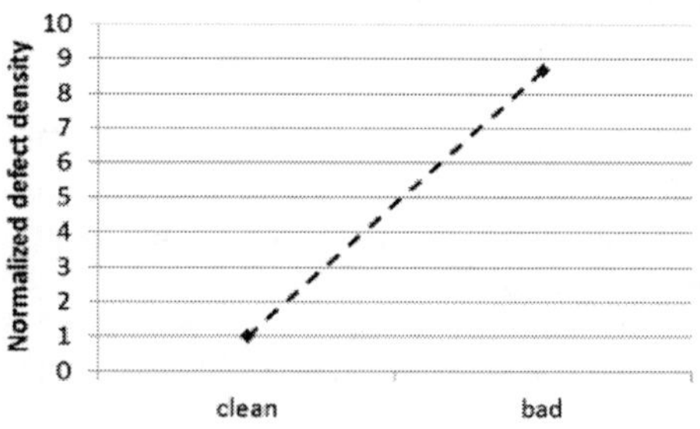

FIGURE 3. Defect densities after HCl etch

As a next step, two wafers were inspected after fin patterning: One had been processed with a known good process, the other with a process known to create many defects. A sample of defects from each wafer were reviewed in the SEM and classified. Based on the classified sample the defect density was normalized for defects that look like etch pits; defects caused by other effects were excluded. Fig. 3 shows the defect density for etch pit defects; as expected the bad wafer has a higher defect density. This confirms that detection and counting works; thanks to the BFI it provides quantitative results for defect density on full wafers.

CONCLUSION

A method has been developed to highlight dislocations in strained SiGe and count them using brightfield inspection. The advantages of this method are:

- Both steps, HCl highlighting and defect inspection can be run in automation on tools that are available in the production line. This enables regular monitoring during production.
- The automatic detection gives quantitative results from a large area. Even wafer maps are possible.

The same method can be evaluated on patterned pads or fins. In a fin the defect would not look like a square because it is bounded by the liner. But the etch can progress along the fin forming a rectangle that can be detected by the BFI tool.

REFERENCES

1. D. Guo et al, "FINFET Technology Featuring High Mobility SiGe Channel for 10 nm and Beyond", *Proc. VLSI* (2016)

2. S. W. Bedell, D. K. Sadana, K. Fogel, H. Chen, A. Domenicucci, "Quick turnaround technique for highlighting defects in thin Si/SiGe bilayers", *Electrochemical and Solid State Letters*, **7**, G105-G107 (2004)
3. Y. Bogumilowicz, J. M. Hartmann, R. Truche, Y. Campidelli, G. Rolland, T. Billon, "Chemical vapour etching of Si, SiGe and Ge with HCl; applications to the formation of thin relaxed SiGe buffers and to the revelation of threading dislocations", *Semicond. Sci. Technol.* **20**, 127-134 (2004)
4. A. Abbadie, F. Allibert, F. Brunier, "Defect delineation and characterization in SiGe, Ge and other semiconductor-on-insulator structures", *Solid-State Electronics* **53**, 850-857 (2009)
5. J. M. Hartmann, A. Abbadie, "HCl revelation in 200 mm SiGe virtual substrates: A systematic study", *Thin Solid Films* **557**, 110-114 (2014)

KEYWORDS

SiGe, strain, detection, dislocations, brightfield inspection, SECCO, HCl

Inline X-Ray Metrology of Epitaxial Thin Film for Gate-All-Around Structures

[1]B. Mendoza, [2]P. Gin, [2]K.M. Matney, [2]M. Wormington [3]J. Wall, [3]P. Ryan, [1]R. Chao, [1]J. Gaudiello

[1]IBM STR Albany, 257 Fuller Road, Albany, NY 12203 USA
[2]Bruker Semi Division, 112 Robin Hill Road, Santa Barbara, CA 93117 USA
[3]Bruker JV UK Ltd., Belmont Business Park, Durham, DH1 1TW United Kingdom

ABSTRACT

As the size of transistors continues to decrease with advancing CMOS technology nodes, we have seen a shift from planar to multi-gate transistors. This is due in large part to reduce short-channel effects such as gate leakage, which is a major contributor to higher power consumption and diminished device performance [1]. As the name suggests, multi-gate devices offer better electrical control over the channel by surrounding it with several gates. In addition, the more compact multi-gate design allows for denser packing of transistors compared to planar devices leading to increased functionality on smaller die. Currently the most popular variant is the finFET device that employs a 3-dimensional thin Si fin surrounded by multiple gates. This technology has been adopted since the 22 nm technology node [2]. However, at sub-7nm, variations in channel width makes FinFET technology susceptible to short channel effects, making CMOS scaling extremely challenging. Alternatively, gate all around (GAA) FET devices, such as Si nanowire transistors with gates completely surrounding all four sides of the Si channel provide better electrostatistics relative to finFETs.GAA is a promising choice from not only an electrostatics perspective, but from an integration perspective. Because of its capability to stack vertically, GAA devices in a horizontal configuration can enable higher transistor density. One of the precursor steps to forming the horizontal GAA transistors is the formation of fully strained periodic multilayers of thin films, such as Si/SiGe, hereafter defined as nanosheets. The Si/SiGe nanosheet is epitaxially grown in a single process, allowing high quality single crystal material to be produced. However, subsequent processing of the stack, such as etching or annealing may create defects or free surfaces around the nanosheet layers, causing the built-in strain to relax, which diminishes the performance of the device. It is therefore, crucial that each nanosheet layer can be characterized in-line in a non-destructive manner during various manufacturing steps.

High-resolution x-ray diffraction (HRXRD) is an established and well regarded technique for epitaxy characterization and metrology including SiGe/Si in advanced CMOS processes [3-4]. In this study, we describe how HRXRD provides a powerful tool for in-line monitoring of nanosheet stacks consisting of multiple strained Si/SiGe epitaxial layers. In addition, we describe the use of specular x-ray reflectivity (XRR) for accurate thickness measurements of each layer and interfacial roughnesses even when the epi quality is poor or relaxed.

We illustrate the use of these techniques by presenting representative data and results from the development of nanosheet technology at IBM (Albany, NY). In-line metrology at key processing steps is a necessity for Si/SiGe nanosheet development, as in-line techniques reduce the time needed for each learning cycle. As an example, Ge concentration and relaxation of the individual SiGe layers must be monitored and controlled in order to maintain majority carrier mobility and concentration in the silicon layers. Similarly, high interfacial roughness between the layers can adversely affect device performance by altering the uniformity of the gate dielectric, thereby degrading device switching response. This information is unattainable through inline AFM, as the technique only provides surface roughness for the top layer. As such, a combination of in-line HRXRD and XRR metrology was selected for this application.

We will introduce the capabilities and methodology of the HRXRD and XRR techniques. For blanket samples, HRXRD ω-2θ scans proved to be a quick and reliable way to measure Ge composition and layer thickness, as seen in Fig 1. For improved thickness precision, as well as to collect information on interfacial roughness, XRR was utilized (Fig 2). In addition, fast reciprocal space maps (RSMs) were used to characterize layer strain and changes in stress states during the formation of the GAA structures, as seen in Fig 3. We will detail the recent advancements, which allowed for quick and automated in-line RSM measurements. In the past RSMs have been shown to provide a wealth of valuable information but it's utilization for in-line metrology has been limited due to slow measurement speeds and lack of automated data extraction capability. Now, by utilizing a linear (1D) detector, RSMs can be acquired with reasonable throughputs and together with new software developments for automatic data extraction, it is possible to monitor the nanosheet superlattices and other complex epitaxial structures in-line [4]. Furthermore, for patterned samples, we show that the HRXRD capabilities are scalable down to 50x50 um using a micro-XRD source. This small beam size is critical especially in the advanced technology nodes as pad sizes continually shrink. We will discuss the impact of the micro-XRD source on data quality. In this paper, we will show typical HRXRD and XRR data from nanosheet Si/SiGe structures and describe how it is used to characterize and control key aspects of the process as an important step in gate GAA FET development.

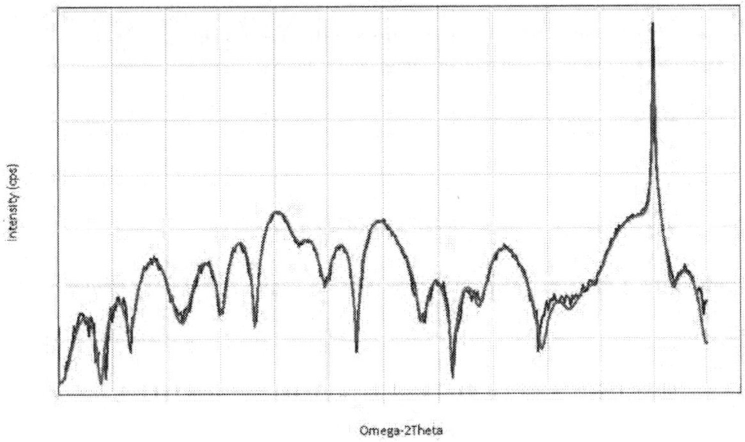

FIGURE 1. HRXRD ω-2θ scans around the symmetric (004) Bragg reflection from a SiGe/Si nanosheet on a bulk Si substrate. Analysis of the diffracted intensity distribution gives information on the lattice parameter of the individual layers with respect to the silicon substrate. Ge concentration and layer thickness are calculated from the best fit simulations (red) of the measured (blue) intensity profiles.

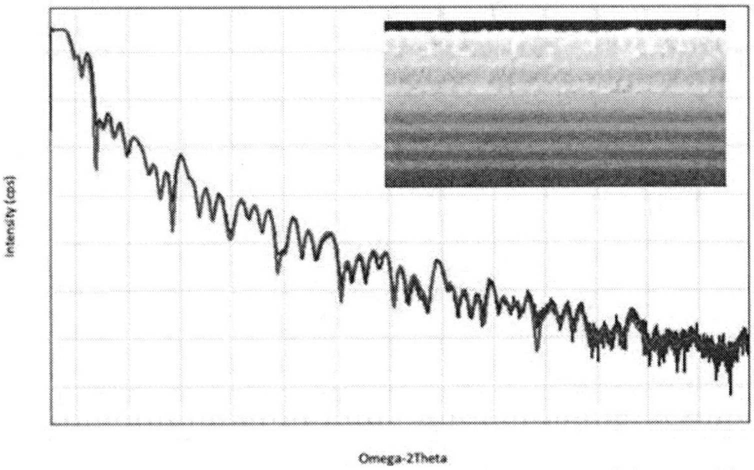

FIGURE 2. XRR from SiGe/Si nanosheet on a bulk Si substrate. Analysis of the measured data provides information on the individual layer thickness, density, and interfacial width (roughness). The measured intensity profile (blue) is overlapped with the best fit simulation (red). Comparison with a destructive SEM cross-section (inset) shows good thickness correlation.

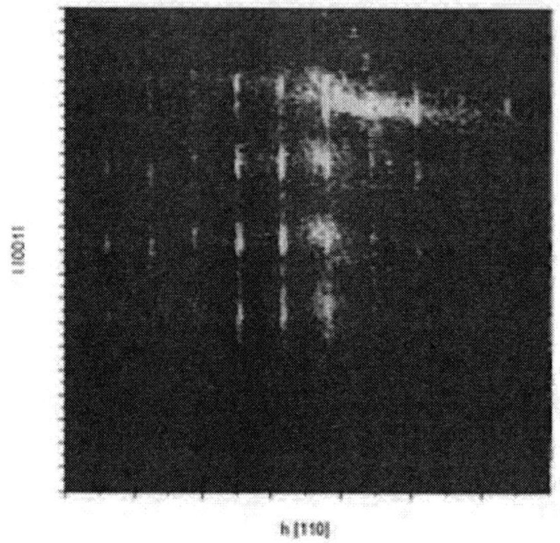

FIGURE 3. HRXRD reciprocal space map (RSM) in the vacinity of the 113ge Bragg reflection of fins etched into a nanosheet stack on Si(001) substrate.

REFERENCES

1. Oh, Sang-Hyun, Don Monroe, and J. M. Hergenrother. "Analytic description of short-channel effects in fully-depleted double-gate and cylindrical, surrounding-gate MOSFETs." IEEE electron device letters 21.9 (2000): 445-447.
2. Farkhani, Hooman, et al. "Comparative study of FinFETs versus 22nm bulk CMOS technologies: SRAM design perspective." 2014 27th IEEE International System-on-Chip Conference (SOCC). IEEE, 2014.3. Le Cunff, D., et al. "HRXRD for in-line monitoring of advanced FD-SOI technology: Use-cases: AM: Advanced metrology." SEMI Advanced Semiconductor Manufacturing Conference (ASMC), 2016 27th Annual. IEEE, 2016.
4. Mendoza, B., et al. "Inline monitoring of SiGe strain relaxed buffers (SRBs) using high-resolution X-ray diffraction: AM: Advanced metrology." SEMI Advanced Semiconductor Manufacturing Conference (ASMC), 2016 27th Annual. IEEE, 2016.

KEYWORDS

Gate-all-around, Multi-gate High resolution X-ray diffraction, X-ray reflectivity, Reciprocal space maps

Nano-Particle Detection, Identification and Concentration Determination in Liquids for sub-10 nm Technology Nodes

Basil G. Eleftheriades, Steve A. Rutledge, Manuchehr Ebrahimi, Emily E. Storey, Marie Tripp,[§] Bhagrava Ravoori,[§] Tuyen K. Tran,[§] and Amr S. Helmy

Department of Electrical and Computer Engineering, University of Toronto, Toronto M5S 3G4, Canada.
[§]Intel Corporation, Hillsboro, OR, USA
e-mail: a.helmy@utoronto.ca

INTRODUCTION

Exposure to aqueous solutions and liquid chemicals is inevitable for semiconductor wafers as they pass through the various stages of a process line. Given the aggressive shrinking in features sizes of upcoming technology nodes, wafers are more prone to experience reduced yields due to process liquid-borne particle contamination. The maximum tolerable contaminant particle size scales roughly as the half-pitch dimension of a given technology node. Particles can be detected with relative ease when they already reside on the semiconductor surface, however their chemical makeup can only be determined using expensive EDS analysis. The most effective approach to preserve process yield is to avoid particles getting onto the wafer in the first place. As such particle detection in process liquids is of great concern to the semiconductor industry as it can play an important role to impact the yield.

There has been a significant body of work outlining the implications of particle inclusion, however there are less viable solutions reported to address these concerns without a prior knowledge of the type/nature of the materials, where they need to be concentrated for a discernable signal to be detected [1,2]. Recently we have been able to demonstrate how enactments in conventional optical spectroscopy such as Raman and Fourier Transform Infrared Spectroscopy (FTIR) can result in a viable solution to these challenges [3]. Optical spectroscopy techniques can be optimal candidates to monitor and quantify nanomaterials in liquids when their sensitivity is enhanced.

The approach described here is different as it utilizes techniques with inherent chemical specificity (Raman and FTIR) to carry out the detection, which does not pose limitations on the size of the contaminants. The limitation of these techniques is their low sensitivity, which the enhancement associated with the optofluidic designs we discuss here alleviate significantly to render the system of practical use with a practical technological approach that provides it with a competitive cost structure.

DETECTION SENSETIVITY ENHANCEMENT OF OPTICAL TECHNIQUES

Optical spectroscopy techniques such as Raman and FTIR can identify the chemical make-up of a contaminant and its concentration in liquids. Standard Raman instrumentation is severely limited in sensitivity for the analysis of liquids gasses and aerosols, requiring large samples, due to modest Raman scattering in these media. Sensitive, commercially available high-resolution and high-sensitivity Raman systems are bulky and, most importantly, extremely expensive. A recently developed approach to increase the received Raman signal (and thus the sensitivity) for all generated Raman modes without altering the bonds of the native analyte is to incorporate optofluidic devices into the Raman spectroscopy setup as the interaction medium [6]. Hollow core optofluidic constructs enable an increased interaction length through the confinement of both the laser light and the liquid of interest into the same cavity, enabling a strong and efficient process of light-matter interaction. This enhancement can enable approaches including Raman and FTIR to serve as a viable approach for environmental monitoring [4] defense [5], biotechnology and pharmaceutical [6], where the majority of samples are gasses, liquids or aerosols.

The enhancement obtained using this technique are expected to enable sufficiently sensitive Raman and FITIR spectra on liquid samples that enable contaminant detection, identification and concentration determination in all

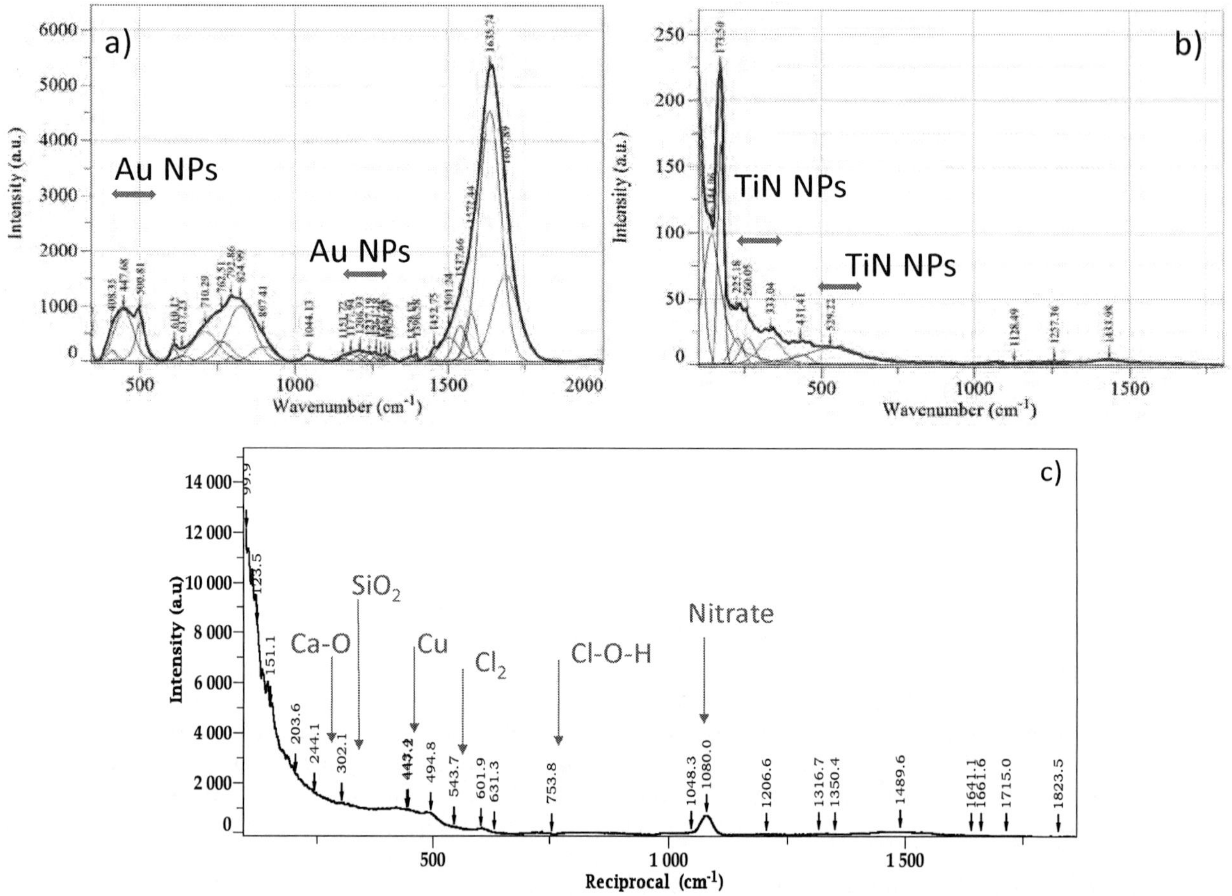

FIGURE 1. The spectra in a) and B) show the Raman modes of Au and TiN, respectively obtained from measurements in ultra-pure water. The known Raman modes for both materials are denoted with red arrows. The Raman spectrum in c) shows the Raman modes obtained from analyzing water as delivered from the municipality source and the contaminants that it contains.

process liquids. Due to the nature of this technique, the limiting factor will be the concentration of the contaminant rather than their size. As such this technique, as we have demonstrated [3] is able to detect nm-scale particles easily.

Figure 1 shows examples of Raman spectra obtained from optofluidic chips as the light-matter interaction medium in the backscattering configuration. In this example, the analyte is filled into the central core of the optofluidic chip from one end through capillary action. At the other end the pump laser light is focused into the same central core through the objective to generate Raman scattering signals from the analyte. Due to the strong confinement of the pump laser light in the chip, the pump laser light maintains a strong power density in the central core. As both the pump laser light and the analyte are confined within the same central core of the optofluidic chip, the strongly confined pump laser can interact with the analyte throughout the entire fiber length. Thus, Raman signals are scattered throughout the entire length of the chip as opposed to just the depth of field of the objective in the conventional scheme. Since Raman signals are mostly shifted by less than several tens of nanometers from the wavelength of the pump laser, they will also be confined inside the liquid core and collected throughout the chip. As a result, the output signal from the fiber end will be collected more efficiently by the objective for detection; thus an enhanced Raman scattering signal could be retrieved at the detector.

In figure 1.a) clear Raman signatures of the existence of nano-scale Au nanoparticles are observed, while figure 1.b) shows clear Raman signatures of the existence of TiN nanoparticles. The spectrum in c) demonstrates the efficacy of this technique in detecting and identifying trace amounts of contaminants in tap-water as delivered.

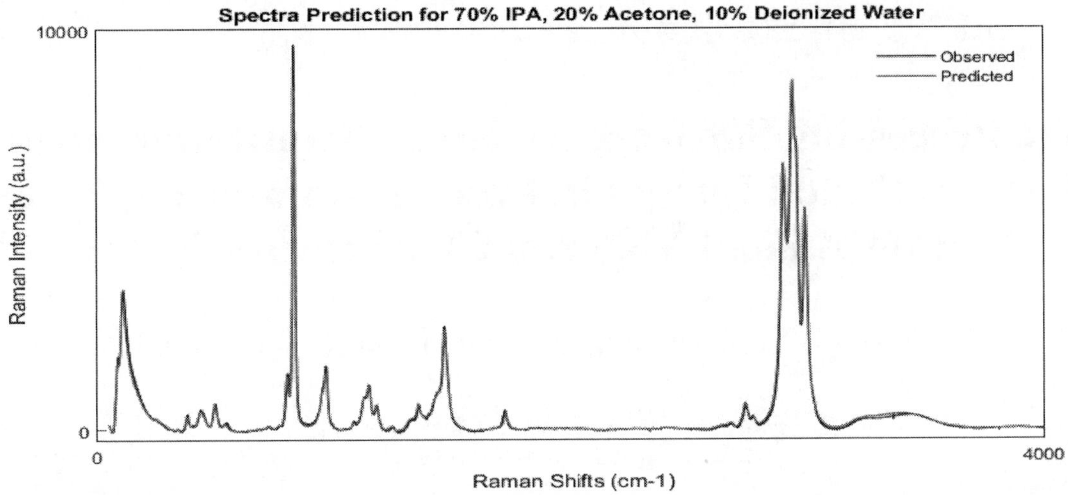

FIGURE 2. Raman spectra Measured for a mixture of IPA, DIW and Acetone. The measured and predicted spectra are compared.

CONCENTRATION DETERMINATION USING RAMAN AND FTIR

Principal component analysis (PCA) is a procedure which reduces the dimensionality of data sets by calculating principal components (PCs), which are functions of the original data and are chosen such that they maximize its variance. This technique is ideal for identifying and quantifying contaminants. Non-Iterative Partial Least Squares (NIPALS) is an algorithm which calculates a predetermined number of PCs. This avoids unnecessary computation in situations where only the PCs with highest variance are needed. A NIPALS program which predicts concentrations within an unknown solution based on its Raman spectra has been developed for the instrumentation described in the previous section. PCs are determined from a collection of spectra with known concentrations. Projecting the unknown data set on to these PCs allows the program to approximate its relative concentrations.

As can be seen in figure 2, the code developed by us was used to estimate concentrations in a solution when supplied with the identities of the components. Actual concentrations were 70% ± 0.5% IPA, 20% ± 0.25% Acetone, and 10% ± 0.25% di-ionized water (DIW). Predicted results were 70.74% IPA, 18.84% Acetone, 10.41% DIW. The measured spectrum compares well with the one predicted by the algorithm.

KEYWORDS

Nanomaterials detection, Chemically Specific Sensing, Optical Spectroscopy, Optofluidics, Handheld Sensors.

REFERENCES

1 Abbas Rastegar, Martin Samayoa, Matthew House, Huseyin Kurtuldu, Sang-Kee Eah, Lauren Morse, and Jenah Harris-Jones, Particle Control Challenges in Process Chemicals and Ultra-pure Water for sub-10 nm Technology Nodes, Extreme Ultraviolet (EUV) Lithography V, Proc. of SPIE, 9048 (2014).
2 Wibowo, J., Shadman, F., Blackford, D., "Measuring and removing dissolved and colloidal silica in ultrapure water". MICRO, Ultrapure Fluids (1997)
3 Janahan Ramanan, Basil G. Eleftheriades, Bhagrava Ravoori, Marie Tripp, Tuyen K. Tran, and Amr S. Helmy, 'Sensitive Nanomaterials Detection and Analysis in Solution,' Frontiers in Nano-electronics Metrology- Dresden (2015).
4 U. Villanueva, J.C. Raposo, K. Castro, A.d. Diego, G. Arana, J.M. Madariaga, Raman spectroscopy speciation of natural and anthropogenic solid phases in river and estuarine sediments with appreciable amount of clay and organic matter, J. Raman Spectrosc. 39 (2008) 1195-1203.
5 M.D. S., S.R. J., Portable Raman explosives detection, Anal. Bioanal. Chem. 393 (2009) 1571-1578.
6 R.M. Abu-Ghazalah, S. Rutledge, L.W.Y. Lau, D.N. Dubins, R.B.M. J., A.S. Helmy, Concentration-dependent structuraltransitions of human telomeric DNA sequences, Biochem. 51 (2012) 7357-7366.

Image Processing Software Assisted Quantitative Analysis of Various Digital Images in Process Monitoring, Process Control and Material Characterization

Woo Sik Yoo, Toshikazu Ishigaki and Kitaek Kang

WaferMasters, Inc., San Jose, California, 95112, USA

INTRODUCTION

A large quantity of digital images is generated daily for process monitoring, process control and material characterization. Each tool provides its own image export and analysis functions. A very small fraction of digital image data is utilized due to the lack of end user friendly image processing software.

It would be beneficial to develop end user friendly, unified image processing software which can support various digital image formats (such as BMP, JPG, GIF, PNG, TIF, DM3 etc.) from various tools. In this paper, a newly developed end user friendly, unified image processing software (PicMan from WaferMasters, Inc.) is introduced, along with a few examples.

APPLICATION EXAMPLES

Figure 1 shows a dimension measurement example from a published article using PicMan software.

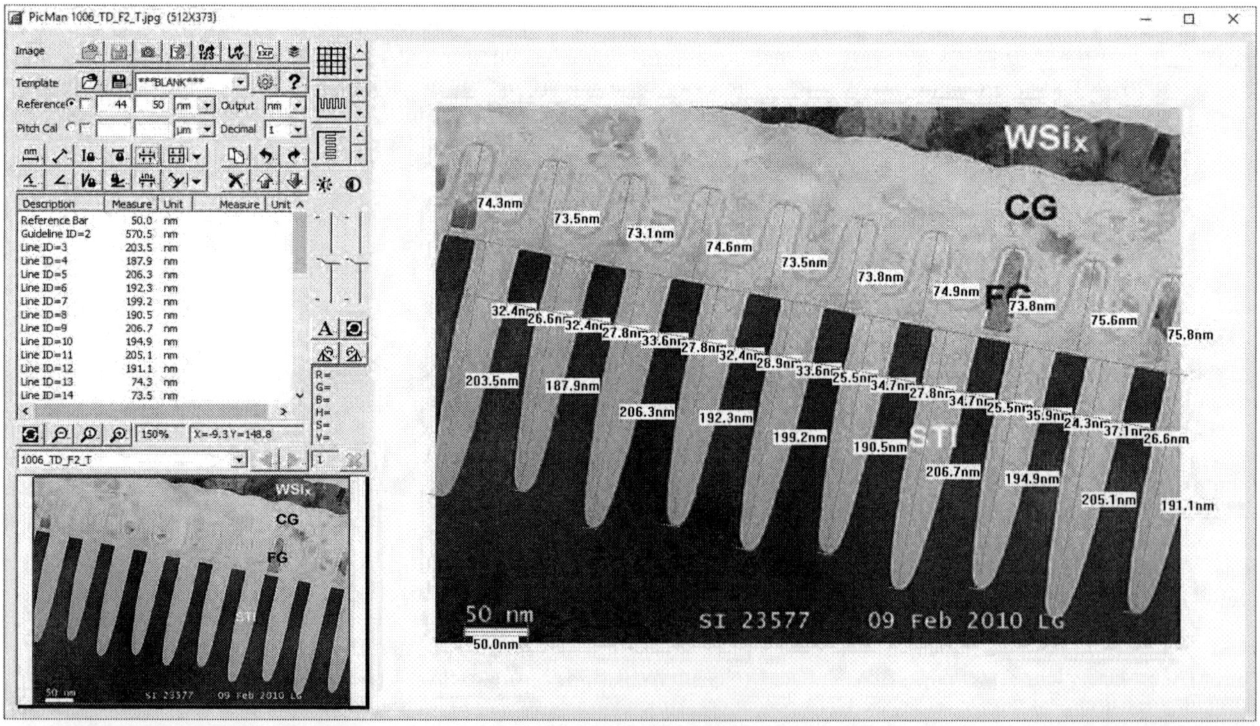

FIGURE 1. Dimension measurement example from a TEM image. [1]

Figures 2~4 show various dimension measurements on TEM images of 3D V-NAND devices on published articles. Automatic dimension measurement and measurement summary are available. Length, angle, area measurement are available. Characteristics of shapes are also classified and recorded with measurement data.

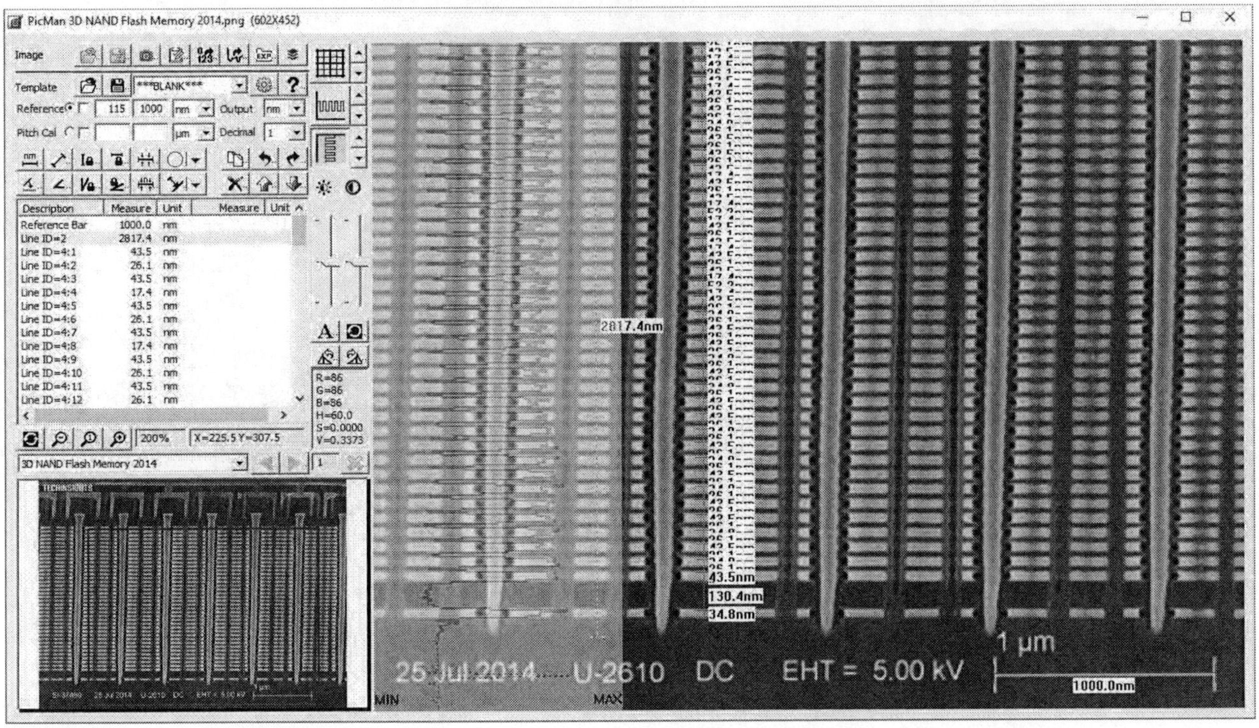

FIGURE 2. Cross section of a 3D V-NAND device (low magnification image). [2]

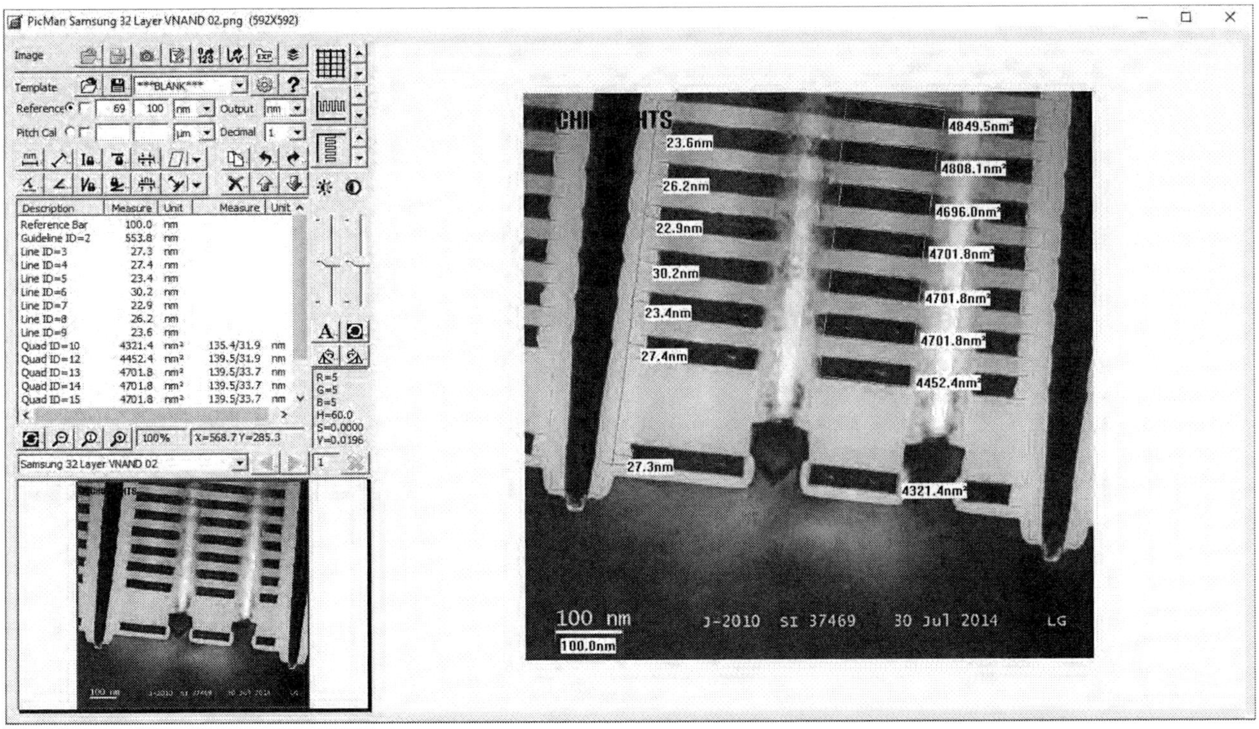

FIGURE 3. Cross section of a 3D V-NAND device (high magnification image). [3]

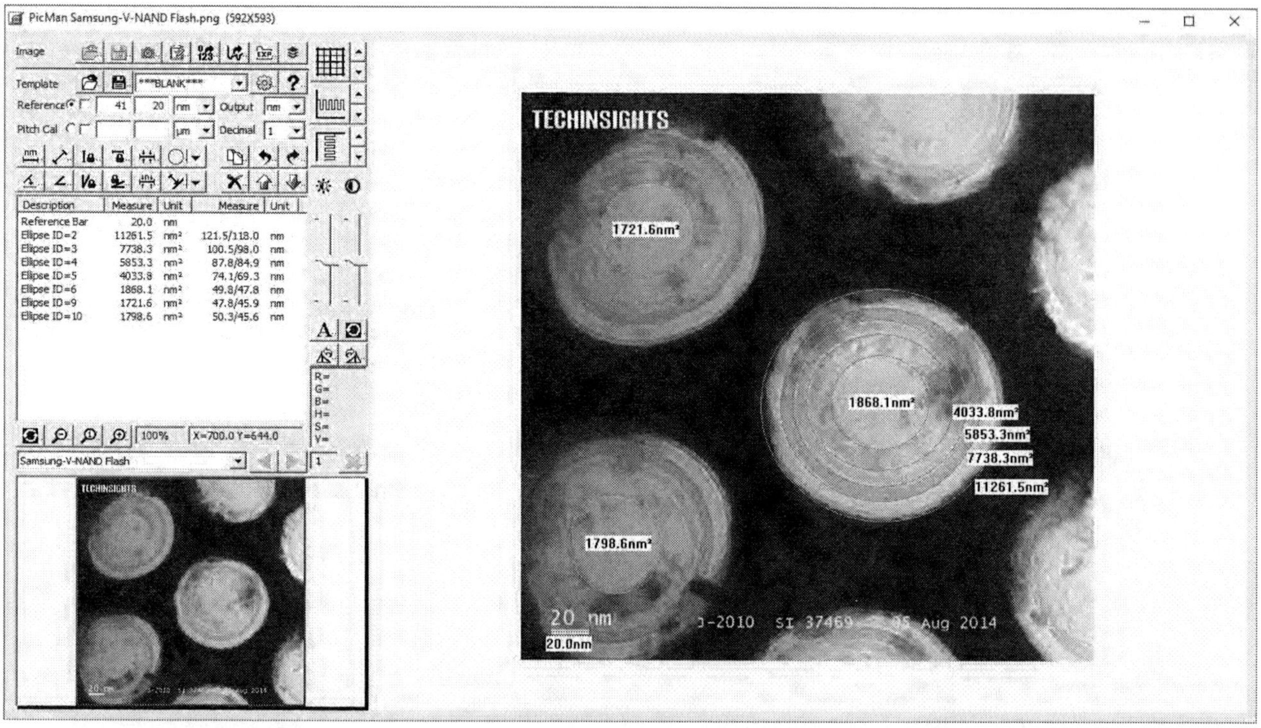

FIGURE 4. Plan view TEM image of 3D V-NAND device for area measurements. [4]

Figures 5 and 6 show line intensity functions using an AFM image and an electron diffraction image for quantitative analysis. The end user can perform offline analysis of any digital images without going back to original tools. Digital images from any source can be analyzed very quickly and accurately using a PC.

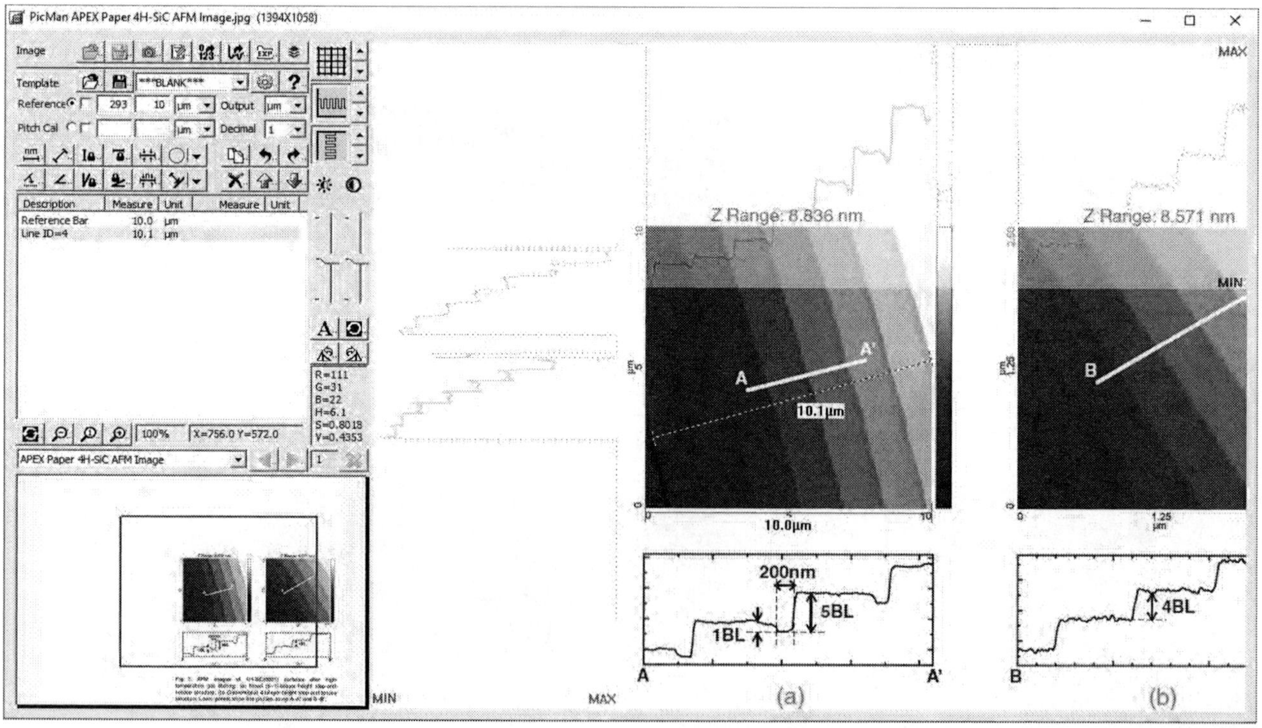

FIGURE 5. AFM image of 4H-SiC surface. [5]

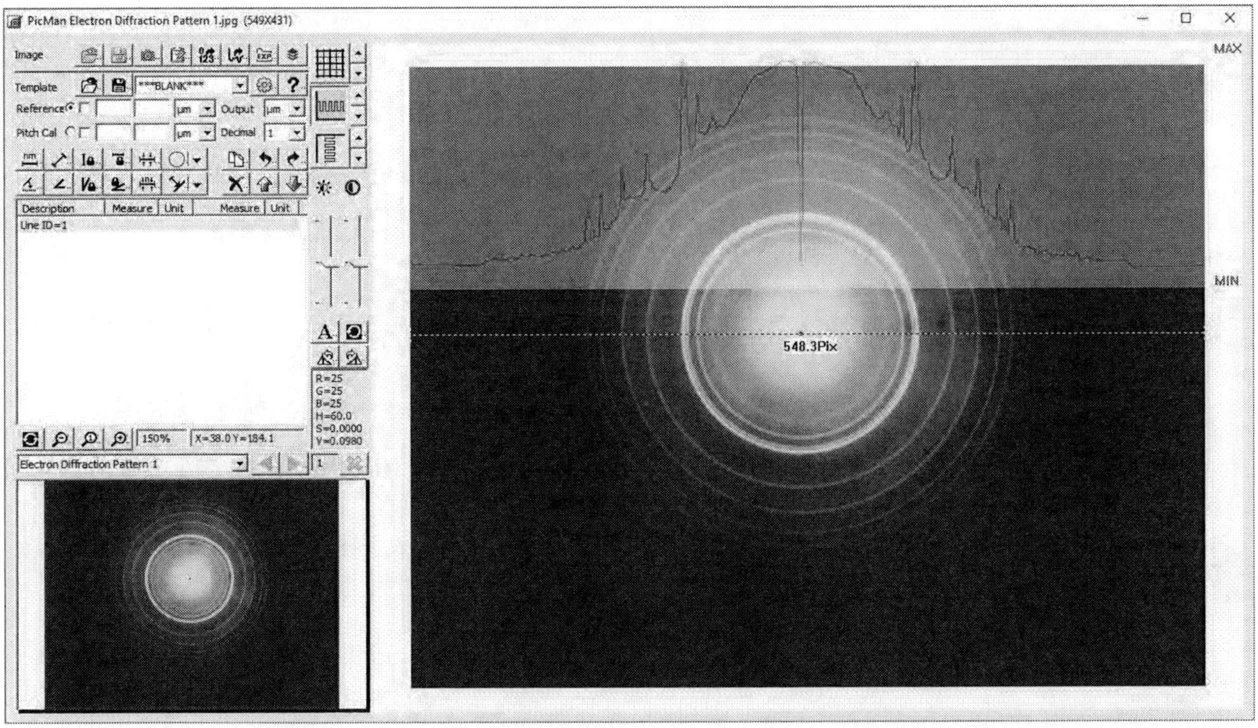

FIGURE 6. Electron diffraction image of microcrystalline specimen. [6]

SUMMARY

Significant efforts have been made to make accurate and precise measurements and characterizations. Very large amount of digital images in the forms of SEM, CDSEM, TEM, AFM, EBIC, optical micrograph are generated and analyzed by specialists every day. Once an image is generated and delivered to end users (requesters) the usefulness of digital image data is less useful than it could be because of the lack of user friendly unified image processing software. Newly developed unified image processing software (PicMan from WaferMasters, Inc.) enables end users to analyze digital images of any format, on their own PCs, easily, accurately and repeatedly. Improved digital image analysis by end users will greatly improve the understanding of characterization results and development cycles.

REFERENCES

1. http://m.eet.com/media/1089741/1578photos_pg44a.jpg
2. http://www.pcper.com/files/imagecache/article_max_width/news/2015-08-11/Picture5.png
3. http://www.3dincites.com/wp-content/uploads/AndyFig5.png
4. http://www.3dincites.com/wp-content/uploads/AndyFIg6.png
5. J. Suda and T. Kimoto, *Applied Physics Express*, **2** 101603 (2009).
6. https://media1.britannica.com/eb-media/28/96828-004-644E46F6.jpg

Optical Critical Dimension Metrology for the 7 nm Node and Beyond Using a Near-field Metalens

Jinlong Zhu[1,2], Aditi Udupa[1], Yating Shi[2], Shiyuan Liu[2] and Lynford L. Goddard[1]

[1]*Photonic Systems Laboratory, Department of Electrical and Computer Engineering, University of Illinois at Urbana-Champaign, Urbana, Illinois 61801, USA*
[2]*State Key Laboratory of Digital Manufacturing Equipment and Technology, Huazhong University of Science and Technology, Wuhan 430074, China*

INTRODUCTION

By the year 2017, the critical dimension (CD) in patterned wafers will shrink down to 7nm. This creates an incredible challenge to balance the requirement of accuracy, non-destruction, and high speed in CD metrology. Atomic force microscopy (AFM) and scanning electron microscopy (SEM) are well-known for high spatial resolution, but inherently suffer a small field of view and cannot be scaled and directly integrated into a manufacturing line. Scanning near-field optical microscopy [1] and multi-photon and fluorescence microscopy [2] offer a resolution down to one-tenth of the illumination wavelength. Unfortunately, the long scanning time and lack of fluorescence, respectively, limit their practical use in the semiconductor industry. Optical scatterometry, which is a far-field based imaging technique, is regarded as one of the most promising candidate methods to meet the demanding requirements in semiconductor CD metrology. However, when the CD of the patterned structure is far smaller than the illumination wavelength, the structure behaves like a thin film. Consequently, its macroscopic optical properties are dominated by its material constants rather than by the lateral CDs. The conventional way to increase sensitivity to small CD features (say, 7 nm) is to reduce the wavelength, e.g. replace visible light with extreme ultra-violet (EUV) light [3, 4]. However, this creates two problems. First, only the superficial CD of a patterned nanostructure can be resolved because semiconductors have strong EUV light absorption. Second, EUV light sources and optics are complex. Despite their resolution advantage over 193 nm excimer lasers, EUV lasers have not significantly penetrated the metrology or photolithography markets. Metrology of sub-10 nm CD remains as a great challenge to all of the existing techniques.

METHODOLOGY

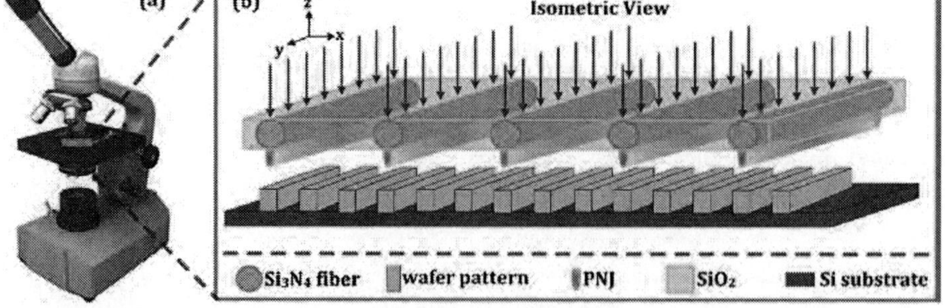

FIGURE 1. (a) Illustrated diagram showing a coherent optical microscope and (b) the mechanism of microfiber array-assisted CD metrology.

In this paper, we propose the metalens-assisted CD metrology system shown in Fig. 1. The metalens consists of periodically arranged Si_3N_4 microfibers that are embedded in a SiO_2 cladding layer. The metalens is situated on top of the patterned wafer in the near-field region. A plane wave illuminates it at normal incidence from the top. The metalens induced super-focusing of the plane wave and refocusing of the scattering field with respect to the underlying nanostructure can result in an enhanced far-field signal that can be imaged. The setup only requires a conventional microscope and a readily available mechanical positioning system (MPS). Current MPSs used in semiconductor industry already achieve nanoscale positioning accuracy.

RESULTS

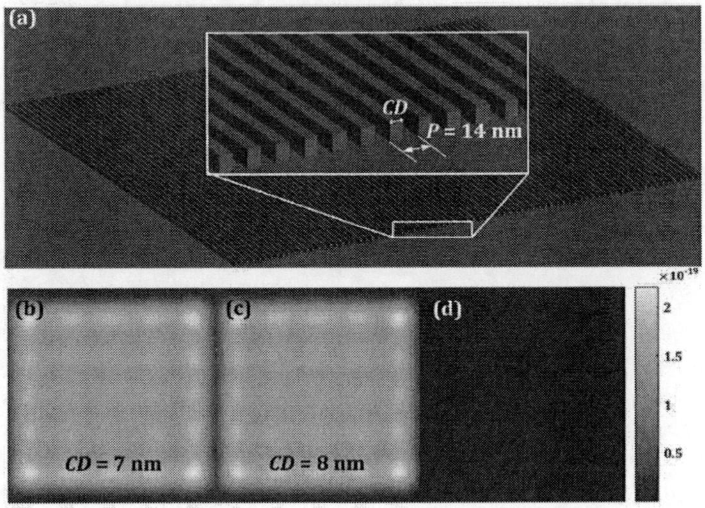

FIGURE 2. Simulations show the extremely low sensitivity of the far-field image to the CD of a line array. (a) Illustrated diagram showing the geometry of the patterned nanostructure. Simulated far-field signals for a (b) 7 nm and (c) 8 nm CD, respectively. (d) Difference signal obtained by subtracting Fig. 2(b) from Fig. 2(c).

Figure 2(a) presents the geometry of a line array with a nominal 7 nm CD. The pitch *P* of the array is set as 14 nm, i.e. a nominal duty cycle of 50%. To analyze how the optical signal varies with the CD, we consider two different patterns, whose CD are set as 7 nm and 8 nm, respectively, while the pitch for each is fixed at 14 nm. A plane wave whose electric field component is parallel to the line impinges on the line array. Next, the far-field signal passing through a high numerical aperture 4f system (NA = 0.9, 206×) is reconstructed using a vectorial imaging modeling algorithm [5, 6]. As can be seen from Fig. 2(b) and Fig. 2(c), it is extremely difficult to distinguish the difference between these two images by the naked eye. We therefore computed the difference signal by subtracting Fig. 2(b) from Fig. 2(c). As shown in Fig. 2(d), the 1 nm bias of the CD on the 7 nm node wafer indeed results in an extremely weak far-field difference signal. Because the signal can be easily covered by noise in the optical system, this far-field based optical CD metrology system is unreliable.

In this talk, we will reveal how the microfiber array metalens can help enhance the sensitivity of the far-field difference signal with respect to tiny variations in the CD.

REFERENCES

1. R. Toledo-Crow, B. W. Smith, J. K. Rogers, and M. Vaez-Iravani, *Proc. SPIE* **2196**, 62-73 (1994).
2. M. Kauranen, H. Husu, S. Kujala, B. K. Canfield, J. Laukkanen, B. Bai, M. Kuittinen, J. Tururen, J. Kontio, J. Viheriala, T. Niemi, E. Chandler, and J. A. Squier, *Proc. SPIE* **7032**, 70321J (2008).
3. H. Gross, A. Rathsfeld, F. Scholze, and M. Bär, *Meas. Sci. Technol.* **20**, 105102 (2009).
4. M. A. Henn, H. Gross, F. Scholze, M. Wurm, C. Elster, and M. Bär, *Opt. Express* **20**, 12771–12786 (2012).
5. P. R. T. Munro and P. Torok, *Opt. Express* **15**, 9293 (2007).
6. J. Zhu, K. Zhang, N. Davoudzadeh, X. Wang, and L. L. Goddard, *Proc. SPIE* **9778**, 97780P (2016).

KEYWORDS

critical dimension metrology, metalens, photonic nanojet, near-field focusing

Enhanced Defect Detection in Patterned Wafers Using a Plasmonic Waveguide Metalens Array

Jinlong Zhu, Sanyogita Purandare and Lynford L. Goddard

Photonic Systems Laboratory, Department of Electrical and Computer Engineering, University of Illinois at Urbana-Champaign, Urbana, Illinois 61801, USA

INTRODUCTION

A single patterning error or defect as small as 10 nm in size can kill the performance of an entire 1 cm^2 chip. To continue the dimensional scaling of Moore's Law, accurately detecting these killer defects has become critical. Further, a fast inspection speed, i.e. the areal throughput, must be maintained to ensure high-volume manufacturing. Optical far-field based inspection platforms, including microscopy combined with through-focus scanning [1, 2] or common-path quantitative phase imaging [3-6], have successfully detected defects around 20 nm in size. Because optical imaging methods are nondestructive, can inspect a large area in a single shot, and can be parallelized, they naturally satisfy most of the rigid requirements of semiconductor inspection. However, the Rayleigh scattering criterion teaches us that the scattering cross-section of a tiny particle, whose diameter is deep sub-wavelength, scales in proportion to CD^6/λ^4. Therefore, tiny defects, especially ones smaller than 10 nm in size, can only generate extremely weak signals. Because the wafer patterns have benign line width variations and the optical imaging system also has random noise, the signal from a killer defect will be hard to identify. This bottleneck, to date, remains unsolvable for far-field optical inspection systems. Hence, it is of vital importance to develop a high throughput optical inspection system that can accurately detect killer defects.

METHODOLOGY

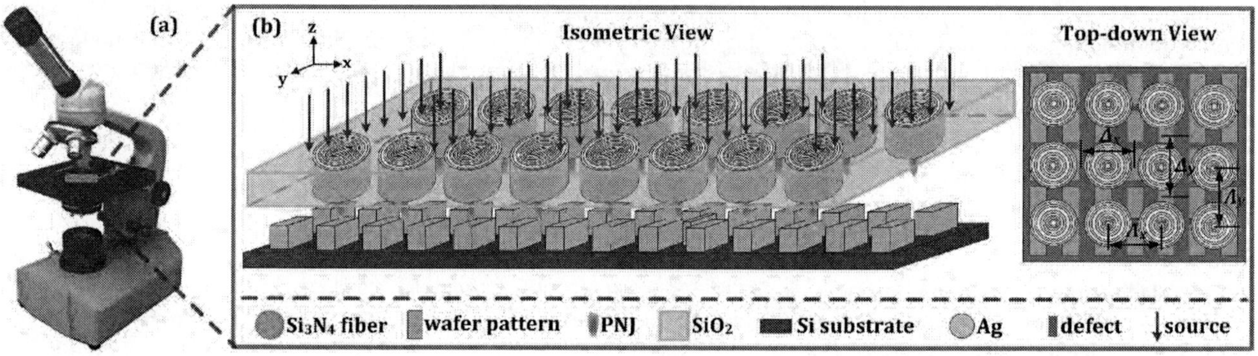

FIGURE 1. (a) Illustrated diagram showing a coherent optical microscope and (b) the mechanism of the metalens array-assisted defect detection system.

In this article, we propose to utilize a periodic metalens array as a near-field high numerical aperture (NA) lens to enhance the signal with respect to the defects on the patterned wafer. As shown in Fig. 1(b), each basic element of the metalens array consists of an Ag cylinder with several annular slits filled by air [7]. The widths and positions of these slits should be carefully engineered to create constructive interference in the desired focal positions. Supposing a linearly x-polarized plane wave is normally incident on the metalens, then the focusing is governed by the constructive interference condition [8]. Note that the slit-in-metal geometry is essentially a planar plasmonic metal-insulator-metal (MIM) waveguide. Thus, we can easily predict the properties of the output electromagnetic field by computing the propagation constant β using a semi-analytic algorithm. By utilizing phase compensation, we can generate sub-diffraction limit focusing with an intensity that is far stronger than that of the illumination plane wave. A conventional optical microscope is utilized to record the images by looking through the metalens, as illustrated in Fig. 1(a). Note that the metalens array would work in scanning mode. We would record a series of images with respect to different xy positions of the array. These images can be stitched together to form a complete panoramic image of the underlying patterned nanostructure. The selection of a metalens array instead of a single metalens, dramatically reduces the necessary scanning area because of the periodicity. See Fig. 1(b).

RESULTS

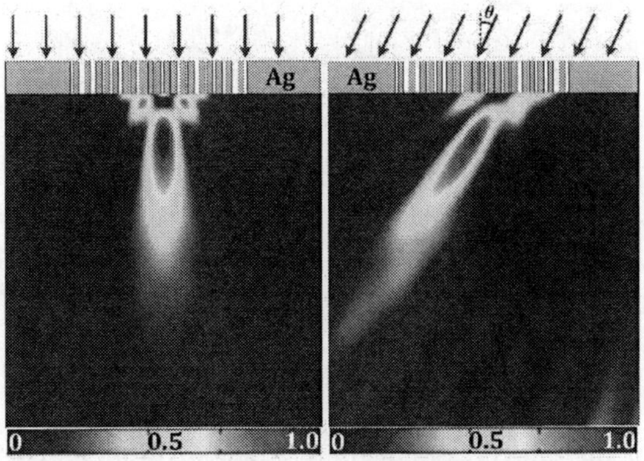

FIGURE 2. Focusing characteristics of the designed metalens. The intensity maps are obtained by setting the incident angles as (a) 0° and (b) 30°, respectively. Metalens height: 215 nm. Slit widths: 10-30 nm. Material in slits: air. All the sub-figures have been normalized by the maximal intensity.

Before answering how the metalens array can help in enhancing the signal associated with a killer defect on the patterned wafer, we first analyze the focusing characteristics of a precisely tailored metalens, as shown in Fig. 2. We performed a 2D full-wave simulation using the wave optics module of COMSOL Multiphysics ® Modeling Software (version 5.2, COMSOL, Inc., Stockholm, Sweden). The simulation domain is surrounded by a perfectly matching layer (PML) with the triangular mesh size set as $\lambda/25$ to approximate an open domain. Triangular mesh elements with sizes equal to $\lambda/35$, $\lambda/65$ and $\lambda/150$ are applied to the ambient, Ag film, and air slit domains, respectively, to ensure accuracy. As shown in Fig. 2(a), a focal spot is generated beneath the metalens for the normal plane wave incident from the top. We then consider using tilted plane wave illumination to understand the variation in the focus. Obviously, the tilt illumination results in a distorted focus spot with a primary axis that is tilted by almost the same angle. See Fig. 2(b). This focusing characteristic indicates that the designed metalens indeed behaves similar to a conventional lens, except that the corresponding beam waist in air is only 0.36 λ, i.e. sub-diffraction limited. The far-field signal enhancement of the defects not only benefits from the super-focusing beam illumination, but also from the refocusing of the scattering field with respect to the defect in near-field region.

In this talk, we will discuss how these results scale for an array and how they build the foundation for wafer defect detection.

REFERENCES

1. B. M. Barnes, F. Goasmat, M. Y. Sohn, H. Zhou, and R. M. Silver, *Proc. SPIE* **8681**, 86810E (2013).
2. B. M. Barnes, M. Y. Sohn, F. Goasmat, H. Zhou, A. E. Vladar, R. M. Silver, and A. Arceo, *Optics Express* **21**, 26219-26226 (2013).
3. R. Zhou, C. Edwards, A. Arbabi, G. Popescu, L. L. Goddard, *Nano Letters* **13**, 3716-3721 (2013).
4. R. Zhou, C. Edwards, G. Popescu, L. L. Goddard, *Proc. SPIE* **9050**, 905017 (2014).
5. R. Zhou, C. Edwards, C. Bryniarski, G. Popescu, and L. L. Goddard, *Proc. SPIE* **9424**, 942416 (2015).
6. J. Zhu, K. Zhang, N. Davoudzadeh, X. Wang, and L. L. Goddard, *Proc. SPIE* **9778**, 97780P (2016).
7. C. Ma and Z. Liu, *Appl. Phys. Lett.* **96**, 183103 (2010).
8. M. Khorasaninejad, W. Chen, R. C. Devlin, J. Oh, A. Y. Zhu, and F. Capasso, *Science* **352**, 1190-1194 (2016).

KEYWORDS

Defect detection, coherent optical microscope, metalens

Combining Nano-Scale Inert-Gas Ion Microscopy and Secondary Ion Mass Spectrometry

Sybren Sijbrandij[a], John Notte[a], David Dowsett[b], Jean-Nicolas Audinot[b] and Tom Wirtz[b]

[a] *Carl Zeiss Microscopy, LLC - Ion Microscopy Innovation Center - 1 Corporation Way - Peabody, MA - USA*
[b] *Advanced Instrumentation for Ion Nano-Analytics (AINA), MRT Department, Luxembourg Institute of Science and Technology (LIST) - 41, Rue de Brill - L-4422 - Belvaux - Luxembourg*

INTRODUCTION

This contribution reports on the ongoing development of a new instrument that combines a scanning helium/neon ion microscope (FIGURE 1), based on a gaseous field-ion source (GFIS), with a secondary ion mass spectrometer (SIMS). The new instrument approaches the limits of spatial resolution achievable in scanning ion microscopy and SIMS analysis, benefitting nano-scale microscopy, metrology and materials characterization.

FIGURE 1. 3rd generation prototype of a GFIS based Microscope & SIMS combined instrument.

THE INSTRUMENT

The Microscope

The helium ion microscope (HIM) was commercially introduced ten years ago[1], and has rapidly become a widely used tool for high-resolution imaging and nano-fabrication. The HIM offers sub-nanometer resolution, excellent depth of focus, and unique image contrast mechanisms and beam-sample interaction characteristics.

Four years ago the capability of using neon was added[2]. The neon ion beam, due to increased sputter rate, expands its use for nano-scale sample modification processes[3,4], and secondary ion imaging and analysis techniques.

The Spectrometer

A custom system was designed for the helium / neon ion microscope, with special care taken to avoid that the ion extractor produces undesirable distortions of the primary ion beam[5].

The goal of characterizing small material volumes necessitates a highly efficient analysis system. To this end a high transmission ion extractor was coupled with a double-focusing magnetic sector spectrometer with a parallel detection scheme. To enhance secondary ion yields from the inert primary beam, O_2 can be flowed over the sample surface to boost positive ion yields, Cs can be flowed to boost negative ion yields[6].

PUSHING THE LIMITS IN ION MICROSCOPY AND ANALYSIS

In current state-of-the-art SIMS instruments for nano-scale analysis, both secondary electron (SE) imaging and secondary ion (SI) elemental mapping analysis are limited by the size of the primary ion beam, to a lower limit of ~50 nm (this situation is depicted in the upper region of FIGURE 2). In a GFIS-based instrument, SE imaging is limited by the size of the focused primary beam to less than 0.5 nm while SI mapping and analysis are limited to less than 10 nm by beam-sample interactions[6] (this situation is depicted in the lower region of FIGURE 2).

In comparison, the GFIS-based SIMS implementation has the following advantages: The effective spot-size for SI elemental mapping and analysis is improved (reduced) by a factor of 5, and in addition, the spot-size for SE imaging is improved (reduced) by 2 orders of magnitude, thereby significantly widening the windows of application towards smaller sample-feature dimensions. This allows for the acquisition of optimized, correlative image and composition data from a single (combined) instrument[7,8].

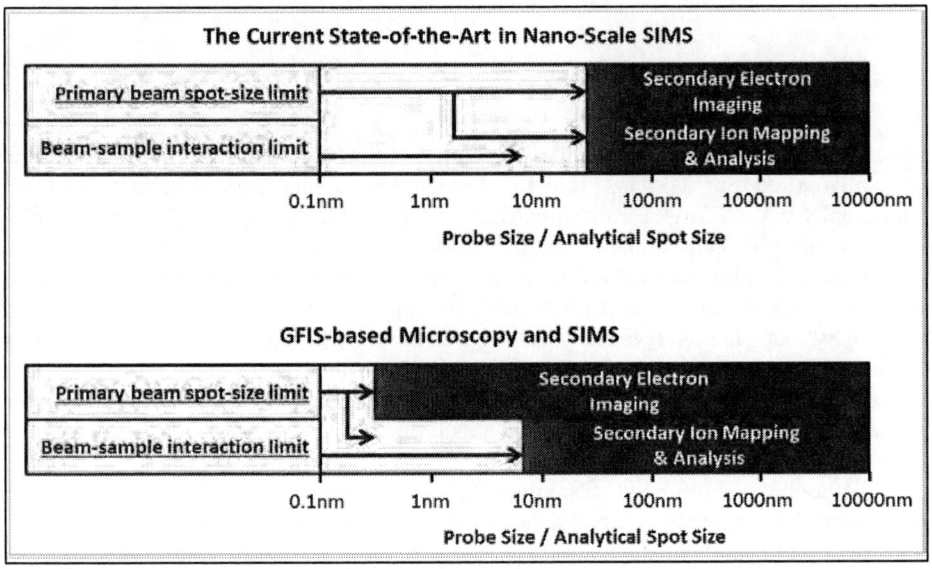

FIGURE 2. The applicable windows for Secondary Electron Imaging and Secondary Ion Mapping & Analysis.

In this contribution the technique and instrument implementation, the benefits for materials research, and recent experimental application results will be presented.

REFERENCES

1. B.W. Ward, J.A. Notte, and N.P. Economou, *J Vac. Sci. & Tech. B* **24**, 2871-2874 (2006).
2. F.H.M. Rahman, J. A. Notte, R.H. Livengood and S. Tan, *Ultramicroscopy* **34**, 10-18 (2013).
3. S. Tan, R. Livengood, P. Hack, R. Hallstein, D. Shima, J. Notte, and S. McVey, *J Vac. Sci. & Tech. B* **29**, 06F604 (2011).
4. D. Winston, V.R. Manfrinato, S.M. Nicaise, L.L. Cheong, H. Duan, D. Ferranti, J. Marshman, S. McVey, L. Stern, J. Notte, and K. Berggren, *Nano Letters* **11**, 4343-4347 (2011).
5. T. Wirtz, D. Dowsett, J.-N. Audinot, and S. Eswara, *Microscopy and Microanalysis*, **22(S3)**, 160–161 (2016).
6. T. Wirtz, N. Vanhove, L. Pillatsch, D. Dowsett, S. Sijbrandij, and J. Notte, *Appl. Phys. Lett.*, **101**, 041601 (2012).
7. T. Wirtz, P. Philipp, J.-N. Audinot, D. Dowsett and S. Eswara, *Nanotechnology*, **26** 434001 (2015).
8. P. Gratia, G. Grancini, J.-N. Audinot, X. Jeanbourquin, E. Mosconi, I. Zimmermann, D. Dowsett, Y. Lee, M. Grätzel, F. De Angelis, K. Sivula, T. Wirtz, and M. Khaja Nazeeruddin, *J. Am. Chem. Soc.,* DOI 10.1021/jacs.6b10049 (2016).

KEYWORDS

Ion, Microscopy, Analysis, Nano-Scale, HIM, GFIS, SIMS

Analyzing Post-CMP Surface Topography from White Light Interference Microscopy

A.F. Bello, Jusang Lee, Shinichiro Kakita, Nicholas Pieniazek

Advanced Module Engineering, GLOBALFOUNDRIES, 400 Stone break Extension, Malta, NY, 12020, USA,

INTRODUCTION

Interference microscopy, also called coherence scanning interferometry or scanning white-light interferometry, is a technique especially suited for CMP [1-3] (chemical mechanical planarization) because of its ability to measure wafer die level topography with its large (~mm) area field of view with high spatial (~um per pixel) resolution. The output is an image topography height map. Statistics such as the height distribution, root-mean-squared and kurtosis are output parameters which could be tracked for statistical process control (SPC). The problem with those output parameters is that they are not descriptive enough for topography images that are usually rich with data. In this study the topography maps are analyzed beyond the traditional calculations for root-mean-squared (RMS) and distribution. Surface map analysis [4-9] traditionally used in other fields such as image recognition, facial recognition, and geographical terrain topography are applied to the post-CMP surface topography measurements. In such analyses, the pixels are classified into types such as type such as peak, ridge, saddle, flat, ravine, pit, hillside according to the surface patch it defines. Statistics on the image according to the types reveals an overall sketch of the topography which may be useful for SPC tracking of the CMP process. Quantitative metrics beyond the simple RMS can thus be used to determine surface characteristics. The mathematical formalism for determining the label of each pixel is described in general and an example is shown to demonstrate its viability.

EXPERIMENTAL DETAILS

Fig. 1 shows the basic concept and the optical path for scanning white-light interferometry. On splitting a broadband light source, one path goes to the sample surface and the other to a reference surface mirror. The reference surface is mounted on a piezoelectric to control and vary the path length to it. The light from the two paths are recombined at a camera sensor where they are mixed to produce the interference signal. In white light interference, where the light source is comprised of a range of wavelengths, the peak intensity in the camera from the interference signal is obtained when the light path is identical. The use of a range of wavelengths has the effect of producing a non-zero signal at a narrow range of the path length difference. That is, there is a strong damping effect as the path length difference becomes large. Scanning the optical system for each pixel produces the height topography map. A commercially available instrument (Nanometrics Unifire) employing white light interferometry was used to scan regions of interest in the wafer die. All measurements were made post-CMP processing.

Topography Surface Image Analysis

A gray level digital image is the input in any computer vision system and one challenge is to determine characteristics of the object surfaces in the image. Classifying the image intensity surface profile into primitive topographic types helps in solving that challenge. In a study by Watson, et al. [4] the image is decomposed into surface patches comprised of a small set of pixels – 9 x 9 in their study. Each surface patch is classified into a label type such as peak, ridge, saddle, flat, ravine, pit, hillside as shown in Fig. 2. The classification is done by first fitting the surface patch to an analytic function, in this study a third order polynomial in the x and y-directions. The directional gradients of the function are calculated, and the values, and whether positive or negative, determine the label. The location and statistics on each label can then be used to characterize the objects, for example, for edge detection. One surface patch label cannot by itself be useful in determining characteristics of the object. For example, one surface patch labeled a flat may not add any meaningful information by itself, but taken in context with how many flat surface patches there are in the image may add value. Fig. 3 is a test of the classification on a generated image (inset) that contains a peak and concentric circles of ridges and ravines defined by a decaying sinusoidal. Note the peak (1) at the center and hillsides (7) around it. There are the concentric ridges (2) and ravines (5), and at the edge of the figure are the flat (4) areas. This validates the algorithm, which is used to analyze real images from the measured surface topography.

In this study, post-CMP wafers were measured for topography maps using high resolution (2um pixel size) in the region of interest within a die. These are samples representative of 14nm technology design rules. Measurements were made post-CMP at a back-end-of-line process where most of the wafer surface was still covered in metallic film. This avoids the artifacts in white light interferometry associated with transparent films where light may not be reflected directly from the top surface. Surface patches of 7 x 7 pixels were calculated for the label. The distribution of labels from the approximately 1800 x 2500 pixels is shown in Fig. 4 for a center and edge die. The labels for ridge, flat, ravine, and hillside are shown because those are the parameters with statistically significant instances. The difference between center and edge is significant. A repeatability (10x) measurement of the center die defines the error bars in the figure, and as shown is much less than the difference to the edge die.

In conclusion we now have a technique to characterize the within-wafer and wafer-to-wafer uniformity of surface characteristics or topography maps post-CMP. The calculated characteristics are much more detailed than the typical root-mean-squared (RMS) use to quantify the image. Beyond RMS, topography images are typically not used in quantifiable terms in a production setting. Using surface patch characterization we now have more quantifiable parameters that could be used for SPC of the CMP process.

REFERENCES

1. De Lega, Xavier Colonna, and Peter De Groot. "Optical topography measurement of patterned wafers." Characterization and Metrology for ULSI Technology 2005 788 (2005): 432-436.
2. de Groot, Peter. "Coherence scanning interferometry." Optical Measurement of Surface Topography. Springer Berlin Heidelberg, 2011. 187-208.
3. Dettoni, Fernando, et al. "Interferometry: a direct die level characterization technique." Planarization/CMP Technology (ICPT 2012), International Conference on. VDE, 2012.
4. Watson, Layne T., Thomas J. Laffey, and Robert M. Haralick. "Topographic classification of digital image intensity surfaces using generalized splines and the discrete cosine transformation." Computer Vision, Graphics, and Image Processing 29.2 (1985): 143-167.
5. Wang, Jun, et al. "3D facial expression recognition based on primitive surface feature distribution." Computer Vision and Pattern Recognition, 2006 IEEE Computer Society Conference on. Vol. 2. IEEE, 2006.
6. Besl, Paul J., and Ramesh C. Jain. "Three-dimensional object recognition." ACM Computing Surveys (CSUR) 17.1 (1985): 75-145.
7. Falcidieno, Bianca, and Michela Spagnuolo. "A new method for the characterization of topographic surfaces." International Journal of Geographical Information System 5.4 (1991): 397-412.
8. Zhang, M. C., J. B. Campbell, and R. M. Haralick. "Automatic delineation of drainage basins within digital elevation data using the topographic primal sketch." Mathematical geology 22.2 (1990): 189-209.
9. Besl, Paul J. Surfaces in range image understanding. Springer Science & Business Media, 2012.

KEYWORDS

CMP, Topography, Image Processing, Metrology, Interference microscopy, surface processing

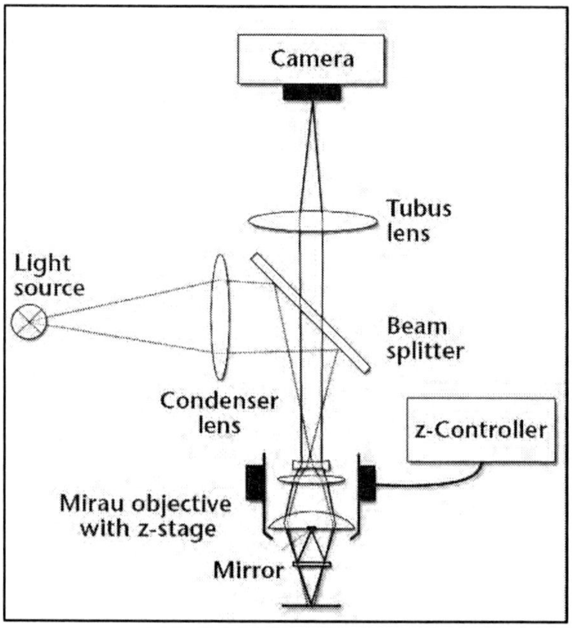

FIGURE 1. Optical path in white-light interference microscopy. (figure attributed to Polytec GmbH)

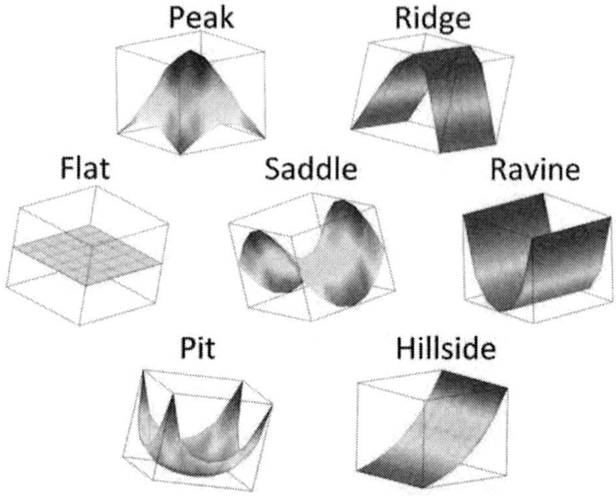

FIGURE 2. Surface patch labels. Pixels of 7x7 are classified according to characteristics.

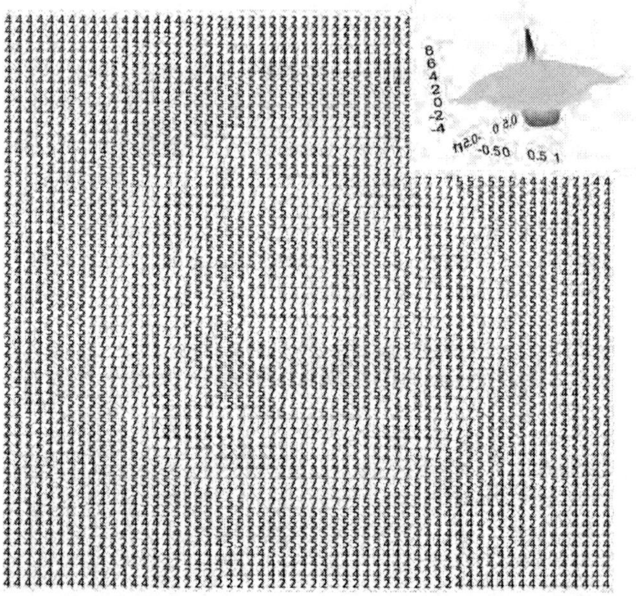

FIGURE 3. Surface patch labels for a test structure (inset) whose characteristics are known. Note the peak (1) at the center and ridges (2) and ravines (5) in concentric regions.

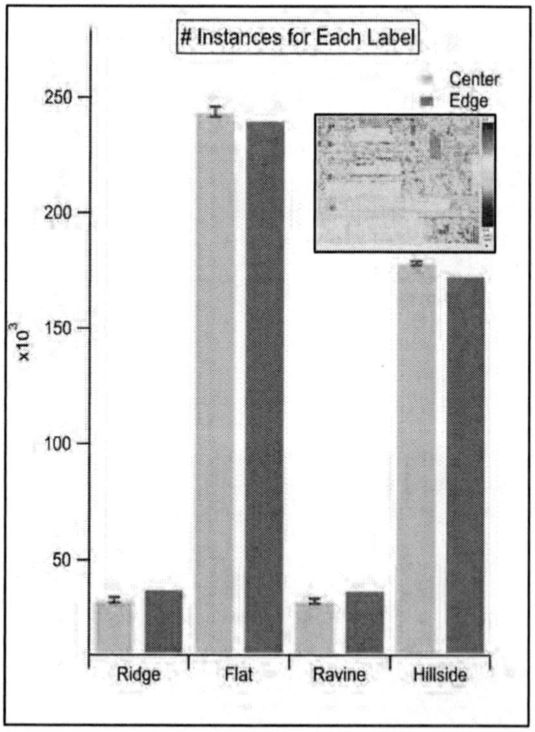

FIGURE 4. Distribution of surface patch labels for a measured region of interest (inset) in a die. Center and edge are compared which show a difference.

Characterization of Doping Distribution in Fin Field Effect Transistor (FinFET) Structures Representative for 14nm CMOS Technology and Beyond

M.J.P. Hopstaken[1,*], M. Saccomanno[1], M. Ebrish[2], O. Gluschenkov[2]

[1] *IBM T.J. Watson Research Center, Yorktown Heights, NY 10598, USA*
[2] *IBM Research Albany Nanotech Center, Albany, NY 12203, USA*
[*] *marco.hopstaken@us.ibm.com*

INTRODUCTION

FinFET devices are rapidly emerging as a very promising candidate to extend CMOS scaling beyond planar technology nodes because of superior electrostatic channel control [1, 2]. One challenge is to achieve conformal doping of source/drain (S/D) regions along the Fin sidewall, which is crucial to minimize access resistance at realistic dimensions [3]. The transition from planar to 3D FinFET architecture is posing challenges to metrology / physical characterization techniques, notably for traditional 1D depth profiling techniques such as Secondary Ion Mass Spectrometry (SIMS) [4].

Here we have successfully employed the so-called 'SIMS through Fins' concept on large arrays of periodic FinFET structures at realistic dimensions (*i.e.* <10 nm Fin width, tight pitch), relevant for 14 nm node and below. This approach enables to delineate S/D doping along the sidewalls from top- and trench bottom contributions at realistic Fin dimensions.

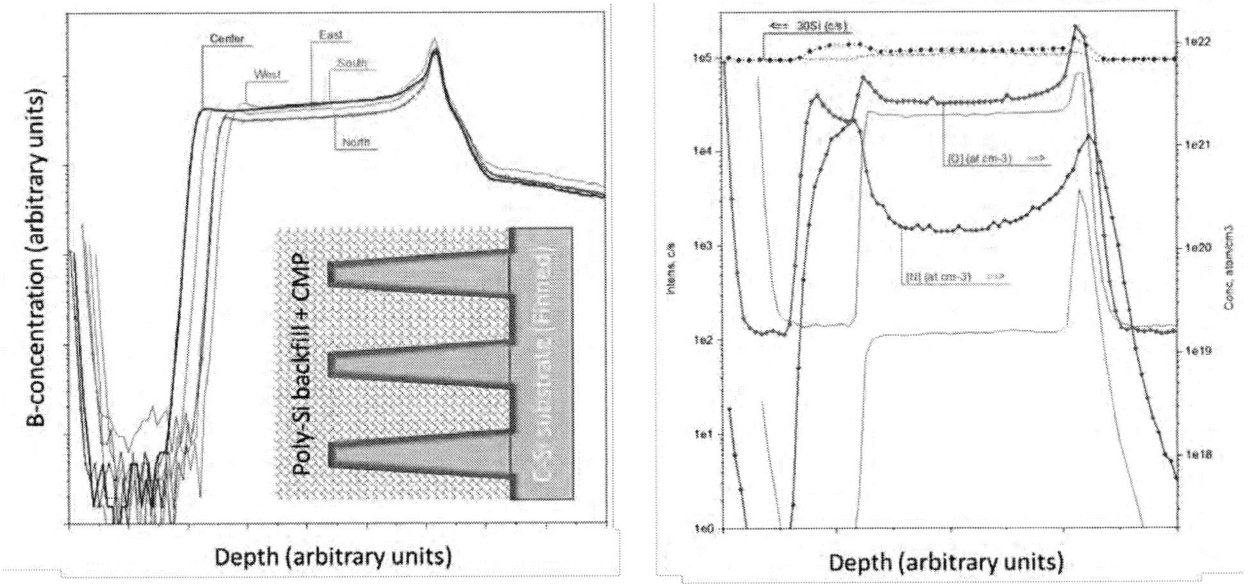

FIGURE 1. **panel a (left):** 1.5D SIMS B-profiles on Fin arrays from different wafer locations after Boron plasma doping and subsequent RTA, cleans / poly-Si backfill / CMP, and **panel b (right):** 1.5D SIMS N and O profiles on Fin array with (symbols) and without (no symbol) SiNx residual hardmask

We demonstrate conformal sidewall doping (B, As) for novel low-damage plasma immersion doping processes on patterned (Fin) wafers (see Figure 1a). We will compare with more conventional 1D depth profiling on blanket plasma doped wafers for fundamental understanding of plasma dopant interactions and dopant penetration. In addition, we propose monitoring of relevant signals for atmospherics (C, N, O) as a quick diagnostic for Fin integrity, presence of (residual) SiNx hardmask. For example, the O-signal –originating from the chemical oxide along the perimeter of the Fins after clean– provides a signature for the Fin geometry and integrity after various post-plasma processing (see Figure 1b).

REFERENCES

1. W. Haensch *et al.,* IBM J. Res. Dev., vol. 50, p. 339, 2006.
2. C.-H. Lin *et al.*, IEDM 2014
3. M. van Dal, (2007).. *2007 IEEE Symposium on VLSI Technology* (Vol. 80, pp. 110–111). IEEE.
4. J. Mody et al., J. Vac. Sc. Technol. **28** (1) C1H5 (2010)

KEYWORDS

FinFET, dopant profiling, dynamic SIMS

Optical Properties of Pseudomorphic $Ge_{1-x-y}Si_xSn_y$ on Ge

Nalin S. Fernando,[1] Ryan Hickey,[2] John Hart,[2] Ramsey Hazbun,[2]
Dainan Zhang,[2] James Kolodzey[2], Stefan Zollner[1]

[1] *Department of Physics, New Mexico State University, Las Cruces, NM 88003.*
[2] *Department of Electrical and Computer Engineering, University of Delaware, Newark DE 19716.*

INTRODUCTION

Ge is an indirect band gap material. The band structure of Ge is a strong function of strain and alloy composition, and a transition from an indirect to a direct band gap has been observed for y~6-10% for relaxed $Ge_{1-y}Sn_y$ indicating the possibility of widespread applications of Ge-based photonic devices[1]. The pseudomorphic nature of the Ge-based alloy layer on a substrate is important to keep dislocation densities low at the interface to improve the performance of the device. Band gap engineering of Ge by controlling strain and alloying with Si and Sn has attracted great interest since $Ge_{1-x-y}Si_xSn_y$ ternary alloy with two compositional degrees of freedom allows decoupling of the lattice constant and electronic structures. Hence the knowledge of the compositional and strain dependence of the $Ge_{1-x-y}Si_xSn_y$ band structure is critical for the design of photonic devices with the desired interband transition energies.

THEORY AND MODEL

We have used elastic theory and deformation potential theory to predict the variation of the direct (E_g^{Γ}), indirect (E_g^{ind}), E_1, and $E_1+\Delta_1$ band gaps of pseudomorphic $Ge_{1-x-y}Si_xSn_y$ on Ge as a function of Si (x) and Sn (y) compositions. The pseudomorphically grown $Ge_{1-x-y}Si_xSn_y$ layer on Ge experiences a biaxial stress due to the lattice mismatch between the alloy layer and the Ge. The in-plane strain $\varepsilon_{\parallel}$ on the epilayer can be calculated by using the relaxed lattice parameters of the $Ge_{1-x-y}Si_xSn_y$ and the out-of-plane strain $\varepsilon_{\perp}$ on the epilayer can be related to $\varepsilon_{\parallel}$ from the elastic constants and can be measured by XRD. The compositional dependence of the hydrostatic strain (ε_H) and shear strain (ε_S) can be calculated from $\varepsilon_{\parallel}$ and $\varepsilon_{\perp}$. The shear strain splits the degeneracy of the valence band at the Γ-point (v_1, v_2 and v_3), and conduction band at the X-point (X2-doublet and X4-quadruplet). The hydrostatic strain shifts the conduction and valence bands relative to their unstrained positions. Deformation potential theory was used to find the strain dependence of the positions of the conduction band and valence band at the Γ, X, and L points to determine the compositional dependence of the E_g^{Γ}, E_g^{ind}, E_1, and $E_1+\Delta_1$ band gaps as a function of Si and Sn compositions.[2,3] As shown in Fig. 1, the smallest band gap becomes direct in the compressively strained region for $y + 0.37x > 0.25$, it will never become direct in the tensile strained region for Si, Sn <20%. Effects of the substrate on the indirect-direct transition will be discussed.

Ellipsometry Results

The predictions of the deformation potential theory are validated for pseudomorphic $Ge_{1-y}Sn_y$ (for Si=0) on Ge through measurements of the optical properties. The complex pseudodielectric functions of pseudomorphic $Ge_{1-y}Sn_y$ alloys grown on Ge by MBE were measured using ellipsometry in the 0.1-6.6 eV energy range for Sn contents up to 11% (see Fig. 2a), to investigate the compositional dependence of the band gaps. Critical point energies and related parameters were obtained by analyzing the second-derivative of the dielectric function. Our experimental results (Fig. 2b) are in good agreement with the theoretically predicted CP energies. The effects of the partial relaxation of the $Ge_{1-y}Sn_y$ epilayers on the dielectric function and critical points have been investigated.

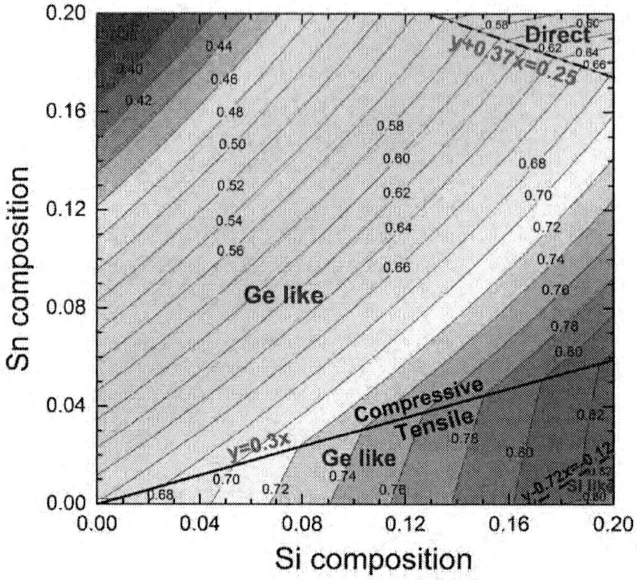

FIGURE 1. Compositional dependence of the lowest band gap (either direct or indirect) of pseudomorphic $Ge_{1-x-y}Si_xSn_y$ on Ge.

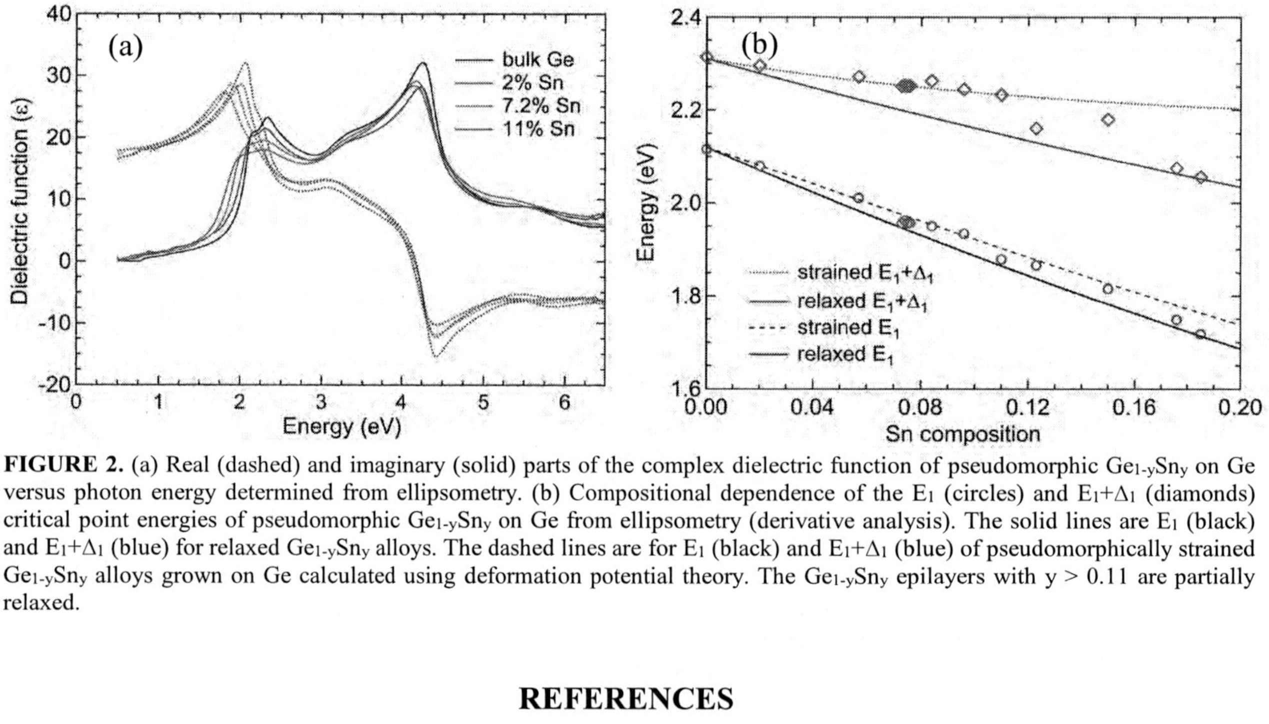

FIGURE 2. (a) Real (dashed) and imaginary (solid) parts of the complex dielectric function of pseudomorphic $Ge_{1-y}Sn_y$ on Ge versus photon energy determined from ellipsometry. (b) Compositional dependence of the E_1 (circles) and $E_1+\Delta_1$ (diamonds) critical point energies of pseudomorphic $Ge_{1-y}Sn_y$ on Ge from ellipsometry (derivative analysis). The solid lines are E_1 (black) and $E_1+\Delta_1$ (blue) for relaxed $Ge_{1-y}Sn_y$ alloys. The dashed lines are for E_1 (black) and $E_1+\Delta_1$ (blue) of pseudomorphically strained $Ge_{1-y}Sn_y$ alloys grown on Ge calculated using deformation potential theory. The $Ge_{1-y}Sn_y$ epilayers with $y > 0.11$ are partially relaxed.

REFERENCES

1. J. D. Gallagher, C. L. Senaratne, J. Kouvetakis and J. Menendez, Appl. Phys. Lett. **105**, 142102 (2014).
2. M. El Kurdi, G. Fishman, S. Sauvage, and P. Boucaud, J. Appl. Phys. **107**, 013710 (2010).
3. A. Attiaoui and O. Moutanabbir, J. Appl. Phys. **116**, 063712 (2014).

KEYWORDS

Spectroscopic ellipsometry, deformation potential, GeSiSn, pseudomorphic, direct band gap, dielectric function, critical points.

In-situ TEM Observation of Nickelide Contact Formation in InGaAs Nanowire Channels

Renjie Chen[1] and Shadi A. Dayeh[1,2,3]

[1] *Department of Electrical and Computer Engineering, University of California San Diego, La Jolla, California 92093, USA.* [2] *Materials Science and Engineering Program, University of California San Diego, La Jolla, California 92093, USA.* [3] *Department of NanoEngineering, University of California San Diego, La Jolla, California 92093, USA*

Corresponding email: sdayeh@ucsd.edu or rec006@ucsd.edu

Semiconductor nanowires and Fins are promising building blocks for next generation ultrascaled devices for electronic and optoelectronic applications.[1-2] An important aspect for the development, maturity, and efficiency of these ultrascaled devices is the detailed understanding of and control over the phase transformation that accompanies the formation of their alloyed contacts for lithography-free self-aligned gate design.[3] Prior pioneering research has revealed the large differences between the nanoscale metallization processes with their planar counterparts, evoking reevaluation of the thermodynamics, kinetics, and resultant phases in alloyed and compound nanoscale contacts.[4] The majority of these studies were carried out by in-situ heating and imaging inside a transmission electron microscopy (TEM) which proved powerful in investigating these thermally driven phase transformations in real-time and at atomic resolution.[5] Though prominent discoveries have been made on nanostructures on elemental semiconductors, i.e. Si and Ge nanowires,[6-7] the dynamics of the phase transformation have not been uncovered in III-V nanowire channels at atomic resolution, possibly due to the multifold elements and complicated phases involved in the phase transformation from a binary/ternary compound semiconductor to a ternary/quaternary compound metallic contact, respectively, even for a single element metal contact such as Ni.

In this work, we utilized the in-situ TEM heating technique to observe the solid-state reactions between Ni contact and $In_{0.53}Ga_{0.47}As$ nanowire channels. Figure 1(a) provides an overview of the interfacial profile between $In_{0.53}Ga_{0.47}As$ nanowire and the reacted nickelide phase ($Ni_2In_{0.53}Ga_{0.47}As$) after applying the in-situ heating at 320 °C. The interface followed the $In_{0.53}Ga_{0.47}As$ (111) || $Ni_2In_{0.53}Ga_{0.47}As$ (0001) atomic planes, and the formed $Ni_2In_{0.53}Ga_{0.47}As$ phase nucleated near the top of the nanowire channel where $In_{0.53}Ga_{0.47}As$ interfaced with SiO_2 layer, as opposed to the $HfO_2/In_{0.53}Ga_{0.47}As$ interface. The $Ni_2In_{0.53}Ga_{0.47}As$ ledge formed in a train of strained single-bilayers, as observed near the top surface of the nanowire channel in each time frame of Fig. 1(d). Once the single-bilayer ledges released their strain energy by forming misfit dislocations, their velocity slowed down and formed coupled single-bilayer ledges. Consequently, these ledges moved with a double-bilayer height that became the step unit height of ledges in this phase transformation.

Our atomic models depicted the ledge propagation behaviors that are related to collective gliding of three $In_{0.53}Ga_{0.47}As$ single-bilayers along three Shockley partial directions to cancel out the shear stress during phase transformation, as shown in Figure 2. The $In_{0.53}Ga_{0.47}As$ lattice has a zinc-blende (face-centered-cubic, fcc) structure, and the atomic planes arrange in a A-B-C-A-B-C... manner along the [111] direction. The $Ni_2In_{0.53}Ga_{0.47}As$ lattice has a NiAs (hexagonal, B8) structure, and the atomic planes arrange in a A-A-A... manner, along the [0001] direction. The $In_{0.53}Ga_{0.47}As/Ni_2In_{0.53}Ga_{0.47}As$ phase transformation not only caused an unusual change in atomic stacking (A-B-C... type to A-A-A... type), but also introduced large elastic strain near the interfaces, resulting in interfacial disconnections (ledge + dislocation).

We will present these results and contrast the phase transformation behavior for III-Vs with those of binary compound contacts to elemental semiconductors. We will present a nucleation model that validates the heterogeneous nucleation at the nanowire top surface to be favored both kinetically and thermodynamically in good agreement with our experimental observations. Our in-situ studies demonstrated for the first time the applicability of interfacial disconnection theory in contact metallization for compound semiconductor nanoscale channels.

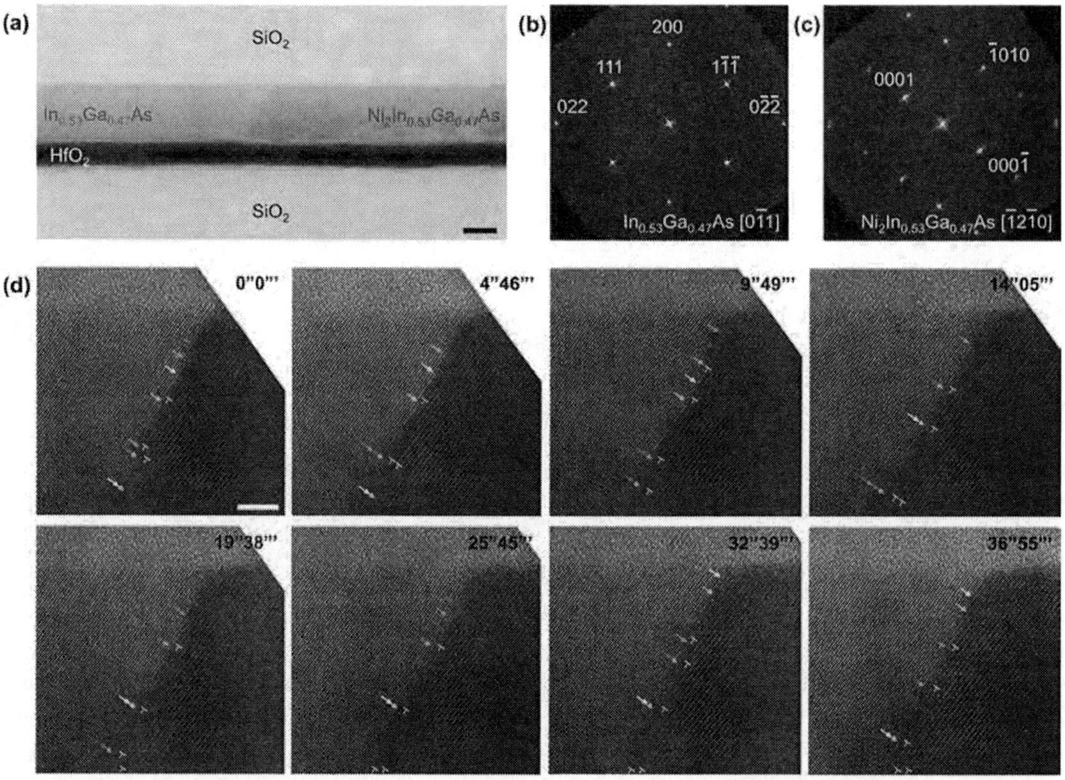

FIGURE 1. In-situ TEM results of the nickelide contact formation on InGaAs nanowire channel. (a) TEM image of the $In_{0.53}Ga_{0.47}As/Ni_2In_{0.53}Ga_{0.47}As$ interface showing a slanted interfacial profile. Scale bar is 20 nm. (b) and (c) Fast Fourier transform (FFT) images of the $In_{0.53}Ga_{0.47}As$ and $Ni_2In_{0.53}Ga_{0.47}As$ segments, respectively. The diffraction patterns indicate that the slanted interface follows the $In_{0.53}Ga_{0.47}As$ (111) || $Ni_2In_{0.53}Ga_{0.47}As$ (0001) atomic planes. (d) HRTEM sequences extracted from the in-situ recorded video, showing the ledge formation at the top-surface of nanowire channel and ledge propagations into the nanowire body. All these HRTEM images share the same scale bar of 5 nm.

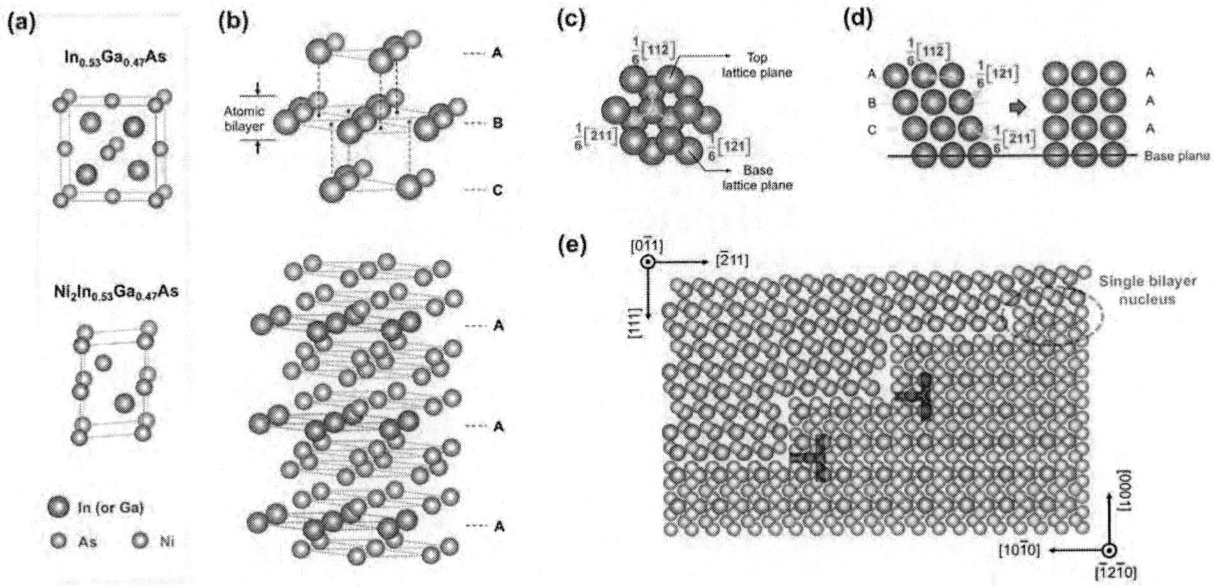

FIGURE 2. Atomic models that reveal the formation of interfacial disconnections during $In_{0.53}Ga_{0.47}As$ to $Ni_2In_{0.53}Ga_{0.47}As$ phase transformation. (a) Lattice structures of $In_{0.53}Ga_{0.47}As$ (zinc blende) and $Ni_2In_{0.53}Ga_{0.47}As$ (simple hexagonal). (b) Atomic arrangements in $In_{0.53}Ga_{0.47}As$ (with [111] direction pointing upwards) and $Ni_2In_{0.53}Ga_{0.47}As$ (with [0001] direction pointing upwards) respectively. (c) Top-view of two lattice planes in an A-B-C type of stacking, indicating that the top lattice plane has three possible in-plane gliding vectors in order to be overlapping with the bottom lattice plane. (d) Side-view of how the A-B-C stacking of lattice planes shuffle into an A-A-A stacking manner by collective gliding of three lattice planes along three different gliding vectors as shown in (c). (e) The reconstructed atomic structures at the interface in between $In_{0.53}Ga_{0.47}As$ (top left) and $Ni_2In_{0.53}Ga_{0.47}As$ (bottom right). The nature of single-bilayer nucleus and double-bilayer step height is reflected in this schematic, together with the labeled misfit dislocations.

REFERENCES

1. Y. Cui, and C. M. Lieber, *science* **291**, 851-853 (2001).
2. M. Ieong, B. Doris, J. Kedzierski, K. Rim, and M. Yang, *Science* **306**, 2057-2060 (2004).
3. Y. Wu, J. Xiang, C. Yang, W. Lu, and C. M. Lieber, *Nature* **430**, 61-65 (2004).
4. W. Tang, B.-M. Nguyen, R. Chen and S. A. Dayeh, *Semiconductor Science and Technology* **29**, 054004 (2014).
5. H. Saka, T. Kamino, S. Ara, and K. Sasaki, *MRS bulletin* **33**, 93-100 (2008).
6. W. Tang, S. A. Dayeh, S. T. Picraux, J. Y. Huang and K.-N. Tu, *Nano Letters* **12**, 3979-3985 (2012).
7. W. Tang, S. T. Picraux, J. Y. Huang, A. M. Gusak, K.-N. Tu and S. A. Dayeh, *Nano Letters* **13**, 2748-2753, (2013).

KEYWORDS

In-situ TEM, nanowire, alloyed contact, nickelide, interfacial disconnection

High Precision Optical Characterization of Carrier Diffusion Length

Will Chism[1] and Michael Current[2]

[1]*Xitronix Corporation, 106 East Sixth Street, Ninth Floor, Austin TX, 78701*
[2]*Current Scientific, 1729 Comstock Way, San Jose, CA 95124*

INTRODUCTION

Photo-modulated reflectance (PMR) is a powerful optical technique to study the bandstructure and interfacial electric fields of semiconductors and semiconductor nanostructures [1]. As a high precision optical technique with direct sensitivity to semiconductor electronic properties, PMR is well suited for process control of semiconductor manufacturing processes involving electrical activation and/or strain [2, 3, 4] The conventional PMR system utilizes an intensity modulated "pump" laser beam to photo-inject carriers in the sample, thereby inducing a small differential reflectance of the sample. A second "probe" light beam is used to measure the modulated reflectance. The pump is typically modulated at a known frequency so that phase-locked detection may be used to suppress unwanted noise, resulting in the ability to measure reflectance changes at the ppm level. The measured PMR signal is the change in amplitude of the reflected probe light as the photo-injected carriers interact within the sample. Therefore, the measurement signal will exhibit a spatial profile dependent upon the diffusion length of the carriers photo-injected by the pump. In particular, the pump will induce a differential reflectance within a effective radius of modulation determined by the diffusion length. In this paper we report the high precision determination of carrier diffusion length by acquiring PMR data with the pump beam waist located at varying distances from the sample surface, followed by a regressive fit to the data to determine the diffusion length. The relative uncertainty in the determined diffusion length is demonstrated at less than 0.5%.

THEORETICAL PRINCIPLES

For probe beam wavelengths that are resonant with semiconductor interband transitions, the PMR signal arises from an electro-modulation effect (i.e. the Franz-Keldysh effect). This provides a means for optical detection of semiconductor bandstructure and internal electric fields through the appearance of sharp, third derivative spectra. Thus, in addition to high precision and spatial resolution, PMR techniques offer excellent sensitivity to semiconductor bandstructures provided the probe wavelength is chosen properly. Moreover, in accord with the Poisson relation, the experimentally measured PMR signal under such conditions is:

$$\frac{\Delta R}{R} = \frac{2qN_e\Delta V}{\varepsilon_S} \times L(\lambda), \quad (1)$$

where q is the electronic charge, N_e is the active dopant concentration, ΔV is the pump induced photo-voltage, ε_s is the static dielectric constant, and $L(\lambda)$ is a line-shape function determined by the semiconductor bandstructure (λ is the probe wavelength). The linearity of the PMR signal with respect to active dopant concentration is particularly convenient for process control of advanced annealing processes [3]. The photo-voltage will depend on the pump beam and the physics of its interaction with the sample. For example, in the 1-d limit, the dependence of the photo-voltage on diffusion length may be obtained from the conventional SPV amplitude analysis.

However, the technique discussed herein provides a means to directly measure the diffusion length by and through its effect on the radial profile of the pump beam. In particular, the PMR probe beam will record the radial profile of the pump beam as broadened by the radial diffusion of carriers within the sample. For example, provided the pump and probe beams have radially symmetric Gaussian profiles and are collinear and co-focused, the electric field of the reflected probe beam at the surface of the sample may be written:

$$E_r = \frac{E_0\omega_0}{\omega(Z)}\exp\left\{-\frac{\rho^2}{\omega^2(Z)}\right\}\left[\tilde{r} + \frac{\partial\tilde{r}}{\partial n}(\Delta n + i\Delta k)\frac{2qN_e\Delta V(\omega_p^2 + L_d^2)}{\varepsilon_s(\omega_p^2(Z) + L_d^2)}\exp\left\{-\frac{2\rho^2}{\omega_p^2(Z) + L_d^2}\right\}\right], (2)$$

where $|E_0|^2$ is the intensity of the probe beam at focus, ω_o is the probe beam waist, Z is the location of the sample surface as measured from the probe beam waist, $\omega^2(Z) = \omega_o^2\{1+(Z/z_o)^2\}$, where $z_o = \pi\omega_o^2/\lambda$ is the Rayleigh range of the probe beam, ρ is radial distance as measured from the probe beam axis, $\tilde{r}$ is the complex reflectance coefficient, n is the sample refractive index, Δn and Δk are the induced changes in the sample refractive index and the extinction coefficient, respectively, ω_p is the pump beam waist, $\omega_p^2(Z) = \omega_p^2\{1+(Z/z_p)^2\}$, where $z_p = \pi\omega_p^2/\lambda_p$ is the Rayleigh range of the pump beam (λ_p is the pump beam wavelength), and L_d is the diffusion length. The $\tilde{r}$ term in Eqn. 2 gives rise to the linear (DC) portion of the reflected field whereas the $\partial\tilde{r}/\partial n$ term gives rise to the modulated (AC) component. Note that while the amplitude of the AC term generally depends on the diffusion length (including through the photo-voltage), the L_d term in the denominator controls the Z behavior of the PMR signal. Eqn. 2 thus demonstrates the effect of carrier diffusion on the radius of modulation as seen by the PMR probe. The PMR signal upon transmittance through an aperture of radius a may be identified from the expression:

$$\frac{R + \Delta R}{R} \equiv \frac{\int_0^a |E_r|^2 \rho d\rho}{\int_0^a |E_{DC}|^2 \rho d\rho}, (3)$$

where E_{DC} refers to the linear component of the reflected field. In the open aperture case ($a = \infty$) the integrals in Eqn. 3 may be evaluated at the sample surface. Neglecting the Δn^2 and Δk^2 terms yields:

$$\frac{\Delta R}{R} \cong \frac{4\Delta n}{n^2 - 1} \times \frac{2qN_e\Delta V}{\varepsilon_s} \times \frac{\omega_p^2 + L_d^2}{\omega^2(Z) + \omega_p^2(Z) + L_d^2}, (4)$$

where $n^2 \gg k^2$ and the dependence of the PMR amplitude on carrier diffusion length and the focal parameters of the pump and probe has been made explicit. (Thus, the open aperture PMR signal from a substrate satisfying $n^2 \gg k^2$ arises solely from the modulated refraction.) Eqn. 4 demonstrates the PMR signal as a function of sample position measured with respect to the focal plane may be parameterized by the expression:

$$\frac{\Delta R}{R} = \frac{A}{B\left\{1 + \left(\frac{\lambda_p Z}{\pi B}\right)^2\right\} + C + D\left\{1 + \left(\frac{\lambda Z}{\pi D}\right)^2\right\}}, (5)$$

where $A = 4\Delta n/(n^2-1)\times 2qN_e\Delta V/\varepsilon_s\times(\omega_p^2+L_d^2)$, $B = \omega_p^2$, $C = L_d^2$, and $D = \omega_o^2$. (Note the dependence of the amplitude on diffusion length is absorbed into the amplitude parameter A.) The carrier diffusion length and its measurement uncertainty may be established by performing a regressive (Levenberg-Marquardt) fit to the PMR amplitude using Eqn. 5. In particular, once the best fit is established (preferably holding B and D fixed since these are system parameters), all fit parameters may be held constant with the exception of C (the parameter corresponding to the carrier diffusion length) in order to determine the diffusion length and its uncertainty [5].

EXPERIMENTAL RESULTS

Using a tightly focused laser probe beam in our PMR setup, we measured the PMR signal as a function Z. Our basic setup has been described elsewhere [4]. The modulated pump (λ_p = 488 nm, Ω_m = 750 kHz) and CW probe beams (λ = 375 nm) were co-focused to a ~2 μm spot on silicon USJ samples activated by flash anneal. The reflected probe beam was collected by the focusing objective and integrated (spatially) at the detector. The sample set comprised n-type (> 2 Ω-cm) silicon (100) substrates implanted with a 40keV As halo implant at a dose of ~4×10^{13}/cm^2, followed by a 500eV B implant at a dose of ~1×10^{15}/cm^2, and subsequently annealed using various millisecond timescale flash anneals. Certain samples were annealed following the halo implant in order to achieve better As activation without causing B diffusion. This is referred to as the halo pre-anneal. Certain sample splits were additionally implanted with a 40keV Ge pre-amorphization implant (PAI) at a dose of ~1×10^{15}/cm^2. SIMS data indicated post-activation B doping levels of ≈1×10^{19}/cc at ≈20 nm for the majority of the samples. The detailed process conditions have been described elsewhere [3]. The physics of samples the under the influence of the modulated pump is essentially that of a p/n junction with various defect levels. For each sample tested, relaxation during the pump-off portion of the modulation cycle was confirmed by checking the linearity of the PMR phase with respect to modulation frequency over the range Ω_m = 600-900 kHz. Figs. 1 and 2 show representative experimental *Z*-scan PMR data and fits obtained from samples with and without PAI.

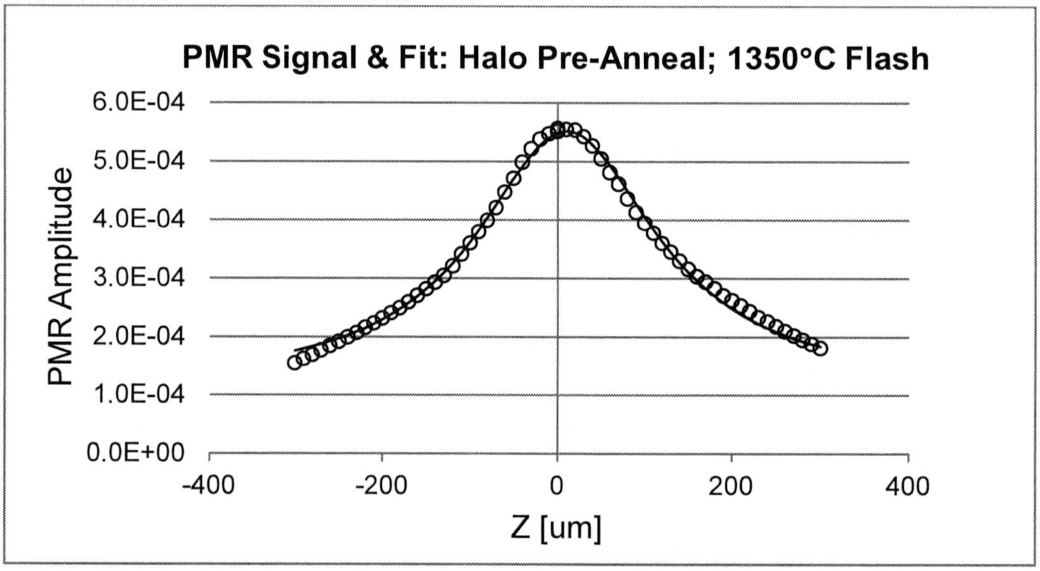

FIGURE 1. Exemplary PMR signal and fit for sample w/o PAI as a function of pump focal plane from sample surface.

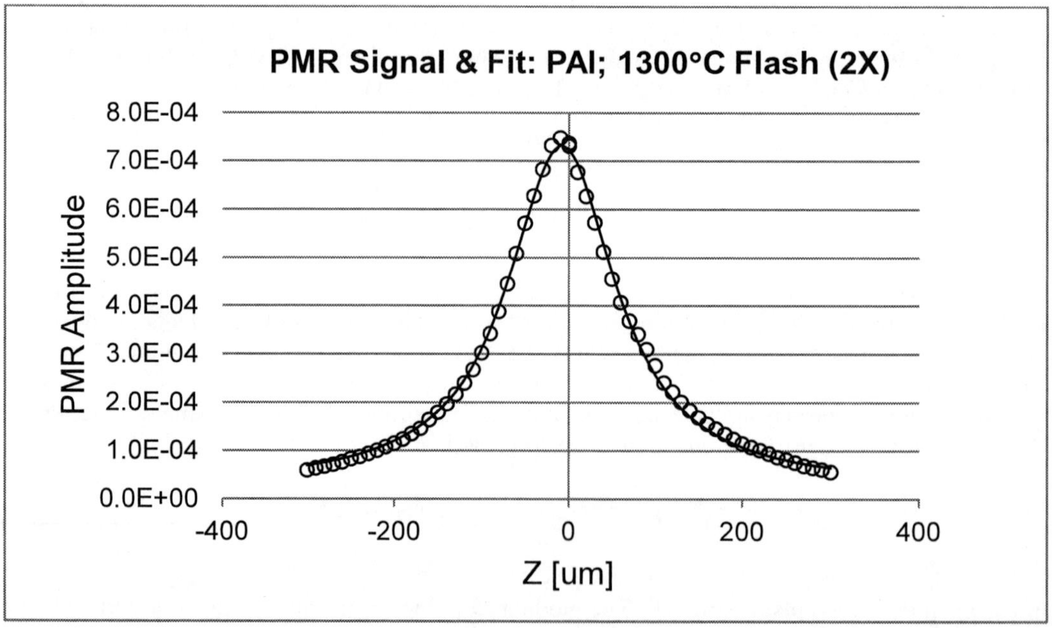

FIGURE 2. Exemplary PMR signal and fit for sample with PAI as a function of pump focal plane from sample surface.

The Z-scan PMR amplitude shown in Figs. 1 and 2 is symmetric with respect to Z, as expected from Eqn. 4. The more sharply peaked PMR response (as a function of Z) seen on the samples with PAI evidences a shorter diffusion length. This is expected since the PAI damage greatly enhances recombination. Table 1 lists extracted values of diffusion length for the samples tested, assuming a measurement uncertainty of 5 ppm for the PMR amplitude. Generally, the measured diffusion lengths are shorter for the samples with PAI. The diffusion lengths measured on the samples with PAI show little sensitivity to the halo pre-anneal. This is also expected since the PAI occurred after the halo pre-anneal. Systematic variations in diffusion length with annealing treatment may also be observed across the sample set. The measured diffusion lengths are in good agreement with values expected from the doping levels. Moreover, although the measurement uncertainties input to the fit procedure exceed the actual measurement uncertainties by a least a factor of 2, the output uncertainties in diffusion length remain less than 0.5% in all cases.

TABLE 1. Measured Carrier Diffusion Lengths.

Flash Temp:	**No PAI**		**With PAI**	
Peak/Ramp [°C]	**With Halo Pre-Anneal**	**Flash Only**	**With Halo Pre-Anneal**	**Flash Only**
1250/550	–	16.20 ± .02 μm	–	–
1300/550	–	16.40 ± .05 μm	5.82 ± .01 μm	5.87 ± .01 μm
1300/550 (2X)	16.72 ± .04 μm	27.33 ± .02 μm	8.50 ± .01 μm	8.89 ± .01 μm
1300/600	–	–	9.21 ± .04 μm	8.73 ± .04 μm
1350/600	14.59 ± .02 μm	13.50 ± .03 μm	10.91 ± .03 μm	9.50 ± .04 μm

CONCLUSION

The carrier diffusion length is sensitive to key device properties such as dopant activation, leakage current, and semiconductor material quality. Generally the combination of low junction sheet resistance and leakage current with defect-free semiconductor material, corresponding to a long carrier diffusion length, is a favorable condition for high quality electrical device functions. The PMR based technique discussed herein has been used to characterize carrier diffusion length with high precision. As an optical technique which is sensitive to dopant activation, junction leakage and other key properties of electrical device structures it is generally suitable for non-contact process monitoring in high volume semiconductor manufacturing. Moreover, the technique offers direct sensitivity to semiconductor electronic properties as may be required for development of manufacturing processes for band engineered semiconductor nanostructures.

REFERENCES

1. N. Bottka et al., *J. Electron. Mater.* **17**, 161-170 (1988).
2. W. Chism, D. Pham, and J. Allgair, *AIP Conf. Proc.* **931**, 64-68 (2007).
3. W. Chism, M. Current, and V. Vartanian, *J. Vac. Sci. Technol. B* **28**, C1C15-C1C20 (2010).
4. W. Chism and J. Cartwright, *Thin Solid Films* **520**, 6521-6524 (2012).
5. W.H. Press, et al., *Numerical Recipes in Fortran 77*, 2nd Ed., Cambridge University Press, 1996, Chapter 15.

KEYWORDS

Photoreflectance, carrier diffusion length

Crack Evolution In Cu/Low-K Stacks And Crack Stop Evaluation Using In-Situ Micro-DCB In A Nano-XCT Tool

Kristina Kutukova, Jürgen Gluch, Yvonne Standke, Ehrenfried Zschech

Fraunhofer Institute for Ceramic Technologies and Systems, Maria-Reiche-Str. 2, 01109 Dresden, Germany

INTRODUCTION

Thermomechanical stress caused by chip-package interaction (CPI) is an increasing reliability concern for advanced microelectronic products. A sophisticated reliability engineering, that includes knowledge from materials and stress engineering as well as fracture mechanics, is needed to guarenatee the requested product quality. This task is particularly challenging for advanced technology nodes using low-k and ultra low-k materials in backend-of-line (BEoL) stacks [1]. The design of on-chip metal interconnects, fracture toughness of the materials and interface strength (adhesion) in the BEoL stack as well as the CPI stress state have to be considered to evaluate the risk of mechanical failure. To prevent fracture in the BEoL stack, e. g. originated by micro-cracks during dicing, the design and the materials used, including the introduction of so-called crack-stop structures, have to be optimized (Fig. 1). These dense metal structures are supposed to dissipate energy in such a way that the crack propagation is slowed down or even stopped effectively. The direct observation of the crack path evolution, the determination of the weakest structures in the BEoL stacks and the evaluation of the effectiveness of crack-stop structures require nondestructive high-resolution 3D imaging of cracks in on-chip interconnect stacks. Transmission X-ray microscopy (TXM) and nano X-ray computed tomography (nano-XCT) are techniques that allow to detect the opened cracks in dielectrics and along interfaces of on-chip interconnect stacks nondestructively with sub-100nm resolution.

FIGURE 1. SEM cross-section images with crack stop structures of an Intel processor with 12 Cu metallization layers

In this study, we are using a novel micro Double Cantilever Beam (micro-DCB) test performed in an X-ray microscope [2] to study crack opening and propagation in materials with high resolution. Cracks in the Cu/low-k interconnect stack of an Intel processor manufactured in cutting edge technology are visualized (in-situ) in the X-ray microscope during mechanical loading. The pathways of cracks, and consequently the weakest structures for cracking, are nondestructively identified applying the micro-DCB test in the X-ray microscope and subsequent analysis of a 3D data set based on a nano-XCT study.

EXPERIMENTAL SETUP

A commercial laboratory X-ray microscope (Xradia/Zeiss, NanoXCT-100) equipped with a rotating anode X-ray source (Cu- Kα radiation, 8.05 keV photon energy) and two exchangeable Fresnel zone plates (FZPs) as focusing lenses was used for this study. The two FZPs applied for overview imaging and high resolution imaging, respectively, have field of views (FOVs) of 66.5 μm x 66.5 μm and 16.5 μm x 16.5 μm, respectively, resulting in pixel sizes of the digital radiograph of 65 nm (10x FZP) and 16 nm (40x FZP). A miniaturized micro Double Cantilever Beam (micro-DCB) test setup was integrated in the X-ray microscope (Fig. 2) [1]. The sample of interest (silicon substrate and BEoL stack) was bonded to a dummy sample (bare silicon wafer) with similar dimensions to obtain a sandwich structure in such a way that the layer stack to be studied was located in its center. A notch was introduced into the specimen, to have a defined starting point for the crack evolution. The stable crack propagation in the BEoL stack occurs in a controlled way, which allows to determine the pathway of the crack along weak interfaces (adhesive failure) or within the soft dielectrics (cohesive failures).

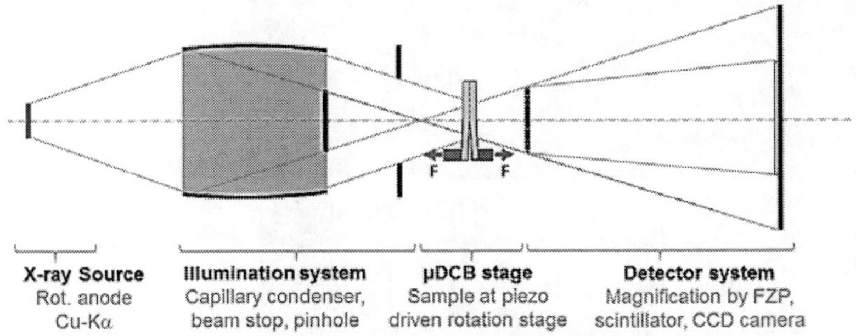

a)

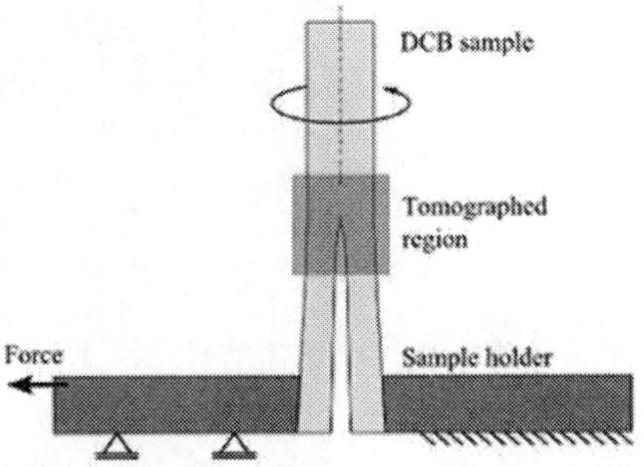

b)

FIGURE 2. a) Scheme of microDCB test setup inside a nano-XCT tool, b) schematic drawing of microDCB test setup [1]

EXPERIMENTAL PROCEDURE AND RESULTS

During the experiment, the crack was opened by the piezo-driven stage of the micro-DCB test setup. A stitched radiograph of the micro-DCB specimen under load is shown in Fig. 3 a. The layer stack of the sandwich-type specimen consists of the original test wafer (Si substrate, BEoL stack), a thin layer of epoxy, and the silicon dummy [3]. The sample studied includes two crack stop structures. The crack propagated from the inner chip side to the scribe line and vice versa, and it was studied within the BEoL stack, along weak interfaces or in interlayer dielectrics towards the crack-stop structure. The crack propagation through the BEoL stack towards the crack-stop structure was visualized. The virtual horizontal cross-sections, i. e. parallel to the metallization layers, allow to track the crack path through the BEoL structure (Fig. 3 b).

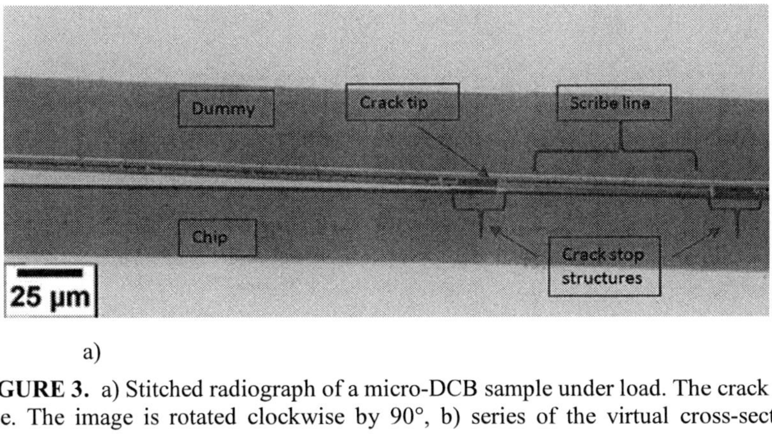

a)

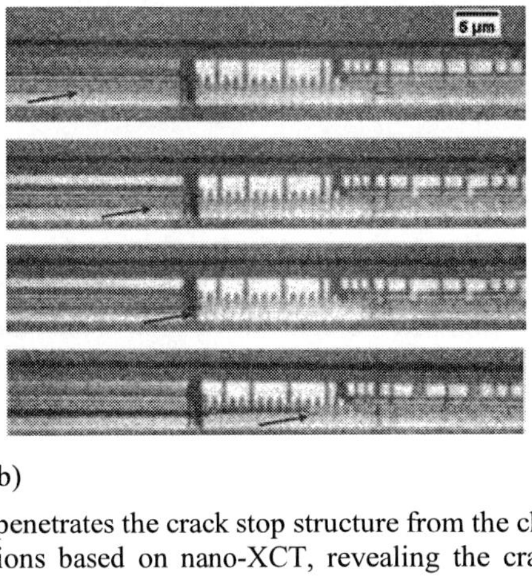

b)

FIGURE 3. a) Stitched radiograph of a micro-DCB sample under load. The crack penetrates the crack stop structure from the chip side. The image is rotated clockwise by 90°, b) series of the virtual cross-sections based on nano-XCT, revealing the crack propagation at the Cu/low-k crack stop structure.

The crack propagated horizontally inside the BEoL stack until it reached the crack-stop structure. At the crack-stop structure, energy was dissipated and the crack changed the direction of propagation. Eventually the crack propagated into the substrate. The tomography data of the superimposed cross-sectional views (perpendicular to the metallization layers) allow to evaluate the number of the damaged layers (Fig. 3 b). Scanning electron microscopy (SEM) was applied after the nondestructive micro-DCB / nano-XCT experiment to prove the applied approach (Fig. 4). The SEM images showing the crack front at the crack stop structure allow to detect the number of the damaged layers (Fig. 4 b and c). This data confirm the results from the tomography data based on the in-situ nano-XCT experiment.

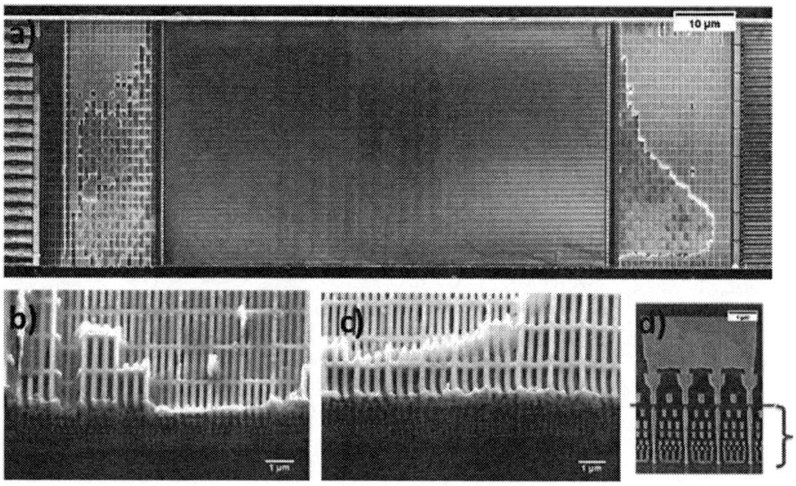

FIGURE 4. SEM images: a) overview of the fracture surface with two fractured crack stop structures, and at higher magnification b) and c); d) comparison of the number of the damaged layers with non-damaged structure.

CONCLUSION

The combination of high-resolution X-ray imaging with in-situ mechanical testing provides an unprecedented level of details on the fracture behavior of microchips. The application of the novel micro-DCB test in an X-ray microscope for (in-situ) visualization of crack opening and propagation in Cu/low-k interconnect stacks was demonstrated. The experimental approach was proven for the mechanical characterization of the BEoL stack, and its robustness against CPI, and for the evaluation of the effectiveness of crack stop structures. Summarizing, the applied approach provides valuable information for reliability engineering and design-for-reliability (DFR) in semiconductor industry.

ACKNOWLEDGEMENT

The authors thank Rüdiger Rosenkranz, Christoph Sander and Martin Gall (all with Fraunhofer IKTS Dresden, Germany) and Sven Niese (AXO Dresden GmbH, Germany) for their support during the experiment. Han Li (with Intel Inc., Hillsboro/OR, US) is acknowledged for fruitful discussions.

REFERENCES

[1] R. Radojcic, M. Nowak, M. Nakamoto, “TechTuning: Stress management for 3D Through-Si-Via stacking technologies”, AIP Conf. Proc. 1378, 5 - 20 (2011)

[2] S. Niese, “Lab-based in-situ X-ray microscopy – Methodical developments and applications in materials science and microelectronics“, PhD thesis, BTU Cottbus (2015)

[3] K. Kutukova, J. Gluch, Y. Standke, E. Zschech, Crack path localization in Cu/low-k stacks during micro-DCB tests using nano-XCT”, IEEE TDMR, in press (2017)

KEYWORDS

Chip-package interaction (CPI), low-k dielectric materials, crack-stop structures, micro double cantilever beam test (micro-DCB), nano-XCT.

New Frequency Domain Fiber Optic Interferometry for Advanced Wafer, Micro- and Nanostructure Metrology

Wojtek J. Walecki, Alexander Pravdivtsev, Jae Ryu, Nikos D. Jaeger, Yuen Lim, Ann Koo

Frontier Semiconductor, 2127 Ringwood Ave., San Jose CA 95131 USA
Tel: 408 432 8838, email: fsm100@frontiersemi.com

INTRODUCTION

Low coherence interferometric imaging known also as optical coherence tomography (OCT) was initially developed for medical applications, however in the last decade it found many applications in many other fields including semiconductor metrology [1]. Frontier Semiconductor has developed many semiconductor applications and deployed over hundred high resolution OCT tools in last decade. In this paper, we report on newest developments in high resolution OCT metrology for modern semiconductor wafers and structures.

The real space OCT system used in semiconductor metrology is based on fiber optic Michelson interferometer design [2]. The analysis of the interferogram measured by the Michelson interferometer leads to establishment of distances of interest within measured structures as shown in Figure 1. Unlike free space interferometers the fiber optic interferometers suffer very large thermal drift caused by thermal dependence of refractive index of glass transmitting light beam. The invention of the auxiliary reference mirrors [3] effectively eliminated this thermal drift and enabled absolute distance measurement [4]. Later the multiprobe synchronized systems were also developed enabling measurements of nontransparent, opaque wafers [5]. The typical static repeatability of the thickness measurement is about 50 nm for Si wafers while the repeatability of the position measurements is about 200 nm. The time needed to perform single shot measurement is limited by speed of scanning mirror and is typically in a range between 50 – 100 ms, however systems operating at higher than 1 kHz rate were also developed in past [6].

Since the invention these systems found applications for measurement of complex wafer structures [7],[8], micro electromechanical systems (MEMs) [9], deep trenches [10], bumps, and vias [11]. The introduction of auxiliary reference mirror [3] enabled measurements of the wafer topography and stress [12]. This technology has also been shown that it can be easily integrated with other measurements techniques [13], and to have great potential for in-situ applications [14]. The applications of this technique are not limited to silicon devices. The same technology can be used to measure many other materials (see for example application to III-V semiconductors [15]).

Recently we observe need for faster and higher resolution measurements driven by more demanding customer applications. This need is addressed by the frequency domain system described below.

FREQUENCY DOMAIN SYSTEM FOR THICK LAYER METROLOGY

Thin slabs of material are often inspected to determine thickness using known methods of observation and analysis of Fabry Perot (FP) interference fringes. In the case of a simple single slab of material, these known methods of inspection are based on the observation of interference fringes in an etalon formed by the parallel interfaces of the sample. However, when the thickness of a slab of material is greater than about 50 μm, employing these known methods is difficult due to limited resolution of small spectrometers. These known methods are not effective when the thickness of a slab of material is greater than 50 μm.

We describe performance and applications of our new patent pending system shown in Figure 2. The system employs an additional etalon filter. The low coherence light reflected from the sample is filtered by a wafer and an additional etalon before entering slit of grating spectrometer enabling observation of closely spaced FP fringes. The resulting spectrum observed by the spectrometer is a product of the spectrum of the light source, the reflectivity of the wafer, and the transmission of the etalon. Both the etalon and the wafer transfer functions exhibit fast varying due to FP effect. In our paper we show that when the optical thicknesses of the wafer and the etalon are similar then the light impinging the slit of the spectrometer exhibits slow varying oscillations in frequency domain corresponding to the difference of the optical thickness of these two components. These slow varying oscillations can be easily measured using a small grating spectrometer. This phenomenon is somewhat analog to acoustic beats.

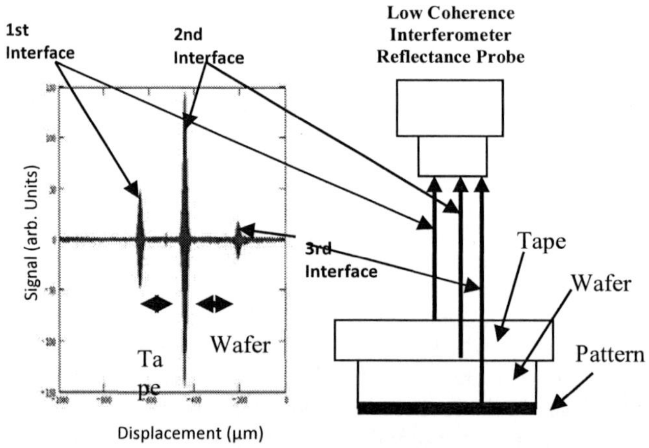

FIGURE 1. Measured asymmetric interferogram exhibits strong peaks corresponding to boundaries in the measured structure.

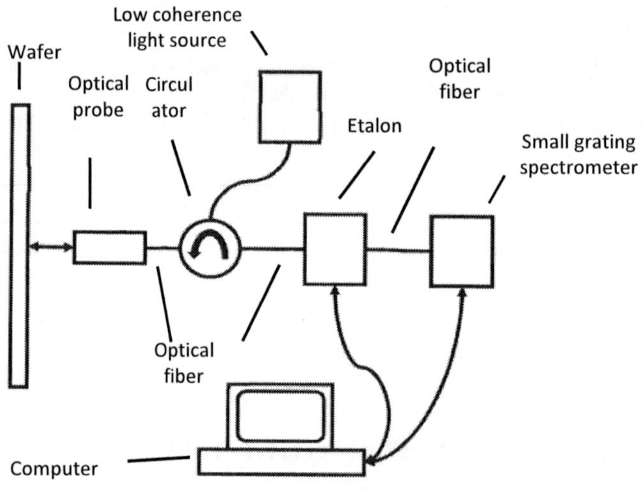

FIGURE 2. System enabling observation of Fabry Perot fringes in thick samples.

We have built a system based on this concept, and we found that it has static reproducibility of the order of 5 nm for 0.75 mm and thicker wafers, and the acquisition time of the order 10 ms . This is a significant improvement over real space system described in [1], while addressing most or all its predecessor applications.

CONCLUSIONS

We presented a novel probe for measuring of the thickness of thick layers with few nm repeatability employing an etalon – small grating spectrometer scheme. This new probe can be used for metrology of MEMs, and other novel structures and materials with greater speed and accuracy than conventional real space techniques.

REFERENCES

1. Stifter, D., 2007. Beyond biomedicine: a review of alternative applications and developments for optical coherence tomography. Applied Physics B, 88(3), pp.337-357.
2. Walecki, Wojciech J., and Phuc Van. "Determining thickness of slabs of materials by inventors." U.S. Patent No. 7,116,429. 3 Oct. 2006.
3. Walecki, Wojciech, and Phuc Van. "Temperature insensitive low coherence based optical metrology for nondestructive characterization of physical characteristics of materials." U.S. Patent No. 7,502,121. 10 Mar. 2009.
4. Walecki, Wojciech J., et al. "Low-coherence interferometric absolute distance gauge for study of MEMS structures." MOEMS-MEMS Micro & Nanofabrication. International Society for Optics and Photonics, 2005.
5. Walecki, W.J., Pravdivtsev, A., Santos II, M. and Koo, A., 2006, August. High-speed high-accuracy fiber optic low-coherence interferometry for in situ grinding and etching process monitoring. In SPIE Optics+ Photonics (pp. 62930D-62930D). International Society for Optics and Photonics.
6. Walecki, W.J., Pravdivtsev, A., Lai, K., Santos II, M. and Koo, A., 2005, August. Synchronized low coherence interferometry for in-situ and ex-situ metrology for semiconductor manufacturing. In Optics & Photonics 2005 (pp. 58800H-58800H). International Society for Optics and Photonics.
7. Walecki, Wojciech J., Vitali Souchkov, Kevin Lai, Phuc Van, Manuel Santos, Alexander Pravdivtsev, S. H. Lau, and Ann Koo. "Novel Noncontact Thickness Metrology for Partially Transparent and Nontransparent Wafers for Backend Semiconductor Manufacturing." In MRS Proceedings, vol. 829, pp. B9-31. Cambridge University Press, 2004.
8. Walecki, W.J., Pravdivtsev, A., Lai, K., Santos, M., Mikhaylov, G., Mihaylov, M. and Koo, A., 2005, September. Fiber-optics Low-coherence Integrated Metrology for In-Situ Non-contact Characterization of Novel Materials and Structures. In D.G. Seiler, A.C. Diebold, R. McDonald and C.R. Ayre eds.,, AIP Conference Proceedings (Vol. 788, No. 1, pp. 338-342). AIP.
9. Walecki, W.J., Azfar, T., Pravdivstev, A., SantosII, M., Wong, T., Feng, A. and Koo, A., 2006, January. Novel combined low-coherence interferometry spectrally resolved reflectometry compatible with high-resolution Raman spectroscopy for nondestructive characterization of MEMS structures. In MOEMS-MEMS 2006 Micro and Nanofabrication (pp. 61110M-61110M). International Society for Optics and Photonics.
10. Walecki, Wojciech, Frank Wei, Phuc Van, Kevin Lai, Tim Lee, S. H. Lau, and Ann Koo. "Novel low coherence metrology for nondestructive characterization of high-aspect-ratio microfabricated and micromachined structures." In Micromachining and Microfabrication, pp. 55-62. International Society for Optics and Photonics, 2004.
11. Walecki, Wojciech J., Talal Azfar, Alexander Pravdivstev, Manuel Santos II, and Ann Koo. "Combined low-coherence interferometry and spectrally resolved reflectometry for nondestructive characterization of small-diameter high-aspect ratio micro-fabricated and micro-machined structures and multilayer membranes." In MOEMS-MEMS 2006 Micro and Nanofabrication, pp. 61090J-61090J. International Society for Optics and Photonics, 2006.
12. Walecki, W.J., Pravdivtsev, A., Lai, K., Santos, M., Mikhaylov, G., Mihaylov, M. and Koo, A., 2005. Fiber-optics low-coherence integrated metrology for in-situ non-contact characterization of wafer curvature for wafers having non-uniform substrate and thin film thickness. In MRS Proceedings (Vol. 875, pp. O9-4). Cambridge University Press.
13. Walecki, W.J., Azfar, T., Pravdivstev, A., SantosII, M., Wong, T., Feng, A. and Koo, A., 2006, January. Novel combined low-coherence interferometry spectrally resolved reflectometry compatible with high-resolution Raman spectroscopy for nondestructive characterization of MEMS structures. In MOEMS-MEMS 2006 Micro and Nanofabrication (pp. 61110M-61110M). International Society for Optics and Photonics.
14. Walecki, W.J., Pravdivstev, A., Santos, M. and Koo, A., 2006, August. Fast and accurate non-contact in situ optical metrology for end-point detection. In SPIE Optics+ Photonics (pp. 629211-629211). International Society for Optics and Photonics.
15. Walecki, W.J., Lai, K., Souchkov, V., Van, P., Lau, S.H. and Koo, A., 2005. Novel noncontact thickness metrology for backend manufacturing of wide bandgap light emitting devices. physica status solidi (c), 2(3), pp.984-989.

KEYWORDS

Novel Measurement Methods, Breakthroughs, Wafer metrology, Low coherence interferometry, Spectral domain interferometry, MEMS/NEMS Metrology Applications, thickness measurement